Mohammad Hossein Keshavarz, Thomas M. Klapötke
Energetic Materials

Also of interest

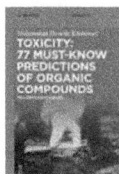

Toxicity: 77 Must-Know Predictions of Organic Compounds.
Including Ionic Liquids
Mohammad Hossein Keshavarz, 2023
ISBN 978-3-11-118912-3; e-ISBN (PDF) 978-3-11-118967-3;
e-ISBN (EPUB) 978-3-11-119092-1

Combustible Organic Materials.
Determination and Prediction of Combustion Properties
2nd edition
Mohammad Hossein Keshavarz, 2022
ISBN 978-3-11-078204-2; e-ISBN (PDF) 978-3-11-078213-4;
e-ISBN (EPUB) 978-3-11-078225-7

Energetic Compounds.
Methods for Prediction of their Performance
2nd edition
Mohammad Hossein Keshavarz, Thomas M. Klapötke, 2020
ISBN 978-3-11-067764-5; e-ISBN (PDF) 978-3-11-067765-2;
e-ISBN (EPUB) 978-3-11-067775-1

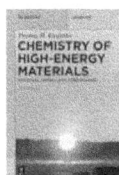

Chemistry of High-Energy Materials.
Explosives, Propellants, Pyrotechnics
7th edition
Thomas M. Klapötke, 2025
ISBN 978-3-11-144698-1; e-ISBN (PDF) 978-3-11-144708-7;
e-ISBN (EPUB) 978-3-11-144713-1

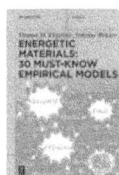

Energetic Materials: 30 Must-Know Empirical Models
Thomas M. Klapötke, Sabrina Wahler, 2025
ISBN 978-3-11-109602-5; e-ISBN (PDF) 978-3-11-109702-2;
e-ISBN (EPUB) 978-3-11-109782-4

Energetic Materials Encyclopedia
3th edition
Thomas M. Klapötke, 2026
ISBN 978-3-11-146836-5; e-ISBN (PDF) 978-3-11-146969-0;
e-ISBN (EPUB) 978-3-11-147006-1

Mohammad Hossein Keshavarz,
Thomas M. Klapötke

Energetic Materials

Sensitivity, Physical Properties, Thermodynamic
Properties

3rd, Completely Revised and Extended Edition

DE GRUYTER

Authors
Prof. Dr. Mohammad Hossein Keshavarz
Department of Chemistry
Faculty of Applied Sciences
Malek-Ashtar University of Technology
Ferdowsi Street
8315713115 Shahin-shahr
Iran

Prof. Dr. Dr. h.c. Thomas M. Klapötke
Department of Chemistry
Ludwig-Maximilians-Universität München
Butenandtstr. 5-13 (D)
81377 Munich
Germany
tmk@cup.uni-muenchen.de

ISBN 978-3-11-914703-3
e-ISBN (PDF) 978-3-11-220676-8
e-ISBN (EPUB) 978-3-11-220702-4

Library of Congress Control Number: 2025941266

Bibliographic information published by the Deutsche Nationalbibliothek
The Deutsche Nationalbibliothek lists this publication in the Deutsche Nationalbibliografie;
detailed bibliographic data are available on the Internet at http://dnb.dnb.de.

www.degruyterbrill.com
Questions about General Product Safety Regulation:
productsafety@degruyterbrill.com

Preface

For a chemist who is concerned with the synthesis of new energetic compounds, it is essential to be able to assess physical and thermodynamic properties, as well as the sensitivity of possible new energetic compounds before synthesis is attempted. Various approaches have been developed to predict important aspects of the physical and thermodynamic properties of energetic materials including (but not exclusively): crystal density, heat of formation, melting point, enthalpy of fusion and enthalpy of sublimation of an organic energetic compound. Since an organic energetic material consists of metastable molecules capable of undergoing very rapid and highly exothermic reactions, many methods have been developed to estimate the sensitivity of an energetic compound with respect to detonation-causing external stimuli such as heat, friction, impact, shock, and electrostatic discharge. This book introduces these methods and demonstrates those methods which can be easily applied.

https://doi.org/10.1515/9783112206768-202

Preface to the third edition

The core principles outlined in the second edition's preface remain valid and largely unchanged. However, this revised third edition incorporates updates and recent advancements in energetic materials research, including:

1. **Three New Chapters**
   - **Chapter 12: Estimation of Properties of Metal-Containing Energetic Complexes and Energetic MOFs**
   Metal-containing energetic complexes and energetic metal-organic frameworks (EMOFs) represent cutting-edge approaches to balancing energy density and stability. While traditional metal complexes offer high performance, their sensitivity remains a challenge; EMOFs provide superior structural control and thermal resilience. This chapter presents predictive models for density, heat of formation, and decomposition temperature to guide the design of safer, high-performance materials.
   - **Chapter 13: Estimating Properties of Energetic Polymers**
   Energetic polymers are critical for optimizing the performance and safety of explosives, propellants, and pyrotechnics. Moving beyond conventional binders like HTPB, newer polymers (e.g., GAP, BAMO) incorporate nitrogen-rich groups for enhanced energy output and cleaner combustion. Predictive models for solubility, viscosity, glass transition, and refractive index are introduced to streamline material development.
   - **Chapter 14: Computational Prediction of Energetic Material Properties**
   Computational methods now play a pivotal role in energetic materials research, reducing reliance on costly and hazardous lab synthesis. Early empirical models have evolved into sophisticated quantum chemistry and machine learning approaches (e.g., CBS-QB3, RoseBoom), though challenges persist – particularly for complex salts. This chapter explores the current capabilities and limitations of predictive tools.
2. **Expanded and Updated Content**
   - **Chapter 1**: New sections on high-nitrogen organic compounds, high-nitrogen salts/ionic liquids (HNCSILs), and machine learning applications.
   - **Chapter 2**: Added discussion on machine learning advances in predicting formation enthalpies.
   - **Chapter 3**: Extended model for melting point prediction in ionic liquids.
   - **Chapter 4**: New sections on predictive models for enthalpy of fusion in hydrocarbons and group contribution methods for ionic liquids.
   - **Chapter 6**: Advances in impact sensitivity modeling, including phonon-vibration coupling and multiplicative incremental theory.
   - **Chapter 7**: New section on electrostatic discharge (ESD) sensitivity in nitrogen-rich heterocyclic compounds.

https://doi.org/10.1515/9783112206768-203

    –   **Chapter 10**: Predictive model for thermal decomposition onset in heterocyclic compounds and salts.

3. **Corrections and References**

    Errors identified in the second edition have been corrected, and references have been updated to reflect the latest research.

We hope these updates will further support researchers and practitioners in the field.

Mohammad Hossein Keshavarz
Thomas M. Klapötke

Preface to the second edition

Everything said in the preface to the first edition still holds and essentially does not need any addition or correction. In this revised second edition, we have updated the manuscript and added some recent aspects of energetic materials:

1. Some errors which unfortunately occurred in the first edition have been corrected and the references have been updated where appropriate.
2. Recent works have been reviewed and discussed in each chapter. Moreover, new sections have been inserted including:
   (a) Chapter 1 – The use of group additivity methods for prediction of crystal density of energetic neutral and ionic liquids or salts
   (b) Chapter 2 – The condensed phase heat of formation of energetic ionic liquids and salts
   (c) Chapter 3 – Melting points of ionic liquids
   (d) Chapter 4 – Group additivity method for prediction of enthalpy and entropy of fusion
   (e) Chapter 5 – Group additivity method for prediction of the heat of sublimation
   (f) Chapter 6 – Impact sensitivity of quaternary ammonium-based energetic ionic liquids or salts
   (g) Chapter 7 – Simple prediction of electrostatic spark sensitivity based on the new ESZ KTTV instrument
   (h) Chapter 8 – Critical diameter of solid pure and composite high explosives
   (i) Chapter 9 – Friction sensitivity of quaternary ammonium-based energetic ionic liquids
   (j) Chapter 10 – Thermal stability of selected classes of energetic ionic liquids and salts
   (k) Chapter 11 – A general correlation between electric spark sensitivity and impact sensitivity of nitroaromatics and nitramines as well as the relationship between shock sensitivity of nitramine energetic compounds based on small-scale gap test and their electric spark sensitivity

Mohammad Hossein Keshavarz
Thomas M. Klapötke

https://doi.org/10.1515/9783112206768-204

About the authors

Mohammad Hossein Keshavarz, born in 1966, studied chemistry at Shiraz University and received his BSc in 1988. He also received an MSc and PhD at Shiraz University in 1991 and 1995. From 1997 until 2008, he was Assistant Professor, Associate Professor, and Professor of Physical Chemistry at the University of Malek-ashtar in Shahin-shahr of Iran. Since 1997, he is a lecturer and researcher at the Malek-ashtar University of Technology, Iran. He is the editor of two research journals in the Persian language. Keshavarz has published over 400 scientific papers in international peer-reviewed journals, 7 book chapters, and nine books in the field of energetic materials assessment (four books in Persian and five in English language). His English-language books include *Toxicity: 77 Must-Know Predictions of Organic Compounds* (2023), *Combustible Organic Materials: Determination and Prediction of Combustion Properties* (2018 and 2022 editions), *The Properties of Energetic Materials* (2017 and 2021 editions, with T.M. Klapötke), *Energetic Compounds: Methods for Prediction of Their Performance* (2017 and 2020 editions, with T.M. Klapötke), and *Liquid Fuels as Jet Fuels and Propellants* (2018). He has also contributed chapters to several scientific volumes, including *Materials Informatics III* (Springer, 2025), *An Introduction to Propellants* (2020), and multiple books on explosive materials and hazardous compounds published by Malek-ashtar University of Technology.

Thomas M. Klapötke received his PhD in 1986 (TU Berlin), post-doc in Fredericton, New Brunswick, habilitation in 1990 (TU Berlin). From 1995 until 1997 Klapötke was Ramsay Professor of Chemistry at the University of Glasgow in Scotland. Since 1997 he has held the Chair of Inorganic Chemistry at LMU Munich. In 2009 Klapötke was appointed a Visiting Professor at CECD, University of Maryland and in 2014 he was appointed a Adjunct Professor at the University of Rhode Island. In 2023 Klapötke received an honorary doctoral degree from the Military Technical Academy Ferdinand I from Bucharest. Klapötke is a Fellow of the RSC (C. Sci., C. Chem. F. R. S. C.), a member of the ACS and the Fluorine Division of the ACS, a member of the GDCh, and a Life Member of both the IPS and the National Defense Industrial Association. Most of Klapötke's scientific collaborations are between LMU and the German Federal Army as well as with ARDEC (Armament Research Development and Engineering Center) in Picatinny, NJ. He is the executive editor of the Journal of Engineering Science and Military Technologies, the Subject Editor in the area of explosives synthesis of the Central European Journal of Energetic Materials and an editorial board member of Propellants, Explosives and Pyrotechnics (PEP), Journal of Energetic Materials, the Chinese Journal of Explosives and Propellants and the International Journal of Energetic Materials and Chemical Propulsion (IJEMCP). Klapötke has published over 950 papers, 38 book chapters and 18 books.

https://doi.org/10.1515/9783112206768-205

Contents

1 Crystal density

Organic compounds containing energetic groups such as nitro, nitramine, and nitrate ester functional groups have wide applications in military and civilian applications as propellants, explosives, and pyrotechnics because they can release their stored chemical energy upon external stimuli such as heat, impact, shock, friction, and electrostatic discharge [1–7]. Ionic molecular energetic materials containing high nitrogen content are attractive for scientists and industries. They may be used as energetic compounds because they have high density, positive heats of formation, and thermal stability [8]. They frequently consist of high nitrogen content cations such as substituted imidazole, triazole, and tetrazole derivatives and bulky anions containing energetic groups, for example, $-NO_2$, $-N_3$, and $-CN$. They can be considered as eco-friendly, low-melting, and thermally stable ionic compounds [2]. Considerable efforts have been done in recent years to introduce new organic and ionic molecular energetic materials with high density because their higher density is always desirable for packing more energy per unit volume.

The crystal density and condensed phase heat of formation of an energetic compound are two important physicothermal properties, and are essential values in order to be able predict the detonation performance using a thermodynamic equilibrium code such as CHEETAH [9], or through empirical methods [1, 10–17]. The performance characteristics of energetic compounds are proportional to their densities; for example, the Chapman–Jouguet pressure is proportional to the square of the initial density [1, 10]. Thus, it is essential to use suitable methods such as gas pycnometry or low-temperature single crystal X-ray diffraction to determine the crystal density of an energetic compound. New molecules which are candidates for possible use as energetic materials can be synthesized, characterized, and formulated by reliable predictive methods, and theoretical molecular design may be used to develop new energetic materials before synthesis is attempted. The synthesis of molecules with significantly increased energy in comparison with current materials, as well as the synthesis of very insensitive materials which have reasonable energies, have been two important goals for scientists in recent years. Since it is essential to have reliable methods to predict the density of energetic compounds, different approaches have been developed to assess the crystal density of an energetic compound at 25 °C.

Attempts have been made to predict the crystal densities of proposed new energetic compounds with satisfactory accuracy. A predicted crystal density that differs by less than 0.03 g/cm^3 from the experimentally obtained value should be defined as "excellent." A value that deviates between 0.03 and 0.05 g/cm^3 from the experimental value is still "informative" [18]. Quantum-mechanically determined molecular volumes [19–22], group additivity [23–25], empirical methods [26–29], quantitative structure-property relationship (QSPR) based on complex descriptors [30, 31] and MD [32] are usual different approaches which have developed to predict the crystal densities

https://doi.org/10.1515/9783112206768-001

of different types of $C_aH_bN_cO_d$ energetic compounds. The group additivity method is a simple approach because it requires only a set of atoms and group volumes that can be summed to obtain an estimate of the effective molecular volume using a simple computer code [24]. Tedious investigations have been undertaken over the last 30 years or so to expand the list of atom and functional group volumes [24, 25]. Although group additivity methods are simple to use with low cost, the predicted value which is obtained may show very large deviation from the experimentally obtained value for some energetic compounds. Moreover, such methods can only be used for those energetic compounds for which the values of all groups contained in the compound have been specified. Quantum mechanical and empirical methods (or QSPR) based on the structures of energetic compounds are more reliable approaches to estimate the density of an energetic compound. The QSPR methods are based on complex descriptors which develop a mathematical relationship connecting a macroscopic property of a series of compounds to microscopic descriptors derived from their molecular structures using an experimental data set. They require computer codes and expert users, as well as complex descriptors. The descriptors used in QSPR models can be empirical, or computed on the basis of the molecular structure. Various statistical tools including multilinear regression (MLR), nonlinear regression (NLR), partial least squares (PLS), artificial neural network (ANN), genetic algorithm or support vector machine (SVM) are frequently used to derive the mathematical equations (or algorithms) linking the property and descriptors [33, 34]. MD is a computer simulation of the physical movements of atoms and molecules in the context of N-body simulation. Due to the higher reliability of quantum mechanical and empirical methods (or QSPR) based on molecular/ionic structures, these approaches have been developed in recent years for neutral and ionic liquid energetic compounds, which are described in this chapter. Some efforts have been made to assess the detonation performance of newly designed explosives and ionic molecular energetic materials with high detonation performance in recent years [13, 15, 35–46]. Since high reliability is an important parameter in selecting predictive methods for different classes of energetic compounds, several of the best available methods are introduced and described in this chapter.

1.1 Group additivity method

The group additivity method sums the volume of atoms, molecular fragments, and functional groups to estimate the molecular volume of an organic energetic compound. It cannot explain the effect of parameters including the molecular conformation, isomerism, and packing efficiency in the crystals on density. Moreover, it provides the same results for organic explosives with different isomers and conformations. Thus, the reliability of group additivity methods is low as compared to the other common methods. Ammon [47] has introduced the latest group additivity method by including a larger database of groups, that is, 96 different groups and atoms from more than 26,000 crystals. Ye and Shreeve [48] introduced a suitable group additivity method that only involves 38 atom/group parameters and three corrections. This approach considers the quantitative impact of strong hydrogen bonding on the densities of energetic materials. Table 1.1 shows volume parameters for atoms, groups, and fragments. The sum of the contributions of these volume parameters can be used to estimate density at room temperature as follows:

$$\rho = \frac{Mw}{0.6022V}, \tag{1.1}$$

where Mw is the molecular weight of the desired explosive in g/mol and V is the total volume in Å^3, respectively.

Table 1.1: Volume parameters for atoms, groups, and fragments based on the method of Ye and Shreeve [48].

Species	Volume (Å^3)	Species	Volume (Å^3)
Neutral			
Imidazole	84	1,2,4-Triazole	79
Tetrazole	75	s-Triazine	90
1,2,4,5-Tetrazine	87	Pyrimidine	100
Cubane	135	Furazan	77
Benzene	110	Pyridine	105
Groups			
H (bonded to N)	7	H (bonded to C)	5
CH_3	30	CH_2 (acyclic)	24
CH_2 (three- or four-membered ring)	22.5	CH_2 (five- or six-membered ring)	22
CH_2 (eight-membered ring)	21	CH (in isowurtzitane)[a]	13.5
$-C=C-$	26.5	$-C=N-$	25
CN	30	$-N=N-$	26
NH_2	20	NO_2	36
NH	15	N_3	41
N (in tetraazapentalene)	9.5	N (in other cases)	10
C=O (not in a ring)	25	C=O (in a ring)	22
COOH	41	OH	15
O (in ether or $-O-NO_2$)	11.5	O (in other cases)	10
F	12.5	CF_2	37.5
NF_2	37	SF_5	82
Corrections			
Strong hydrogen bonds for each NH_2 or NH	-8		
Each sp^3 C or sp^3 N in two or more rings	-1		
Each sp^2 C in two or three rings	-2		

[a] Volumes of other CH moieties were derived from respective CH_2: $V(CH) = V(CH_2) - 5/\text{Å}^3$; while $V(C) = V(CH_2) - 10/\text{Å}^3$.
[b] Except in tetrazole where sp^2 C does not need to be corrected.

Example 1.1: *N,N*-Bis(2-fluoro-2,2-dinitroethyl)nitramide has the following structure:

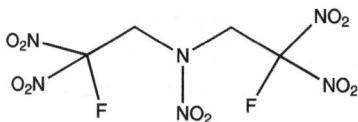

Molecular Weight: 334.11

The use of eq. (1.1) and Table 1.1 gives:

$$V = V(NO_2) \times 5 + V(F) \times 2 + V(CH_2, \text{acyclic}) \times 2 + V(C, \text{acyclic}) \times 2 + V(N)$$

$$= 36 \times 5 + 12.5 \times 2 + 24 \times 2 + (24 - 10) \times 2 + 10 = 291\text{Å}^3$$

$$\rho = \frac{Mw}{0.6022V} = \frac{334.10}{0.6022 \times 291} = 1.907 \, g/cm^3.$$

The measured X-ray density of this compound is 1.917g /cm^3 [48].

Ye and Shreeve [48] considered corrections of sp^3 C or sp^3 N in two or more fused rings as well as sp^3 C in two or more caged rings and sp^2 C in two or more rings, which are important for the design of high-energy-density materials (HEDMs).

Example 1.2: Consider the following three HEDMs:

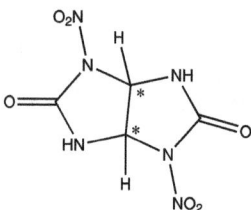

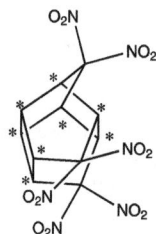

Molecular Weight: 232.11   Molecular Weight: 416.21   Molecular Weight: 388.21

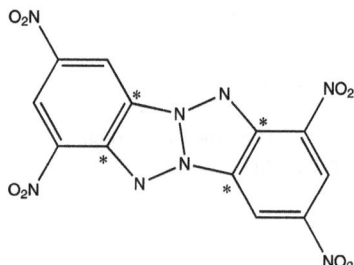

(a) GEMZAZ, aka DINGU   (b) JAJBEB   (c) PUTCEM, aka z-TACOT

Necessary corrections for sp^3 C, sp^3 N, and sp^2 C in these compounds are marked with an asterisk. The use of eq. (1.1) and Table 1.1 for three compounds gives:

(a) Volume of each asterisk CH needs to be corrected by −1 Å^3 in GEMZAZ:

$$V = V(NO_2) \times 2 + V(C=O, \text{in ring}) \times 2 + V(NH) \times 2 + V(N) \times 2 + V(CH) \times 2 - 1 \times 2$$

$$= 36 \times 2 + 22 \times 2 + 15 \times 2 + 10 \times 2 + (22 - 5) \times 2 - 2 = 198 \text{ Å}^3$$

$$\rho = \frac{Mw}{0.6022V} = \frac{232.11}{0.6022 \times 198} = 1.947 \, g/cm^3.$$

The measured X-ray density of GEMZAZ is 1.99 g/cm^3 [48].

(b) There are eight asterisk –CH groups in three rings of JAJBEB. Thus, the volume of each –CH must be corrected by -1 Å^3:

$$V = V(NO_2) \times 2 + V(C=O, \text{ in ring}) \times 2 + V(NH) \times 2 + V(N) \times 2 + V(CH) \times 2 - 1 \times 2$$

$$= 36 \times 2 + 22 \times 2 + 15 \times 2 + 10 \times 2 + (22 - 5) \times 2 - 2 = 380 \text{ Å}^3$$

$$\rho = \frac{Mw}{0.6022V} = \frac{416.21}{0.6022 \times 380} = 1.819 \text{ g/cm}^3.$$

The measured X-ray density of JAJBEB is 1.828 g/cm^3 [48].

(c) There are four sp^2 C in two rings of PUTCEM. Each requires correction of -2 Å^3 in volume:

$$V = V(\text{benzene}) \times 2 - V(H) \times 8 + V(NO_2) \times 4 + V(N, \text{ in tetraazapentalene}) \times 4 - 2 \times 4$$

$$= 110 \times 2 - 5 \times 8 + 36 \times 4 + 9.5 \times 4 - 2 \times 4 = 354 \text{ Å}^3$$

$$\rho = \frac{Mw}{0.6022V} = \frac{388.21}{0.6022 \times 354} = 1.821 \text{ g/cm}^3.$$

The reported X-ray density of PUTCEM is 1.830 g/cm^3 [48].

Ye and Shreeve [48] corrected the volume of each NH_2 or NH group by -8 Å^3 for strong bonding in three categories: (i) both carbons vicinal to the NH_2 or NH group have nitro groups or one C–NO$_2$ and one N-oxide (N–O). Due to electronic and/or steric effects, this condition cannot be applied if the NH_2 group was substituted by an alkyl, phenyl, or other electrondonating groups, that is, N-methyl-2,4,6-trinitrobenzenamine (JUPROB) or N-methyl-2,6-dinitro-4-(trifluoromethyl) benzenamine (FMANIL).

Example 1.3: The two $-NH_2$ groups of N^1-isopropyl-2,4,6-trinitrobenzene-1,3,5-triamine with the following molecular structure require correction for hydrogen bonding but there is no need to consider correction of NH group because it is attached to a isopropyl group (electron donating) on –NH.

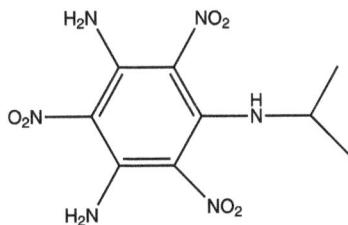

Molecular Weight: 300.23

The use of eq. (1.1) and Table 1.1 provides:

$$V = V(\text{benzene}) - V(H) \times 6 + V(NO_2) \times 3 + V(NH_2) \times 2 + V(NH) + V(\text{isopropyl})$$

$$- V(\text{hydrogen bonding}) \times 2$$

$$= 110 - 5 \times 6 + 36 \times 3 + 20 \times 2 + 15 + (30 \times 2 + 24 - 5) - 8 \times 2 = 306 \text{ Å}^3$$

$$\rho = \frac{Mw}{0.6022V} = \frac{300.23}{0.6022 \times 306} = 1.629 \text{ g/cm}^3.$$

The reported value of the X-ray density of this compound is 1.604 g/cm^3 [48].

(ii) There is no need to correct the volume of the amino group remains at 20 Å^3 in Table 1.1 when only one vicinal carbon bears a nitro group. Meanwhile, if the molecule has $C2$ symmetry, the volume correction of NH_2 groups by $-8\,\text{Å}^3$ should be considered. At least two nitro and two NH_2 groups exist in the molecule and in the vicinal position here. Three examples for this situation are 1,1-diamino-2,2-dinitroethene (FOX-7), 2,4,6-triamino-3,5-dinitropyridine (TIBMUM), and 2,6-diamino-3,5-dinitropyrimidine (CIWMAW01).

Example 1.4: FOX-7 has the following molecular structure:

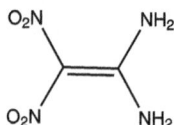

Molecular Weight: 148.08

The use of eq. (1.1) and Table 1.1 provides:

$$V = V(-C{=}C-) + V(NO_2) \times 2 + V(NH_2) \times 2 - V(\text{hydrogen bonding})$$

$$= 26.5 + 36 \times 2 + 20 \times 2 - 8 = 130.5 \text{Å}^3$$

$$\rho = \frac{Mw}{0.6022V} = \frac{148.08}{0.6022 \times 130.5} = 1.884\,\text{g/cm}^3.$$

The measured value of density of this compound is 1.883 g/cm^3 [48].

(iii) The volume of the molecule also needs a correction of $-8\,\text{Å}$ if some heterocycles (triazole, pyrazole, etc.) include an $-NHNO_2$ group and the heterocycle also has an acidic N–H on the ring, that is, 5-nitramino-1,2,4-triazole (NRTZ) and 3-nitramino-4,5-dinitro-pyrazole.

Example 1.5: 5-Nitramino-1,2,4-triazole (NRTZ) with the following molecular structure follows this condition.

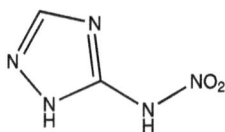

Molecular Weight: 129.08

The use of eq. (1.1) and Table 1.1 provides:

$$V = V(1,2,4-\text{triazole}) - V(H) + V(NO_2) + V(NH) - V(\text{hydrogen bonding})$$

$$= 79 - 5 + 36 + 15 - 8 = 117 \text{Å}^3$$

$$\rho = \frac{Mw}{0.6022V} = \frac{129.08}{0.6022 \times 117} = 1.832\text{g/cm}^3.$$

The reported value of density of NRTZ is 1.83 g/cm^3 [48].

Two further new group additivity models were introduced recently for some specific classes of energetic compounds, which have been illustrated here.

1.1.1 The method of atomic contributions

The method of atomic contributions (MACs) uses the specific molar volume as a sum of volumes of individual atoms constituting the crystal [49]. It can improve the accuracy of the calculations of the molecular crystal density, which contains both explosive and nonexplosive compounds. Thus, the density of molecular crystals is given as a ratio of molar mass to molar volume:

$$\rho = \frac{Mw}{\left(\sum_{A=C,N,O...} \sum_{i=1}^{n_i} B_i \frac{M_i}{\rho_i} + \sum_{j=4}^{n_2} L_j C_j N_j + \sum_{k=1}^{n_3} R_k D_k N_k\right)}, \tag{1.2}$$

where ρ is the density of the compound, Mw is the molecular weight of the compound, B_i is the number of ith atoms in one mole, M_i is the mass of ith atom, C_j is a correction for interactions in cycles, N_j is the number of atoms in a cycle of jth order, L_j is the number of cycles of jth order in one mole, D_k is a correction for atomic functional groups of kth type, N_k is the number of atoms in the group of k type, and R_k is the number of groups of k type in one mole. This method is more complex than the other group additivity methods. Smirnov et al. [49] used this approach for the calculation of the density of several nitramine compounds and energetic compounds containing the oxadiazole ring.

1.1.2 Benzene-derived energetic compounds using atomic volumes

Hofmann [50] reported that the crystal density of a neutral or ionic compound can be calculated by using the average atomic volumes of various elements as well as thermal expansion as follows:

$$\rho = \frac{Mw}{0.01387a + 0.00508b + 0.0118c + 0.01139d} \times 0.00164, \tag{1.3}$$

where a, b, c, and d are the number of carbon, hydrogen, nitrogen, and oxygen atoms; Mw is the molecular weight of the desired explosive. Ghule et al. [51] indicated that eq. (1.2) should be revised for those energetic compounds with strong H-bonding or with strong van der Waals or electrostatic interactions. It has been suggested to account for H-bonding between amino and oxygen-containing groups in group additivity methods because they underestimate the density [29, 52, 53]. Ghule et al. [51] found that the existence of two or more $-NH_2$ or $-NH_3^+$ groups in the molecule increases. They considered the contributions of these groups to improve the reliability of eq. (1.2) for those benzene-derived energetic compounds containing $-NH_2$ and/or $-NH_3^+$ groups. Thus, they introduced the following equation for estimation of densities of neutral nitrobenzenes, energetic salts, and cocrystals:

$$\rho = \frac{Mw}{0.01387a + 0.00508b + 0.0118c + 0.01139d} \times 0.00175. \tag{1.4}$$

This equation is the simplest approach for the calculation of densities of the mentioned classes of energetic compounds.

Example 1.6: 2-[Nitro(2,4,6-trinitrophenyl)amino]ethyl nitrate, 2,4,6-trinitrobenzene-1,3-diamine (DATB), and 2-isopropyl-6-methylpyrimidin-1,3-diium-4-olate picrate have the following structures:

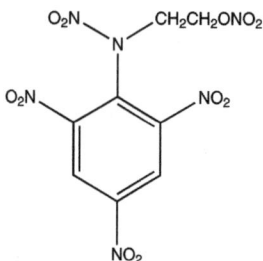

Chemical Formula: $C_8H_6N_6O_{11}$
Molecular Weight: 362.17

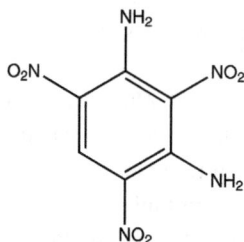

Chemical Formula: $C_6H_5N_5O_6$
Molecular Weight: 243.13

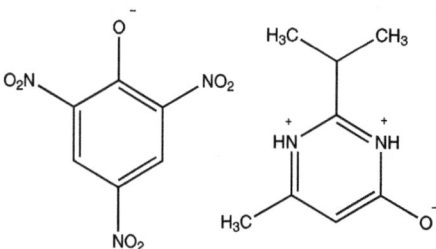

Chemical Formula: $C_{14}H_{15}N_5O_8$
Molecular Weight: 381.30

Equation (1.2) should be used for 2-[nitro(2,4,6-trinitrophenyl)amino]ethyl nitrate and 2-isopropyl-6-methylpyrimidin-1,3-diium-4-olate picrate because there is no $-NH_2$ and/or $-NH_3^+$ groups in their benzene rings:

$$\rho = \frac{Mw}{0.01387a + 0.00508b + 0.0118c + 0.01139d} \times 0.00164$$

$$= \frac{362.17}{0.01387 \times 8 + 0.00508 \times 6 + 0.0118 \times 6 + 0.01139 \times 11} \times 0.00164$$

$$= 1.760 g/cm^3$$

$$\rho = \frac{Mw}{0.01387a + 0.00508b + 0.0118c + 0.01139d} \times 0.00164$$

$$= \frac{381.30}{0.01387 \times 14 + 0.00508 \times 15 + 0.0118 \times 5 + 0.01139 \times 8} \times 0.00164$$

$$= 1.487 g/cm^3$$

The reported crystal densities for 2-[nitro(2,4,6-trinitrophenyl)amino]ethyl nitrate and 2-isopropyl-6-methylpyrimidin-1,3-diium-4-olate picrate are 1.75 [54] and 1.44 g/cm^3 [55], respectively.

Equation (1.3) should be used for DATB because there are two $-NH_2$ groups in its benzene ring:

$$\rho = \frac{Mw}{0.01387a + 0.00508b + 0.0118c + 0.01139d} \times 0.00175$$

$$= \frac{243.14}{0.01387 \times 6 + 0.00508 \times 5 + 0.0118 \times 5 + 0.01139 \times 6} \times 0.00164$$

$$= 1.803 \text{ g/cm}^3$$

The reported crystal density for DATB is 1.83 g/cm^3 [23].

1.1.3 The method of group additivity for estimating densities of energetic ionic liquids and salts

Ye and Shreeve [48] introduced a group additivity method for energetic ionic liquids and energetic ionic salts. They applied the method of the closed packed volume of single ions to estimate the density of 59 ionic compounds containing energetic ionic liquids and energetic ionic salts at room temperature. They developed volume parameters, which depend on the packing of the component ions as well as the size and shape of the ions and the ion–ion interactions. This approach can be used only for limited classes of energetic ionic liquids and energetic ionic salts because many volume parameters of groups and fragments were not defined. The existence of energetic groups such as $-NO_2$, $-N_3$, and $-CN$ can increase strong H-bond interactions in a desired energetic ionic liquid or energetic ionic salt. Ye and Shreeve [48] introduced the effective volumes of cations and anions for calculation of molecular volume (V) of room-temperature energetic ionic liquids and energetic ionic salts with general formula M_pN_q as follows:

$$V = pV^+ + qV^-, \tag{1.5}$$

where V^+ and V^- are the effective volumes of cations and anions, respectively. Equation (1.1) can be used to estimate the density. Table 1.2 shows volume parameters of groups and fragments for ionic liquids and salts at room temperature. As shown, the volume parameters for ionic liquids and salts are different for some cations and anions.

Table 1.2: Volume parameters of groups and fragments for ionic liquids and salts at room temperature.

Species	Volume (Å³)	Species	Volume (Å³)
Cations			
1,3-2H-Imidazolium (+)[a]	79	1,4-2H-1,2,4-Triazolium (+)	73
1,4-2H-Tetrazolium (+)	66	N-H-Pyridinium (+)[a]	95
Guanidinium (+)	69	Triaminoguanidinium (+)	105
Me_4N^+ [a]	113	Me_4P^+ [a]	133
HMTA-H(+)[b]	155	NH_4^+ [c]	21
NH_2NH_3 (+)	30	Azetidinium (+)	76
Anions			
Imidazolate (−)	93	1,2,4-Triazolate (−)	87
1-Tetrazolate (−)	80	$(NO_2)_3C$ (−)	141
Picrate (−)	218	NTO (−)[d]	123
NO_3^-	64	ClO_4^-	82
$N(NO_2)_2^-$	98	N_3^-	60
NO_2^-	55	CN^-	50
CF_3CO_2 (−)	108	CH_3SO_3 (−)	99
TfO^-	129	PF_6^-	107
BF_4^-	73	Tf_2N^- [a]	230
Br^-	56	Cl^-	47
Groups			
CH_3 [a]	30	C=O (not in a ring)	24
NH_2 [e]	20	N	10

[a]For ionic liquids: 1,3-dimethylimidazolium: 154; N-methylpyridinium: 146; Me_4N^+: 136; Me_4P^+: 163; 1,1-dimethylpyrrolidinium: 169; Tf_2N^-: 248; CH_2: 28; CH_3: 35; H 7 Å³. [b]HMTA, hexamethylenetetramine. [c]$V(NH_4^+)$ 15 Å³ for inorganic salts. [d]NTO: 3-nitro-1,2,4-triazol-5-one. [e]For very strong hydrogen bonds, $V(NH_2)$ 12 Å³.

Example 1.7: Consider the following energetic salt:

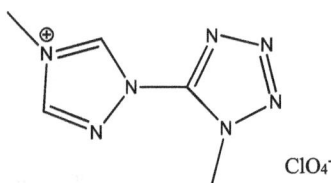

ClO₄⁻

The structure of the cation should be considered as follows:

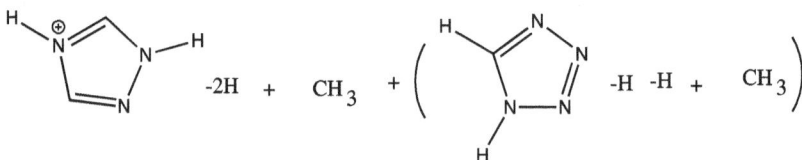

The use of eqs. (1.1) and (1.5) as well as Tables 1.1 and 1.2 gives:

$$V^+ = V(1,2,4-\text{triazolium}) - 2 \times V(\text{H, bonded to N}) + V(CH_3) + V(\text{tetrazole})$$
$$- V(\text{H, bonded to C}) - V(\text{H, bonded to N}) + V(CH_3)$$
$$= 73 - 2 \times 7 + 30 + 75 - 5 - 7 + 30 = 182\,\text{Å}^3$$
$$V^- = V(ClO_4^-) = 82\,\text{Å}^3$$
$$V = V^+ + V^- = 182 + 82 = 264\,\text{Å}^3$$
$$\rho = \frac{Mw}{0.6022V} = \frac{265.61}{0.6022 \times 264} = 1.671\,\text{g/cm}^3.$$

The reported crystal density for this energetic salt is 1.678 g/cm^3 [48].

1.2 Quantum mechanical approach

1.2.1 Quantum mechanical approach for neutral energetic compounds

Quantum mechanical computations require high-speed computers to conduct complicated calculations, and can be used for energetic molecules with simple molecular structures [18, 20–22, 56]. Molecular surface electrostatic potential (MESP) was introduced by Politzer et al. [22, 57] and accounted for the intermolecular interactions in the crystal. It has been widely used in recent years to estimate the density of many classes of $C_a H_b N_c O_d$ energetic compounds. The effectiveness of different MESP-based methods has been advanced to improve the reliability of estimations [18, 20–22, 56, 58–63]. Rice et al. [62] reviewed different quantum mechanical approaches for the prediction of crystal densities of neutral molecular and ionic molecular crystals. Qiu et al. [21] introduced a very direct approach to estimate the crystal densities of nitramines based on the ratio of the molecular mass and the volume of the isolated gas phase molecule. Rice et al. [20] developed the method of Qiu et al. [21] for primarily nitroaromatics, nitramines, nitrate esters, and nitroaliphatics. They presented a method for estimating the densities of neutral and ionic molecular crystals using the quantum-mechanically determined molecular volume of an isolated molecule or formula unit within the crystal as

$$\rho = \frac{M}{V_{\mathrm{m}}}, \tag{1.6}$$

where ρ is the crystal density of energetic compound; M is the molecular mass of the molecule in g/molecule and V_{m} is the volume inside the 0.001 a. u. isosurface of electron density surrounding the molecule which is calculated using density functional theory (DFT) at the B3LYP/6–31G** level with the Gaussian program package. Politzer et al. [22] improved the procedure presented by Rice et al. [20] by adding corrections for electrostatic interactions, in order to better represent the intermolecular interactions in both neutral and ionic molecular crystals. Rice et al. [62] used a suitable DFT method based on the Politzer et al. method [22] for calculating the density of energetic compounds through the interaction index $v\sigma_{tot}^2$ as follows:

$$\rho = \alpha_1 \left(\frac{M}{V_m} \right) + \beta_1 \left(v\sigma_{tot}^2 \right) + \gamma_1, \tag{1.7}$$

where ρ is in g/cm^3; σ_{tot}^2 and v are the total variance of the electrostatic potential on the 0.001 a. u. molecular surface and the degree of balance between the positive and negative potentials on the molecular surface, respectively. The parameter v calculates the degree of balance between the positive and negative potentials on the molecular surface. The parameter $v\sigma_{tot}^2$ has a significant contribution to analytical relations of condensed phase properties which depends on intermolecular interactions. For the model of Politzer et al. [22], the values of three parameters α_1, β_1, and γ_1 are 0.9183, 0.0028, and 0.0443, respectively. Meanwhile, the values of these parameters are $\alpha_1 = 1.0462$, $\beta_1 = 0.0021$, and $\gamma_1 = -0.1586$ for the model of Rice–Byrd [62]. Politzer et al. [22] optimized the structure and the surface properties at the B3PW91/6–31G(d,p) level whereas Rice–Byrd [62] applied the B3LYP/6–31G** level for calculations. Wang et al. [64] compared the computed densities for 31 aliphatic nitrates at room temperature using the DFT method (B3LYP) in combination with 6 basis sets (3–21G, 6–31G, 6–31G*, 6–31G**, 6–311G*, and 6–31+G**) and the semiempirical molecular orbital method (PM3). They recommended the B3LYP/6–31G* method because it provided more reliable results to predict the crystalline densities of organic nitrates as compared to those obtained by the QSPR-based method of Keshavarz and Pouretedal [28] as well as MESP-based method Rice–Byrd [62]. Wang et al. [65] used experimental data of 3694 nitro compounds to introduce a molecular morphology descriptor and a hydrogen-bond descriptor as correction items of eq. (1.5) to build three new density-functional theory (DFT)-QSPR models. For about 91–93 % of nitro compounds, the percent of deviations of the predicted results are less than 5 % at two levels of B3PW91/6–31G(d,p) and B3LYP/6–31G**. Details of the coefficients α_1, β_1, and γ_1 (which can be found by least square fitting with experimental data), are given elsewhere [62]. For example, the calculated crystal densities of some newly designed derivatives of tetrazole, namely, 5,5'-((1Z,5Z)-3,4-dinitrohexaaza-1,5-diene-1,6-diyl)bis(1-nitro-1H-tetrazole),5,5'-((1Z,5Z)-3,4-diaminohexaaza-1,5-diene-1,6-diyl)bis(1-nitro-1H-tetrazole),5,5'-((1Z,5Z)-3,4-dinitrohexaaza-1,5-diene-1,6-diyl) bis(1H-tetrazol-1-amine),3,3'-dinitro-3,3a,3',3'a-tetrahydro-7H,7'H-[6,6'-bitetrazolo[1,5-e]pentazine]-7,7'-diamine,3,3'-,7,7'-tetranitro-3,3a,3',3'a-tetrahydro-7H,7'H-6,6'-bitetrazolo[1,5-e] pentazine are 1.95, 1.84, 1.86, 1.90, and 1.92 g/cm^3, respectively [35], which have high energy content [66]

Nirwan et al. [67] compared the reliability of the best available MESP-based methods to calculate the density of seven groups such as energetic compounds containing nitrate-esters, nitramines, and azides groups as well as energetic materials including benzene, caged, strained, heterocyclic backbone, and fused rings. They computed densities for 221 $C_aH_bN_cO_d$ explosives of different chemical nature and functional groups by these MESP-based methods and compared their outputs with the measured values. They indicated that Politzer et al. [22] as well as Rice and Byrd (Rice–Byrd) [62] methods can provide more reliable predictions as compared to the other MESP-based methods.

The crystal packing method accounts for the molecular interactions and conformation at a modest computational cost. It can also be used to predict the density of energetic compounds. A particular force field cannot be handled for all classes of energetic compounds. The option of force field depends on the type of atoms, hybridization, and chemical bonds in the desired molecule. Ghule and Nirwan [68] used various force fields in the crystal packing to evaluate their role for calculations of 68 energetic materials including aromatic and nonaromatic backbone with explosophoric groups. They predicted densities with the Dreiding force field for 84% energetic compounds with the nonaromatic backbone that gave deviation within 5%. For 83% of energetic materials including nonaromatic backbone, the estimated densities with the polymer consistent force field were found within 4% deviation. Thus, the selection of force field is essential for precise density estimation.

Lal et al. [69] introduced a streamlined and efficient method that improves density prediction, enabling faster computational discovery of new materials. By refining Politzer's approach with validated B3LYP quantum calculations and Multiwfn's precise structural analysis [70], the protocol significantly outperforms traditional methods. This advancement supports high-fidelity virtual screening, accelerating the development of novel, high-performance energetic materials.

Lal et al. [69] categorized compounds into two groups based on chemical characteristics – Group I (with strain and nitro functionalities):

$$\rho = 1.0330 \left(\frac{M}{V_m}\right) + 1.836 \times 10^{-3} \left(\sigma_{tot}^2 v\right) - \frac{v}{6}, \tag{1.8}$$

and Group II (without strain and nitro functionalities):

$$\rho = 0.9568 \left(\frac{M}{V_m}\right) + 1.836 \times 10^{-3} \left(\sigma_{tot}^2 v\right) - \frac{v}{6}, \tag{1.9}$$

where M and V_m are calculated using Multiwfn [70].

This refined computational framework not only enhances accuracy but also establishes a versatile platform for discovering next-generation energetic materials. Future efforts could focus on expanding datasets to further improve robustness and efficiency, paving the way for even more precise and innovative material design.

1.2.2 Energetic ionic liquids and salts as room-temperature energetic materials

For ionic molecular crystals, Rice et al. [62] reparameterized the equation of Politzer et al. [22] as follows:

$$\rho = \alpha \frac{M}{V_m} + \beta \sum \left(\frac{\bar{V}_s^+}{A_s^+}\right) + \gamma \sum \left(\frac{\bar{V}_s^-}{A_s^-}\right) + \delta, \tag{1.10}$$

where the contributions from every ionic molecule in the formula unit are summed over the two ionic contribution terms. In this equation, $\bar{V}_s^+$ is the average of the positive values of V_s and A_s^+ is the portion of the cation surface that has a positive electrostatic potential. Furthermore, $\bar{V}_s^-$ is the average of the negative values of V_s and A_s^- is the portion of the anion surface that has a negative electrostatic potential. Rice et al. [62] obtained the values of α, β, γ, and δ through best fit parameters and have the following values: 1.1145, 0.02056 g Å^2/(cm^3 kcal/mol), −0.0392 g Å^2/(cm^3 kcal/mol), and −0.1683 g/cm^3, respectively.

1.3 Empirical methods for the calculation of the crystal density of different classes of energetic materials

Empirical or QSPR methods based on molecular structures are important because they are easy, effective, and precise methods for the density prediction of $C_a H_b N_c O_d$ explosives. The computer code EMDB_1.0 [71] can calculate densities of organic compounds containing common energetic functional groups including nitro (−NO$_2$), nitrate (−ONO$_2$), and nitramine (−NNO$_2$) using the best available QSPR methods. For some classes of energetic compounds, the reliability of these simple QSPR methods is higher than the best available group additivity approaches [5].

Some QSPR models based on complex molecular descriptors have also been developed in recent years. The molecular structures of 26 energetic cocrystals have been correlated with their densities by the ANN and MLR analysis models [72] through three complex molecular descriptors. Two methods of ANN and MLR with five complex molecular descriptors have also been used to predict the density of 172 polynitroarenes, polynitroheteroarenes, nitroaliphatics, nitrate esters, and nitramines at room temperature [73]. It was found in these works that the ANN model can give a more reliable prediction as compared to the MLR model.

Different simple empirical methods have recently been introduced which enable the reliable prediction of the crystal density of important classes of energetic compounds at room temperature. These methods are reviewed here.

1.3.1 Nitroaromatic energetic compounds

It has been shown that the following general equation is suitable for most nitroaromatic high explosives [27]:

$$\rho' = \frac{10.57a + 0.1266b + 30.38c + 35.18d}{Mw}, \tag{1.11}$$

where ρ' is the uncorrected crystal density in g/cm^3 (it can also be corrected for some compounds in which the molecular structure or intermolecular forces can result in a reduction or expansion of the volume of the compound), a, b, c, and d are the number of carbon, hydrogen, nitrogen, and oxygen atoms respectively, and Mw is the molecular weight of the desired nitroaromatic energetic compound. This equation provides core correlation for estimation of the crystal density for a large number of nitroaromatic explosives. The corrected crystal density correlations can be expressed for some nitroaromatic energetic compounds according to the following:

(a) attachment of $-N_3$ or $-N_2^+$ groups to an aromatic ring,

$$\rho = -0.0238 + 0.9615\rho', \tag{1.11a}$$

(b) presence of positive and negative charges on nitrogen:

$$\rho = 1.3022 + 0.3261\rho', \tag{1.11b}$$

(c) existence of N-oxide in a heterocyclic aromatic structure:

$$\rho = 1.3958 + 0.2657\rho', \tag{1.11c}$$

(d) attachment of an aromatic ring ($-Ar$) to $-OR$ or $-OAr'$ where R and Ar' are alkyl and aromatic groups, respectively:

$$\rho = -0.5139 + 1.2532\rho', \tag{1.11d}$$

(e) the attachment of more than two $-OH$ or $-NH_2$ groups to the aromatic ring:

$$\rho = 1.1024\rho', \tag{1.11e}$$

(f) attachment of one $-OH$ or two $-OH$, or two $-NH_2$ groups to an aromatic ring:

$$\rho = 0.2332 + 0.8872\rho', \tag{1.11f}$$

(g) direct attachment of an aromatic ring to another aromatic ring (Ar–Ar):

$$\rho = -0.2164 + 1.093\rho'. \tag{1.11g}$$

For any nitroaromatic compounds, the order in which the equations should be applied is from eqs. (1.11a) to (1.11g).

Example 1.8: Equation (1.11a) should be used rather than eq. (1.11f) to calculate the crystal density of diazodinitrophenol (with the following molecular structure) because of the presence of the $-N_2^+$ group:

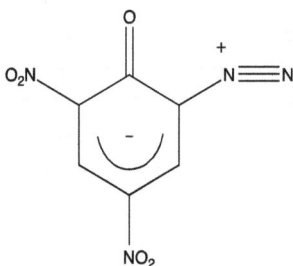

Thus, the crystal density of diazodinitrophenol ($C_6H_2N_4O_5$) is calculated as follows:

$$\rho' = \frac{10.57a + 0.1266b + 30.38c + 35.18d}{Mw}$$

$$= \frac{10.57(6) + 0.1266(2) + 30.38(4) + 35.18(5)}{210.1}$$

$$= 1.720 g/cm^3$$

$$\rho = -0.0238 + 0.9615\rho'$$

$$= -0.0238 + 0.9615(1.720)$$

$$= 1.630 \, g/cm^3.$$

The calculated value is the same as the measured value of 1.63 g/cm^3 [54].

1.3.2 Acyclic and cyclic nitramines, nitrate esters, and nitroaliphatic compounds

A study of nitrate esters and nitroaliphatic systems shows that the atomic composition and number of special functional groups can be inserted into an empirical formula to predict the uncorrected crystal density of these compounds according to the following equation [26]:

$$\rho' = \frac{47.97a - 19.29b + 26.53c + 26.00d - 25.32n_{COO} - 0.6358n_O + 11.54n_{OH}}{Mw}, \tag{1.12}$$

where n_{COO}, n_O, and n_{OH} are the number of ester, ether, and alcohol functional groups, respectively.

The corrected crystal densities for some particular examples of acyclic and cyclic nitramines, nitrate esters, and nitroaliphatic compounds can be obtained based on the molecular structure as follows:

(1) Mononitroalkanes:

$$\rho = 0.4170 + 0.5970\rho'. \tag{1.12a}$$

(2) The attachment of two $-NO_2$ groups to one carbon (which has no additional $-$ COO$-$, $-$O$-$, or $-$OH functional groups):

$$\rho = 0.1233 + 0.8373\rho'. \tag{1.12b}$$

(3) The attachment of three $-NO_2$ groups to one carbon:

$$\rho = \begin{cases} 3.033 - \rho' & \text{if } n_{CH_2} \geq 1.5 n_{NO_2}, \\ -0.3788 + 1.2569\rho' & \text{if } n_{CH_2} \leq 0.6 n_{NO_2}. \end{cases} \tag{1.12c}$$

For those molecules that satisfy both conditions (2) and (3), eq. (1.12c) rather than (1.12b) should be used.

(4) For nitrate compounds (without cyclic ring attachment), if $n_{CH_2ONO_2} + n_{CHONO_2} \geq 4$ then

$$\rho = 0.1745 + 0.9235\rho'. \tag{1.12d}$$

(5) Cage and cyclo nitro compounds in which only one $-NO_2$ (not more) is attached to a carbon atom:

$$\rho = -0.0515 + 0.9142\rho'. \tag{1.12e}$$

For cyclic nitramines and nitramine compounds in which N$-NO_2$ is attached to an aromatic ring (such as in tetryl), as well as polycyclic energetic compounds that contain no more than one oxygen atom in the ring, the crystal density can be calculated by

$$\rho = \frac{13.15a - 5.303b + 39.72c + 29.34d}{Mw}. \tag{1.13}$$

For acyclic nitramines, the following correlation is used:

$$\rho = \frac{66.86a - 27.37b + 52.96c + 12.81d}{Mw}. \tag{1.14}$$

Example 1.9: The calculated density of octanitrocubane ($C_8N_8O_{16}$) can be obtained as follows:

$$\rho' = \frac{47.97a - 19.29b + 26.53c + 26.00d - 25.32n_{COO} - 0.6358n_O + 11.54n_{OH}}{Mw}$$

$$= \frac{47.97(8) - 19.29(0) + 26.53(8) + 26.00(16) - 25.32(0) - 0.6358(0) + 11.54(0)}{464.1}$$

$$= 2.181 \, g/cm^3$$

$$\rho = -0.0515 + 0.9142\rho'$$

$$= -0.0515 + 0.9142(2.181)$$

$$= 1.943 \, g/cm^3.$$

The calculated value is close to the measured value which is reported to be $1.979 \, g/cm^3$ [74].

1.3.3 Improved method for the prediction of the crystal densities of nitroaliphatics, nitrate esters, and nitramines

For various nitroaliphatics, nitrate esters, and nitramines, a more reliable general correlation for predicting the densities of acyclic and cyclic nitramines, nitrate esters, and nitroaliphatic compounds than correlations eqs. (1.12), (1.13), and (1.14) was established, and can be written as follows [28]:

$$\rho = 1.521 + \frac{6.946a - 11.53b + 20.10c}{Mw} - 0.1559E_D + 0.1325E_I, \tag{1.15}$$

where E_I and E_D are specific structural parameters that can increase or decrease the value of the crystal density, and the values of which are based on the molecular structure as follows:

(1) Nitroaliphatics and nitrate esters
   (a) $C(H)_{4-n}(NO_2 \text{ or } ONO_2)_n$: For nitro or nitrate derivatives of methane, the values of E_D and E_I are 1.70 and 0.0, respectively.
   (b) $C_nH_{2n+1}(NO_2 \text{ or } ONO_2)$: The values of E_D and E_I depend on the number of carbon atoms in the alkyl substituents of mononitro- or mononitrate-alkanes:
      (i) if $n = 2$, then $E_D = 1$ and $E_I = 0$;
      (ii) if $n = 3$, then $E_D = 0.5$ and $E_I = 0$;
      (iii) if $n = 4$, then $E_D = 0$ and $E_I = 0$;
      (iv) if $n = 5$, then $E_D = 0$ and $E_I = 0.5$;
      (v) if $n \geq 6$, then $E_D = 0$ and $E_I = 0.75$.
   (c) $C_nH_{2n}(NO_2 \text{ or } ONO_2)_2$: The values of E_D depend on the position of attachment of the nitro or nitrate groups in dinitro- or dinitrate-alkanes:
      (i) if two nitro or nitrate groups are attached to the same $-CH_2-$ group, $E_D = 1.5$ and $E_I = 0$;

(ii) if two nitro or nitrate groups are attached to one >CH– group, $E_D = 1.0$ and $E_I = 0$;

(iii) for the other cases, $E_D = 0.75$ and $E_I = 0$.

(d) If the OH group is present in nitro or nitrate compounds: $E_I = 0.4$ and $E_D = 0.0$.

(e) Nitrate compounds without OH groups:

(i) $E_I = 1.0$ and $E_D = 0$ for $(CH_2ONO_2)_4C$;

(ii) $E_I = 0.5$ and $E_D = 0$ for compounds that contain $(CH_2ONO_2)_3$- or two $(CH_2ONO_2)_2$-fragments.

(2) Nitramines

(a) $(C_nH_{2n+1})_2NNO_2$:

(i) if $n = 1$, then $E_D = 1.5$ and $E_I = 0.0$;

(ii) if $n = 2$, then $E_D = 0.5$ and $E_I = 0.0$;

(iii) if $n = 3$, then $E_D = 0.0$ and $E_I = 0.0$;

(iv) if $n \geq 4$, then $E_D = 0.0$ and $E_I = 0.75$.

(b) The presence of specific molecular entities:

(i) if an aromatic ring is present then $E_D = 0.5$ and $E_I = 0.0$;

(ii) if the NH–NO$_2$ functional group is present then $E_D = 1.5$ and $E_I = 0.0$;

(iii) for cyclic ethers $E_I = 1.0$ and $E_D = 0.0$;

(iv) if the –C(NO$_2$)$_3$ group is present then $E_I = 0.5$ and $E_D = 0.0$.

Example 1.10: The value of the calculated density for $O_2NOCH_2CH_2ONO_2$ can be obtained as follows:

$$\rho = 1.521 + \frac{6.946a - 11.53b + 20.10c}{Mw} - 0.1559E_D + 0.1325E_I$$

$$= 1.521 + \frac{6.946(2) - 11.53(4) + 20.10(2)}{152.1} - 0.1559(0.75) + 0.1325(0)$$

$$= 1.457 \, g/cm^3.$$

It should be pointed out that this compound follows part (iii) of condition (1) (c). The estimated value is close to the measured value which is 1.48 g/cm^3 [23].

1.3.4 Reliable correlation for the prediction of the crystal densities of polynitroarenes and polynitroheteroarenes

From studying the crystal densities of various polynitroarenes and polynitroheteroarenes, it has been shown that it is possible to correlate the crystal density with the elemental composition, as well as to establish the positive and negative contributions of some specific structural parameters using the following equation [29]:

$$\rho = -1.609 + \frac{29.20a + 1.515b + 53.06c + 61.30d}{Mw} + 0.0703C_{PG} - 0.0751C_{NG}, \qquad (1.16)$$

where C_{PG} and C_{NG} are the positive and the negative contributions of some specific structural fragments, which can be specified according to the following.

1.3.4.1 Prediction of C_{PG}

(1) The presence of $n_{OH} \geq 1$ or $n_{NH_x} \geq 2$ without extra further substituents such as the methyl group:

$$C_{PG} = \begin{cases} 1.0 & \text{if } n_{NO_2} - n_{OH} = 1 \text{ or } n_{NO_2} - n_{NH_x} \geq 1; \\ 2.0 & \text{if } n_{NO_2} - n_{NH_x} = 0; \\ 0.5 & \text{if } n_{NO_2} - n_{OH} > 1. \end{cases}$$

(2) For polynitrobenzene compounds containing a center of symmetry, or polynitro-heteroarenes with substituent N-oxide or the explosive containing more than two ⟨structure⟩ groups, $C_{PG} = 1.0$.

(3) For explosives that contain positive and negative charges on nitrogens, such as tetranitrodibenzo tetraazapentalene (TACOT), $C_{PG} = 2.0$.

(4) For polynitroarenes containing the ⟨structure⟩ group, $C_{PG} = 3.0$.

1.3.4.2 Estimation of C_{NG}

(1) If the nitrate group is present:

$$C_{NG} = \begin{cases} 1.0 & \text{for } n_{NO_3} = 1 \\ 2.0 & \text{for } n_{NO_3} \geq 2. \end{cases}$$

(2) For polynitroaromatics with $-N_3$ or $-N_2$ substituents, or polynitroarenes containing more than two alkyl substituents, the value of C_{NG} is 1.0.

(3) If a polynitroarene cycle with only nitro substituents is directly attached to another $-Ar$, $-OR$, and $-OAr$ or $-NHAr$ (where R and Ar are alkyl and aromatic groups), the values of C_{NG} are 1.5, 1.0, and 0.75, respectively.

(4) For polynitroheteroarenes containing amino groups and in which more than two heteroatoms are present per ring, the value of C_{NG} is 1.0.

Example 1.11: 1,3,5-Triazido-2,4,6-trinitrobenzene has the following structure:

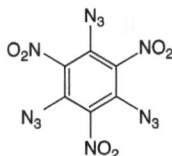

The use of eq. (1.16) gives

$$\rho = -1.609 + \frac{29.20a + 1.515b + 53.06c + 61.30d}{Mw} + 0.0703C_{PG} - 0.0751C_{NG}$$

$$= -1.609 + \frac{29.20(6) + 1.515(0) + 53.06(12) + 61.30(6)}{336.1} + 0.0703(0) - 0.0751(1)$$

$$= 1.826\,g/cm^3.$$

The measured crystal density is 1.805 g/cm^3 [54]. Thus, the percent of deviation (%Dev) of the new method from the measured value is 1.24. The calculated crystal density which is obtained by the group additivity method of Ammon [24] is 1.630 g/cm^3 (% Dev = −9.70).

1.3.5 The extended correlation for the prediction of the crystal density of energetic compounds

It was found that it is possible to establish a general correlation to predict the crystal density of different classes of energetic compounds including various polynitroarenes, polynitroheteroarenes, nitroaliphatics, nitrate esters, and nitramines as follows [75]:

$$\rho = 1.753 + \frac{-10.24b + 9.908c}{Mw} + 0.0992IMP - 0.0845DMP, \tag{1.17}$$

where *IMP* and *DMP* are two correcting functions that depend on intermolecular interactions for increasing and decreasing the second and third terms in eq. (1.17), respectively. For different classes of energetic compounds, the values of *IMP* and *DMP* can be specified based on the molecular structure according to the rules which are outlined in the following subsections.

1.3.5.1 Structural parameters affecting *IMP*

(1) −OH or −ONH$_4$: The value of *IMP* is 1.0. This condition cannot be applied for the attachment of further alkyl groups to polynitroarenes.

(2) The attachment of both −NH$_x$ and nitro groups to carbocyclic aromatic molecular fragments: For those compounds that follow the conditions $1 \leq n_{NO_2} - n_{NH_x} < 0$ and $n_{NO_2} - n_{NH_x} = 0$, the values of *IMP* are 0.9 and 1.8, respectively.

(3) The presence of $\overset{+}{N}-\overset{-}{O}$, [structure] or more than two [structure] groups: If these molecular fragments are present, the value of *IMP* is 1.0, except if the –NH– group is present between two aromatic rings, i. e. Ar–NH–Ar.

(4) Cyclic nitramines: The value of *IMP* is 1.0.

(5) $(CH_2ONO_2)_4C$ and two groups $(CH_2ONO_2)_3$–: if the $(CH_2ONO_2)_4C$ or two groups $(CH_2ONO_2)_3$– are present, the values of *IMP* are 1.5 and 0.5 respectively.

(6) Nonaromatic cyclic compounds containing nitro groups: The value of *IMP* equals 0.9.

1.3.5.2 Structural moieties affecting *DMP*

(1) Nitrobenzenes: For nitrobenzenes which contain only one benzene ring, $DMP = 0.5$, except for those compounds in which $b = 0$. If the two nitrobenzene rings are connected to each other through an –O– or –N=N– group, the values of *DMP* are 2.3 and 1.0, respectively.

(2) $CH_{4-n}(NO_2 \text{ or } ONO_2)_n$ or the presence of the –N_3 group: The value of *DMP* is 3.3.

(3) $C_nH_{2n+1}(NO_2 \text{ or } ONO_2)_2$:
 (a) if two nitro or nitrate groups are attached to the same carbon atom, $DMP = 2.2$;
 (b) if two nitro or nitrate groups are attached to two different carbon atoms, $DMP = 1.5$.

(4) Nonaromatic nitro and nitrate compounds in which –NO_2 or –ONO_2 groups are attached to a –CH_2– group: The value of *DMP* is 0.6 for the following conditions:
 (a) additional oxygen atoms are present in addition to those present in the –NO_2 or –ONO_2 groups in those energetic compounds with general formula R–CH_2–(NO_2 or ONO_2) or R–CH_2–(NO_2 or ONO_2)$_2$;
 (b) more than two –NO_2 or –ONO_2 groups are present.

1.3.5.3 Structural moieties affecting both *DMP* and *IMP*

(1) $C_{n+1}H_{2n+3}(NO_2 \text{ or } ONO_2)$: The values of *IMP* and *DMP* depend on the number of carbon atoms in the alkyl substituents:
 (a) if $n = 1$, then $DMP = 1.9$;
 (b) if $n = 2$, then $DMP = 1.0$;
 (c) if $n = 3$, then $DMP = 0.0$;
 (d) if $n \geq 4$, then $IMP = 1.0$.

(2) $(C_nH_{2n+1})_2NNO_2$: The number of carbon atoms is important for the prediction of *IMP* and *DMP* in acyclic nitramines:
 (a) if $n = 1$, then $DMP = 2.1$;
 (b) if $n = 2$, then $DMP = 0.0$;
 (c) if $n \geq 3$, then $IMP = 1.4$.

This method is more complex than previous methods.

Example 1.12: 1,3,3-Trinitroazetidine (TNAZ) is a melt-cast explosive with the following molecular structure:

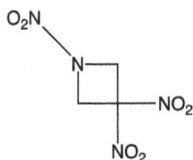

The present method can be applied as follows:

$$p = 1.753 + \frac{-10.24b + 9.908c}{Mw} + 0.0992\text{IMP} - 0.0845\text{DMP}$$

$$= 1.753 + \frac{-10.24(4) + 9.908(4)}{192.1} + 0.0992(1) - 0.0845(0)$$

$$= 1.845 \, g/cm^3.$$

The predicted value is close to the measured value of 1.84 g/cm³ [74].

1.3.6 Energetic azido compounds

It has been shown that the molecular structures of energetic azido compounds can be used to evaluate the density using the following correlation [76]:

$$p = 1.2 + 0.01c + 0.26\frac{d}{a} - 0.1\frac{a}{c} + 0.04\rho^+_{\text{azide}} - 0.52\rho^-_{\text{azide}}. \tag{1.18}$$

The two correcting parameters ρ^+_{azide} and ρ^-_{azide} are defined on the basis of the presence of certain molecular fragments as follows:

(1) The value of ρ^-_{azide} is 0.1 if two substituents of the following form are present:

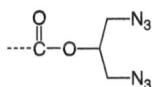

(2) The value of ρ^-_{azide} is 0.5 if the following molecular moiety is present in the compound, where R is an alkyl group:

(3) The value of ρ^+_{azide} is 1.0 if the following fragment is present:

(4) The value of ρ_{azide}^+ is 2.0 if the $-C(NO_2)-CH_2-O-$ fragment is present.
(5) If more than two $-NNO_2$ groups are present, the values of ρ_{azide}^+ are 0.5 and 2.0 for acyclic and cyclic nitramines, respectively.

Example 1.13: Using this method to calculate the density of 1,3-diazido-2-nitro-2-azapropane with the following structure gives

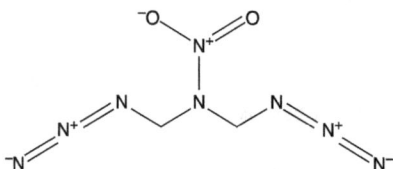

$$\rho = 1.2 + 0.01c + 0.26\frac{d}{a} - 0.1\frac{a}{c} + 0.04\rho_{azide}^+ - 0.52\rho_{azide}^-$$

$$= 1.2 + 0.01(8) + 0.26\frac{2}{2} - 0.1\frac{2}{8} + 0.04(0) - 0.52(0)$$

$$= 1.52\,g/cm^3.$$

If the complex quantum mechanical method is used, a value of 1.45 g/cm^3 is obtained [76], while the experimental determined value is 1.43 g/cm^3 [77].

1.3.7 High-nitrogen-content organic compounds

HEDMs play a critical role in advancing research programs worldwide. These materials are designed to deliver superior performance while maintaining lower sensitivity to external stimuli like impact, shock, friction, and electrostatic discharge [2, 3]. Nitrogen-rich compounds with high density are particularly desirable as propellants because they enhance the efficiency of the combustion process [7].

Many well-known high-performance explosives, such as RDX (hexahydro-1,3,5-trinitro-1,3,5-triazine), HMX (octahydro-1,3,5,7-tetranitro-1,3,5,7-tetrazocine), and CL-20 (2,4,6,8,10,12-hexanitro-2,4,6,8,10,12-hexaazaisowurtzitane), contain up to 38% nitrogen by weight. However, researchers have also explored high-nitrogen-content organic compounds, which are defined as materials containing more than 50% nitrogen by weight in their molecular structure [2, 3]. Some of these compounds can reach nitrogen contents as high as 88%, incorporating nitrogen-rich derivatives of triazole, tetrazole, triazine, tetrazine, furazan, and nitrogen-containing chains.

Most high-nitrogen organic materials are built around five- or six-membered nitrogen-rich rings or chains. These structures often feature various functional groups, such as $-NH_2$, $-NO_2$, $-N_3$, $-C\equiv N$, $-N=N-$, $-C=N-$, and $-N=N^+O^-$, which contribute to their energetic properties. Due to their unique characteristics, these compounds are considered excellent candidates for applications in insensitive high explosives, gun propellants propellants, and clean gas generators used in vehicle airbags [2, 3]. Their combination of high energy, stability, and reduced sensitivity makes them highly valuable in modern energetic materials research.

Two new methods have been devised to predict the density of high-nitrogen-content organic compounds at or near room temperature. These approaches target compounds that feature nitrogen-rich components, including triazoles, tetrazoles, triazines, tetrazines, furazans, and certain organic chains with more than 50% nitrogen by weight. The details and applications of these methods are explained in the following sections.

1.3.7.1 The reliance solely on molecular fragments

It was discovered that the presence of certain specific functional groups and molecular fragments can be utilized to predict the density of high-nitrogen-content organic compounds. This approach is outlined as follows [78]:

$$\rho = 2.278 - 10.58 \left(\frac{a}{Mw}\right) - 4.775 \left(\frac{b}{Mw}\right) - 6.763 \left(\frac{c}{Mw}\right) + 0.1188 \rho^+_{non-add} - 0.10468 \rho^-_{non-add},$$

(1.19)

where $\rho^+_{non-add}$ and $\rho^-_{non-add}$ represent two nonadditive correcting functions. These functions are used to adjust the predicted density of a high-nitrogen compound by either increasing ($\rho^+_{non-add}$) or decreasing ($\rho^-_{non-add}$) it, depending on the specific characteristics of the compound. Different values for these correcting functions are provided in Table 1.3.

Table 1.3: Values of $\rho^+_{non-add}$ and $\rho^-_{non-add}$.

Category			$\rho^-_{non-add}$		
	Molecular moieties	$\rho^-_{non-add}$	Examples	Comments	
1	−N(R)(NO$_2$ or NO)	1.2	O$_2$N—N—CH$_3$	–	
2	−CN	0.5	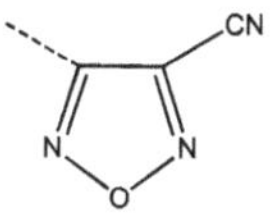 CN	The attachemnt of −CN to heterocyclic ring	
3	−OR	1.1	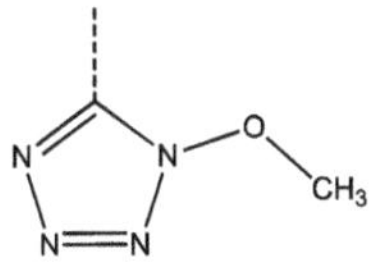 O—CH$_3$	The attachment of −OR to tetrazole ring	

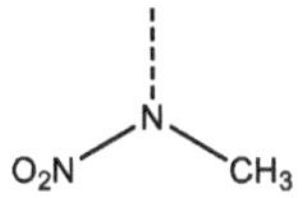

Table 1.3 (continued)

Category				$P^-_{non-add}$	
	Molecular moieties	$P^-_{non-add}$	Examples		Comments
4	$-N_3$	1.3			The attachment of three $-(CH_2)_n-N_3$ to tertiary carbon
5		0.5			The attachment of more than two $-(CH_2)_n-N_3$ to heterocyclic rings

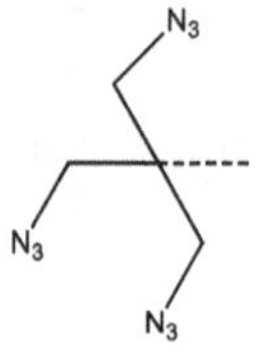

Category				$P^+_{non-add}$	
	Molecular moieties	$P^+_{non-add}$	Examples		Comments
6	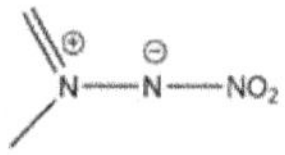	0.6			–
7	$-NH-NO_2$	0.9			The attachment of more than one $-NH-NO_2$ to triazole rings without the other substituents
8	$-NH-$	1.6			The presence of more than two $-NH-$ groups without substituents in ring case
9	$-R$	1.9			The attachment of more than one $-R$ groups to triazole rings
10	$=NH$ (cyclic)	1.0			–

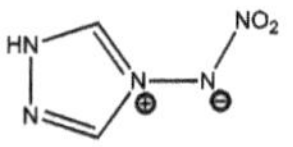

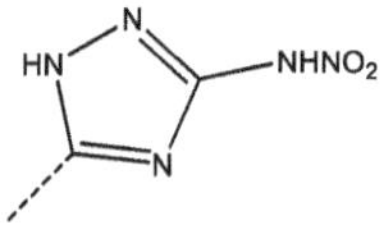

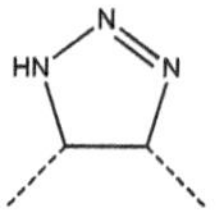

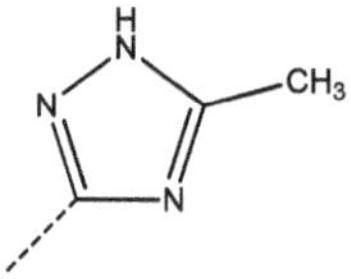

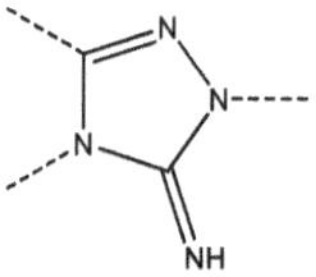

Table 1.3 (continued)

Category			$\overset{+}{p}_{non-add}$		
	Molecular moieties	$\overset{+}{p}_{non-add}$	**Examples**		**Comments**
11	$-NH_2$	1.0			The attachment of more than two $-NH_2$ groups to oxadiazole rings
12		1.4			The attachment of more than one $-NH_2$ groups to triazole rings without the other substituents or $-N=N-$ between two triazole rings

Example 1.14: The energetic compound 3,3'-bis(4-(5-methyl-1H-1,2,4-triazol-3-yl)-4,4'-azofurazan possesses a distinctive molecular architecture, with experimental measurements confirming a crystal density of 1.81 g/cm^3 [79]:

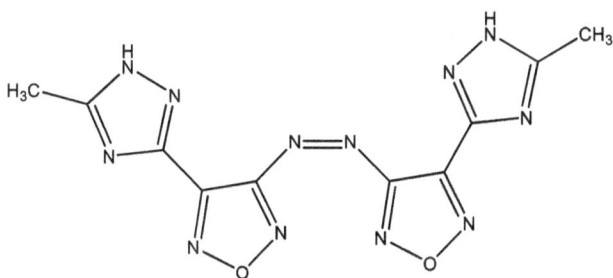

Chemical Formula: $C_{10}H_8N_{12}O_2$
Molecular Weight: 328.26

Applying eq. (1.19) with $\overset{+}{p}_{non-add}$= 1.9 g/cm^3 (Category 9, Table 1.3) yields the following result:

$$p = 2.278 - 10.58\left(\frac{a}{Mw}\right) - 4.775\left(\frac{b}{Mw}\right) - 6.763\left(\frac{c}{Mw}\right) + 0.1188\overset{+}{p}_{non-add} - 0.10468\overset{-}{p}_{non-add}$$

$$= 2.278 - 10.58\left(\frac{10}{328.26}\right) - 4.775\left(\frac{8}{328.26}\right) - 6.763\left(\frac{12}{328.26}\right) + 0.1188(1.9) - 0.10468(0)$$

$$= 1.817\,g/cm^3.$$

Example 1.15: The high-energy density compound 3,3′,5,5′-tetra(azidomethyl)-4,4′-azo-1,2,4-triazole exhibits a unique molecular architecture, with a crystal density of 1.534 g/cm³ [80]:

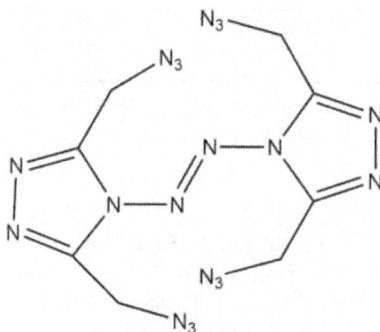

Chemical Formula: $C_8H_8N_{20}$
Molecular Weight: 384.29

Applying eq. (1.19) with the nonadditive density parameter $\rho_{non-add}^- = 0.5$ g/cm³ (Category 5 compounds, Table 1.3) yields the following computational result:

$$\rho = 2.278 - 10.58\left(\frac{a}{Mw}\right) - 4.775\left(\frac{b}{Mw}\right) - 6.763\left(\frac{c}{Mw}\right) + 0.1188\rho_{non-add}^+ - 0.10468\rho_{non-add}^-$$

$$= 2.278 - 10.58\left(\frac{8}{384.29}\right) - 4.775\left(\frac{8}{384.29}\right) - 6.763\left(\frac{20}{384.29}\right) + 0.1188(0) - 0.10468(0.5)$$

$$= 1.553 \, g/cm^3.$$

1.3.7.2 Semiempirical quantum mechanical calculations using the PM6 method

It is possible to predict the densities of high-nitrogen-content organic compounds at or near room temperature by combining structural parameters with semiempirical quantum mechanical calculations using the parameterized model 6 (PM6) method [81]. The PM6 method is a computationally efficient tool for estimating molecular properties such as geometry, electronic structure, and thermodynamic parameters. As part of the semiempirical methods family, PM6 simplifies the complex equations of quantum mechanics by incorporating approximations and empirical parameters. This approach strikes a practical balance between accuracy and computational cost, making it ideal for studying large molecules or systems where more advanced methods, like DFT or post-Hartree-Fock, would be too computationally demanding. The final correlation for predicting density using the PM6 method is expressed as follows [82]:

$$\rho = 1.085 + 0.01298 n_{NH,OH} + 0.03396 n_{NH_2} - 0.02126 n_{N_3} + 0.3729 \rho_{PM6}$$
$$+ 0.1269 \rho^{Inc} - 0.1164 \rho^{Dec},$$

(1.20)

where $n_{NH,OH}$, n_{NH_2}, and n_{N_3} represent the number of >NH (and –OH), –NH₂, and –N₃ groups present in the molecule, respectively. The density calculated using the PM6 method, denoted as ρ_{PM6}, is determined by eq. (1.6) as:

$$\rho_{PM6} = M/V_{PM6}$$

where V_{PM6} is the quantum-mechanically calculated molecular volume of an isolated molecule, obtained using the PM6 method. Additionally, ρ^{Inc} and $\rho^{Dec}c$ are two nonadditive correcting functions that adjust the predicted density. These functions account for specific structural features in the molecule: ρ^{Inc} increases the density, while ρ^{Dec} decreases it. The specific values of ρ^{Inc} and ρ^{Dec} c for various molecular moieties are provided in Table 1.4.

Table 1.4: The values of ρ^{Inc} and ρ^{Dec}.

Category	Molecular moieties	ρ^{Inc}	ρ^{Inc}	Complementary condition (the presence of the other molecular moieties)
1	*(triazole–CH₃ and tetrazole structures)*	1.5	0	–
2	*(triazolone with NH₂ structure)*	0	1.0	–
3	*(O–O ether structure)*	0	1.0	-----N₃
4	*(H₂N–tetrazole structure)*	0	1.0	*(pyrazole)* N-----
5	*(oxadiazole with CN structure)*	0	1.5	–
6	*(triazole with NH₂ and H₂N structure)*	1.5	0	*(NO₂ pyrazole, HN)*
7	*(triazole with NH₂ and NH structure)*	1.0	0	

Table 1.4 (continued)

Category	Molecular moieties	ρ^{Inc}	ρ^{Inc}	Complementary condition (the presence of the other molecular moieties)
8		0	1.0	
9		1.0	0	–
10		0	1.5	
11		0.7	0	–

Example 1.16: The heterocyclic energetic compound 4H-[1,2,5]oxadiazolo[3,4-b][1,2,3]triazolo[4,5-e]pyrazine 6-oxide (F5) exhibits a complex polycyclic architecture, with computational PM6 methods predicting a density of 2.011 g/cm³, while experimental crystallographic measurements yield a slightly lower value of 1.85 g/cm³ [67]:

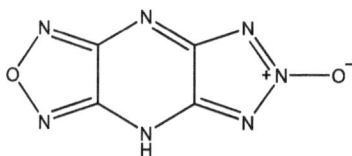

Chemical Formula: $C_4HN_7O_2$
Molecular Weight: 179.10

Application of eq. (1.20) yields the following result:

$$\rho = 1.085 + 0.01298 n_{NH, OH} + 0.03396 n_{NH_2} - 0.02126 n_{N_3} + 0.3729 \rho_{PM6} + 0.1269 \rho^{Inc} - 0.1164 \rho^{Dec}$$

$$= 1.085 + 0.01298(1) + 0.03396(0) - 0.02126(0) + 0.3729\ (2.011) + 0.1269(0) - 0.1164\ (0)$$

$$= 1.847\ g/cm^3.$$

Example 1.17: The nitrogen-rich heterocyclic compound 3,3'-bis(4-amino-1H-1,2,4-triazol-5(4H)-one)-4,4'-azofurazan exhibits a complex fused-ring architecture. Semi-empirical PM6 calculations predicted a density of 1.687 g/cm³, which shows excellent agreement with the experimentally determined crystal density of 1.68 g/cm³, with only 0.4% deviation [79]:

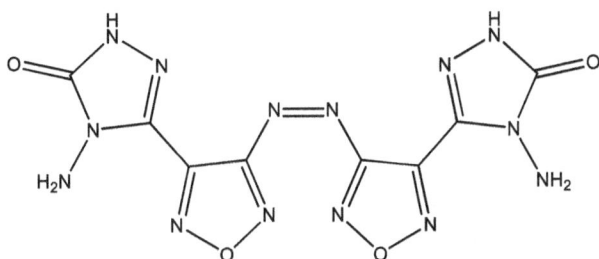

Chemical Formula: $C_8H_6N_{14}O_4$
Molecular Weight: 362.23

Implementation of eq. (1.20) with the decomposition density parameter ($\rho^{Dec} = 1.0$ g/cm³) from Category 2 (Table 1.4) yields the following computational result:

$$\rho = 1.085 + 0.01298 n_{NH, OH} + 0.03396 n_{NH_2} - 0.02126 n_{N_3} + 0.3729 \rho_{PM6} + 0.1269 \rho^{Inc} - 0.1164 \rho^{Dec}$$

$$= 1.085 + 0.01298(2) + 0.03396(2) - 0.02126(0) + 0.3729(1.687) + 0.1269(0) - 0.1164(1.0)$$

$$= 1.683\ g/cm^3.$$

1.4 Empirical methods for the assessment of the crystal density of hazardous ionic molecular energetic materials using the molecular structures

Some QSPR methods based on the structure of ionic compounds have been introduced in recent years for predicting their densities. The computer code EMDB_1.0 [71] uses suitable QSPR methods to calculate densities of ionic molecular energetic materials. There are several empirical methods which can be used to predict the crystal density of ionic molecular energetic materials, and which are demonstrated here.

1.4.1 Two general empirical methods

It was shown that the elemental composition of an ionic molecular energetic compound with general formula $C_aH_bN_cO_d$ can be used to predict its crystal density as follows [83]:

$$\rho = 2.148 - \frac{7.767a + 6.261b + 4.154c}{Mw}. \tag{1.21}$$

The presence of some specific molecular moieties may enhance or reduce the molecular packing in ionic molecular energetic materials. Equation (1.21) can be improved by considering the effects of two correcting functions for increasing and decreasing density (ρ^+ and ρ^-) as

$$\rho = 2.137 - \frac{8.653a + 6.273b + 3.561c}{Mw} + 0.1241\rho^+ - 0.09772\rho^-, \tag{1.22}$$

where ρ^+ and ρ^- are two correcting functions that are equal to 1.0 for the presence of the following specific molecular moieties:

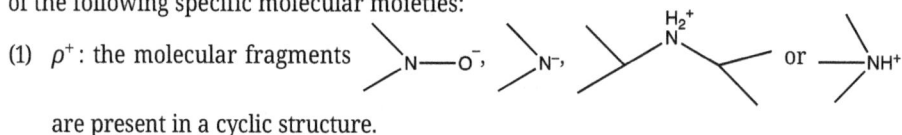

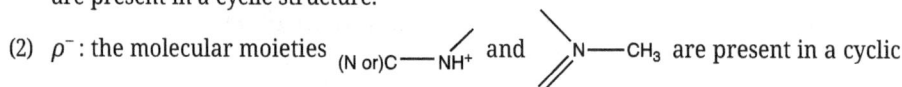

(1) ρ^+: the molecular fragments $\overset{}{\underset{}{>}}N\!-\!O^-$, $\overset{}{\underset{}{>}}N^-$, ⌐ or $-NH^+$

are present in a cyclic structure.

(2) ρ^-: the molecular moieties $_{(N\ or)C}\!-\!NH^+$ and $>N\!-\!CH_3$ are present in a cyclic

structure, as well as $\overset{(H\ or)C}{\underset{(H\ or)C}{>}}N\!-\!NH_3^+$, except in the presence of a bulky

anion such as picrate.

Example 1.18: Triaminoguanidinium 4,5-dicyano-1,2,3-triazolate has the following structure, and the measured crystal density of this compound is 1.48 g/cm³ [62]:

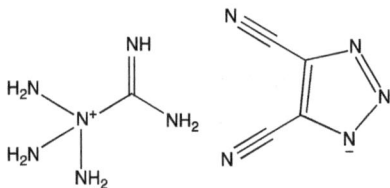

The use of eqs. (1.17) and (1.18) gives

$$\rho = 2.148 - \frac{7.767(5) + 6.261(9) + 4.154(11)}{223.2}$$

$$= 1.517\,g/cm^3\,(\% \, Dev = 2.5);$$

$$\rho = 2.137 - \frac{8.653(5) + 6.273(9) + 3.561(11)}{223.2} + 0.1241(0) - 0.09772(0)$$

$$= 1.515\,g/cm^3\,(\% \, Dev = 2.3).$$

The calculated values which have been obtained by two quantum mechanical methods reported by Rice et al. [20, 62] are 1.503 (% Dev = 1.5) and 1.428 g/cm³ (% Dev = − 3.5), respectively.

1.4.2 The effects of various substituents on the density of tetrazolium nitrate salts

It was found that the effect of various substituents such as N_3, NO_2, NH_2, NF_2, CN, and CH_3 on the density of tetrazolium nitrate salts with general formula $C_aH_bN_cO_dF_e$ can be shown by calculating the density using [84]:

$$\rho = 1.592 + 0.015c + 0.077e + 0.068P_{F_2} + 0.143P_{NO_2} - 0.081P_{CH_3}$$

$$- 0.072\rho^-_{\text{tetrazolium nitrate}},$$

$$(1.23)$$

where P_{F_2}, P_{NO_2}, and P_{CH_3} indicate the presence of the $-NF_2$, $-NO_2$, and $-CH_3$ groups in the salts, respectively. The value of $\rho^-_{\text{tetrazolium nitrate}}$ is 1.0 if the following structures are present:

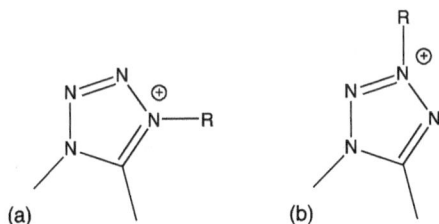

(a)   (b)

Example 1.19: The crystal density of the 1,5-dinitro-3H-tetrazolium nitrate salt containing the following cation:

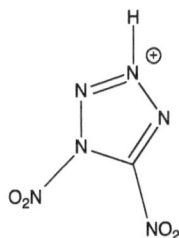

can be calculated as follows:

$$\rho = 1.592 + 0.015c + 0.077e + 0.068P_{F_2} + 0.143P_{NO_2} - 0.081P_{CH_3}$$

$$-0.072\rho_{\text{tetrazolium nitrate}}^-$$

$$= 1.592 + 0.015(7) + 0.077(0) + 0.068(0) + 0.143(1) - 0.081(0)$$

$$- 0.072(0)$$

$$= 1.840 \text{ g/cm}^3.$$

The measured crystal density of this compound is 1.87 g/cm³ [37].

1.4.3 Predicting the density of tetrazole-*N*-oxide salts

A suitable correlation has been introduced to estimate the density of tetrazole *N*-oxide salts using molecular structure descriptors, which has the following form [85]:

$$\rho = 1.514 - 0.047a - 0.025b + 0.028c + 0.073d - 0.039n_{H_2O} + 0.075TET$$

$$+ 0.172\rho_{\text{Tetrazole-}N\text{-oxide}}^+ - 0.118\rho_{\text{Tetrazole-}N\text{-oxide}}^- \tag{1.24}$$

where n_{H_2O} is the number of the H_2O molecules in the crystal of the salt; the values of *TET* parameter are 1.0 and zero for tetrazole salts containing 1*N*-oxide and 2*N*-oxide fragment, respectively; $\rho_{\text{Tetrazole-}N\text{-oxide}}^+$ and $\rho_{\text{Tetrazole-}N\text{-oxide}}^-$ are the positive and negative nonadditive structural parameters, which are defined as [85]:

(a) $\rho_{\text{Tetrazole-}N\text{-oxide}}^+$: The value of $\rho_{\text{Tetrazole-}N\text{-oxide}}^+$ is 1.0 when there is a 2-hydroxytetrazole fragment in a tetrazole-*N*-oxide energetic molecule.

(b) $\rho_{\text{Tetrazole-}N\text{-oxide}}^-$: The value of $\rho_{\text{Tetrazole-}N\text{-oxide}}^-$ is 1.0 for ammonium tetrazolate-1*N*-oxide salts.

(c) For the presence of amino-nitroguanidinium as the cation of the tetrazole-*N*-oxide salt, the value of $\rho_{\text{Tetrazole-}N\text{-oxide}}^-$ is 1.0.

Example 1.20: Bis(oxalyldihydrazidinium) 1H,1H-5,5-bitetrazole-1,1-diolate has the following molecular structure:

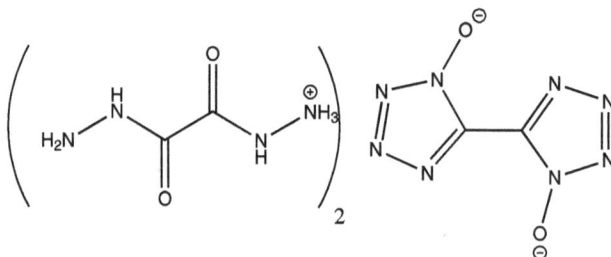

The use of eq. (1.24) gives:

$$\rho = 1.514 - 0.047a - 0.025b + 0.028c + 0.073d - 0.039n_{H_2O} + 0.075TET$$

$$+ 0.172\rho^+_{Tetrazole-N-oxide} - 0.118\rho^-_{Tetrazole-N-oxide}$$

$$= 1.514 - 0.047 \times 6 - 0.025 \times 14 + 0.028 \times 16 + 0.073 \times 6 - 0.039 \times 0 + 0.075 \times 1$$

$$+ 0.172 \times 0 - 0.118\rho^-_{Tetrazole-N-oxide} \times 0$$

$$= 1.843 \, g/cm^3.$$

The measured crystal density of this compound is 1.847 g/cm^3 [86].

1.4.4 High-nitrogen-containing salts and ionic liquids (HNCSILs)

Among HEDMs, energetic ionic salts and ionic liquids with high nitrogen content have gained significant attention [87,88]. These compounds are prized for their exceptionally high heats of formation, which stem from the abundance of energetic NN and CN bonds [89].

High-nitrogen-containing salts and ionic liquids (HNCSILs) are particularly fascinating. These materials are characterized by nitrogen-rich cations and anions, often featuring amino, nitro, or azide groups. To qualify as an HNCSIL, a compound must contain at least 50% nitrogen by weight [2,3]. This high nitrogen content translates into several advantages: higher heat of formation, greater density, and better oxygen balance compared to other classes of salts and ionic liquids [90]. Additionally, because nitrogen gas (N$_2$) is the primary product of detonation, these materials are more environmentally friendly [91].

One standout feature of HNCSILs is the presence of nitrogen-rich species in both the cation and anion. This dual nitrogen enrichment results in lower vapor pressures, higher densities, and improved thermal stability compared to their nonionic counterparts [92].

HNCSILs achieve higher densities due to their ability to pack tightly in the crystal structure, thanks to strong interionic interactions [93]. The density of the crystal structure directly impacts the material's detonation performance: higher crystalline density leads to increased detonation velocity and pressure [94]. It is worth noting that detonation velocity correlates linearly with density, while detonation pressure correlates with the square of density [95].

Previous studies have shown that the elemental makeup and specific polar groups significantly influence ion interactions, which in turn affect crystal density [96]. For HNCSILs, it's essential to understand how different structural parameters can either increase or decrease the density of a nitrogen-rich salt or ionic liquid before synthesis. Notably, hydrogen and nitrogen atoms play a pivotal role in determining density, as indicated by their low P-values (<0.05).

Incorporating $-NH_x$ groups (such as $>NH$, $-NH_2$, NH_3, and NH_4^+) into HNCSILs is a straightforward way to boost density. These groups can act as electron-withdrawing moieties in heterocyclic cations when paired with nitrogen-rich anions like dinitramide anions. Increasing the number of these groups enhances hydrogen bonding and intermolecular attraction [78].

On the flip side, incorporating $-CN$ or $-N_3$ groups tends to reduce density. Branching caused by $-N_3$ can weaken intermolecular interactions, where replacing $-NO_2$ with $-CN$ lowers density. Similarly, a compound with an $-N_3$ group has a lower density compared to one with an $-NO_2$ group.

Using data from the training set, a correlation established between density and structural factors as follows [97]:

$$\rho = 2.000 +$$

$$\frac{-9.165b - 2.832c + 6.510n_{NH} + 5.425n_{NH_2} + 24.15n_{NH_3} + 10.87n_{NH_4} - 11.59n_{N_3} - 12.06n_{CN}}{Mw},$$

$$(1.25)$$

where the terms n_{NH}, n_{NH_2}, n_{N_3}, and n_{CN} correspond to the counts of these specific functional groups; n_{NH_4} indicates the number of ammonium cations present. The equation predicts the density of HNCSILs by considering several key structural factors. In this model, the coefficients for n_{N_3} and n_{CN} groups are negative, meaning that increasing the presence of these groups tends to lower the overall density of the compound. However, the extent of their impact depends on the combined contributions of all other factors, such as hydrogen, nitrogen, and the other functional groups.

Conversely, the coefficients for n_{NH}, n_{NH_2}, n_{NH_3}, and n_{NH_4} are positive, indicating that these groups contribute to increasing the density of HNCSILs. This aligns with their role in enhancing hydrogen bonding and intermolecular attraction, as seen in the trends discussed earlier. Examples 1.21 and 1.22 illustrate this effect: adding an extra amino group to the cation increases density.

Example 1.21: The compound bis(diaminoguanidinium) 4-(carboxylatomethyl)-5-nitroiminotetrazolate has a specific molecular structure, and its crystal density has been measured to be 1.498 g/cm^3 [98]:

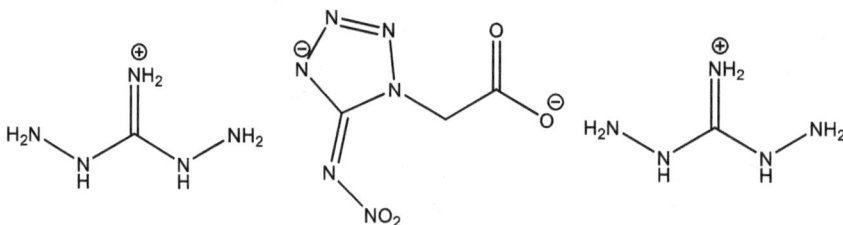

Using eq. (1.25) yields the following result:

$$\rho = 2.000 + \frac{-9.165b - 2.832c + 6.510n_{NH} + 5.425n_{NH_2} + 24.15n_{NH_3} + 10.87n_{NH_4} - 11.59n_{N_3} - 12.06n_{CN}}{Mw}$$

$$= 2.000 + \frac{-9.165(18) - 2.832(16) + 6.510(4) + 5.425(6) + 24.15(0) + 10.87(0) - 11.59(0) - 12.06(0)}{336.39}$$

$$= 1.586 \, g/cm^3.$$

The calculated value which has been obtained by eq. (1.22) is 1.679 g/cm³. Thus, eq. (1.25) gives closer predicted result.

Example 1.22: The compound bis(triaminoguanidinium) 4-(carboxylatomethyl)-5-nitroiminotetrazolate is structured as follows, and its crystal density has been experimentally determined to be 1.529 g/cm³ [98]:

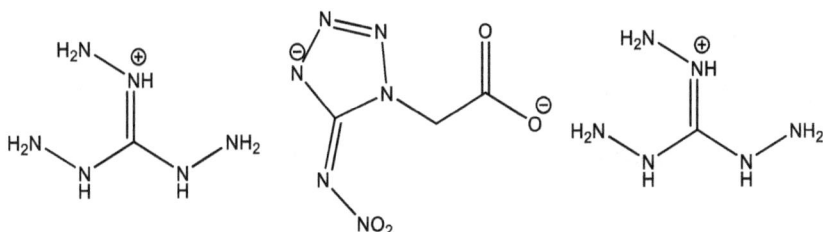

Applying eq. (1.25) results in the following:

$$\rho = 2.000 + \frac{-9.165b - 2.832c + 6.510n_{NH} + 5.425n_{NH_2} + 24.15n_{NH_3} + 10.87n_{NH_4} - 11.59n_{N_3} - 12.06n_{CN}}{Mw}$$

$$= 2.000 + \frac{-9.165(20) - 2.832(18) + 6.510(6) + 5.425(6) + 24.15(0) + 10.87(0) - 11.59(0) - 12.06(0)}{396.43}$$

$$= 1.590 \, g/cm^3.$$

The calculated value obtained using eq. (1.22) is 1.674 g/cm³. Therefore, eq. (1.25) provides a more accurate prediction.

In contrast, the coefficients of n_{N_3}, and n_{CN} are negative, which are consistent with those demonstrated in Examples 1.23 and 1.24.

Example 1.23: The compound guanidinium 2-nitroamino-4,6-diazido[1,3,5]triazinate is structured as follows, and its crystal density has been measured to be 1.60 g/cm³ [99]:

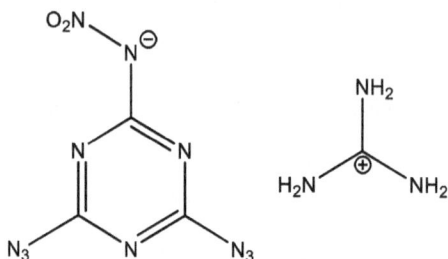

Using eq. (1.25) yields the following result:

$$\rho = 2.000 + \frac{-9.165b - 2.832c + 6.510n_{NH} + 5.425n_{NH_2} + 24.15n_{NH_3} + 10.87n_{NH_4} - 11.59n_{N_3} - 12.06n_{CN}}{Mw}$$

$$= 2.000 + \frac{-9.165(6) - 2.832(14) + 6.510(0) + 5.425(3) + 24.15(0) + 10.87(0) - 11.59(2) - 12.06(0)}{282.24}$$

$$= 1.640 \text{ g/cm}^3.$$

As can be seen, the calculated density value is close to the experimental result.

Example 1.24: The structure of aminoguanidinium 2-nitroamino-4,6-diazido[1,3,5]triazinate is shown below, and its experimentally measured crystal density is 1.68 g/cm³ [99]:

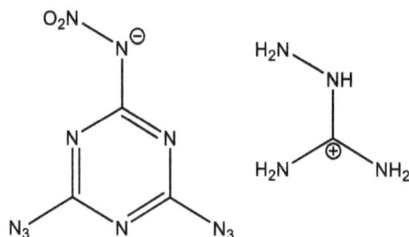

Using eq. (1.25) produces the following result:

$$\rho = 2.000 + \frac{-9.165b - 2.832c + 6.510n_{NH} + 5.425n_{NH_2} + 24.15n_{NH_3} + 10.87n_{NH_4} - 11.59n_{N_3} - 12.06n_{CN}}{Mw}$$

$$= 2.000 + \frac{-9.165(7) - 2.832(15) + 6.510(1) + 5.425(3) + 24.15(0) + 10.87(0) - 11.59(2) - 12.06(0)}{297.26}$$

$$= 1.640 \text{ g/cm}^3.$$

As observed, the calculated density value is in close agreement with the experimental result.

1.5 Machine learning for predicting properties of energetic materials

Chemical theory and calculations have long been used to determine the properties of energetic materials [100,101]. However, traditional methods, like DFT, often involve high computational costs and long estimation cycles, making them less practical for high-throughput needs [102]. To address these challenges, machine learning (ML) has emerged as a powerful tool, enabling fast and accurate predictions of material properties that rival quantum chemical methods [103].

In the age of big data, researchers can leverage massive datasets to uncover relationships between molecular structures and their properties, predicting outcomes for unseen molecules [104]. This approach aligns with the "fourth paradigm" of science, which emphasizes data-driven discovery [105]. In materials science, supervised regression models – trained on labeled data – are widely used to predict properties of energetic materials [106]. Key factors influencing prediction accuracy include dataset size and quality, featurization methods, algorithm choice, and hyperparameter tuning [107].

Data sources typically include experimental results from literature [108], computational outputs from software like Gaussian and Vasp [109], and databases such as CCDC and PubChem [110,111]. Featurization methods, like simplified molecular input line entry system (SMILES) strings or graph-based neural networks (NN), are critical for representing molecular structures in ML models [112]. Algorithms such as linear regression, SVM, k-nearest neighbors, and NN are commonly employed, with performance evaluated using metrics like coefficient of determination (R^2), mean absolute error), mean squared error, and correlation coefficients [113].

Recent studies highlight the growing role of ML in predicting energetic material properties. Some focus on single-property predictions, while others explore multiple properties or compare different models [114]. To avoid redundancy, this review organizes advancements into two categories: single-property and multi-property predictions. It evaluates the accuracy, strengths, and limitations of various methods, emphasizing ML's advantages over traditional approaches while identifying current challenges.

1.5.1 Machine learning for predicting density in energetic materials

Traditionally, scientists have predicted density by calculating molecular or crystal volume using methods like molecular dynamics or group additivity. However, these approaches come with significant limitations. For instance, molecular dynamics heavily depends on the accuracy of force fields, and if the right parameters are missing, predictions can go way off track. Meanwhile, group additivity rules often ignore molecular configurations and intermolecular interactions, making them unable to distinguish between isomers. Worse still, they overlook the effects of temperature and crystallographic forms, which limits their practical use.

Enter ML, a game-changing approach that directly maps molecular structures to density while sidestepping external conditions like temperature. This not only boosts computational efficiency but also reduces prediction errors, offering a clear advantage over traditional methods.

Fathollahi and Sajady [115] took a step forward by using ANNs to predict the density of energetic cocrystals. They started with 26 selected cocrystals, extracting three molecular descriptors based on optimized chemical structures. Their ANN model outperformed MLR, achieving an impressive test precision of 0.9918. They designed a simple yet effective ANN architecture: an input layer with three neurons, a hidden layer with three neurons, and an output layer. This model proved ideal for screening target chemical structures from databases to identify cocrystals with desired densities.

Yang et al. [116] tackled a common challenge in ML: selecting molecular descriptors, which is both time-consuming and requires deep expertise. Instead of relying on hand-picked features, they sought a direct mapping between molecular topology and density. Working with crystallographic data from 2,002 neutral nitro compounds in the CCDC database, they tested three algorithms: random forest (RF), graph NN (GNN), and SVM. Inputs were prepared using tools like Open Babel, Materials Studio, and RDKit. GNN emerged as the top performer, achieving an R^2 value of 0.949 – higher than both RF and SVM, and even outperforming traditional DFT QSPR models ($R^2 = 0.925$). This showed that GNN could deliver high accuracy using just molecular graphs as inputs, proving its superiority over conventional ML models.

Nguyen et al. [117] took a broader approach, collecting 10,521 molecules from the Cambridge Structural Database after rigorous data cleaning. They trained density prediction models using four ML algorithms: support vector regression, RF, PLS regression, and message passing NN, a type of GNN. Interestingly, they found that SMILES strings – a common way to represent molecules – were unsuitable for density prediction. Instead, they opted for three alternative representations: Extended 3D Fingerprints, 2D Molecular Descriptors from RDKit, and graph-based representations with atom and bond descriptors.

These studies collectively highlight the power of machine learning in uncovering complex, nonlinear relationships between molecular structure and density. Compared to traditional empirical methods, ML significantly reduces prediction errors caused by oversimplified parameters [117]. Notably, both Yang et al. [116] and Nguyen et al. [117] found that GNN-based models offered superior accuracy. However, differences in their datasets – such as variations in data distribution – likely influenced model performance and generalization. These findings underscore the importance of choosing appropriate data and representation methods when building predictive models for energetic materials.

Pandey and Roy [118] developed the quantitative read-across structure–property relationship (q-RASPR) method to predict gas-phase heat of formation for energetic compounds, combining QSPR modeling with read-across techniques to improve predictive accuracy over traditional QSPR methods. They first curated and refined datasets from multiple sources to ensure data quality, then represented molecular structures, calculated key descriptors, and split the data into training and test sets before training models using different methodologies and rigorously validating them following Organization for Economic Cooperation and Development (OECD) guidelines while also testing predictive power with multiple ML algorithms.

For density prediction, they used PLS regression to derive the equation:

$$\rho = 0.425 + 0.042 AMW - 0.690 Mp + 0.082 MCD + 0.741 RA \ function(GK) - 0.049 CVsim(GK),$$

$$(1.26)$$

where average molecular weight (*AMW*) contributes positively since density increases with molecular mass, while *Mp* (mean atomic polarizability) contributes negatively because polarizability relates to molecular volume, which inversely affects density. The molecular cyclized degree (*MCD*) descriptor has a positive impact since cyclic molecules exhibit higher density due to stronger London dispersion forces from their rigid ring structures, whereas the *RA function (GK)*, a composite q-RASPR descriptor, also positively influences predictions by incorporating structural similarities from read-across analysis. Finally, *CVsim(GK)* (coefficient of variance of similarity values) shows a negative contribution because higher variation in similarity values among training compounds reduces prediction reliability for test compounds.

This approach demonstrates how integrating read-across techniques with QSPR modeling enhances predictive performance for energetic materials.

1.5.2 Limitations of ML in predicting energetic compound densities

ML is incredibly powerful – yet when it comes to predicting the density of high-energy compounds, it still faces real-world hurdles. To make real progress, we need higher-quality data, models that are easier to interpret, and smarter ways to blend ML with traditional chemistry expertise. Here are some of the key limitations we're up against:

(i) **It does not always work well on new compounds:** ML models often perform great on compounds they have seen before. But when it comes to new or slightly different molecules, they can give poor predictions. That's because the model has not "learned" enough about those kinds of structures [116].

(ii) **Good data is hard to find:** Energetic materials are not exactly easy or safe to work with, so there isn't a ton of high-quality data available – and some of it might be secret or restricted. When the data is limited or contains errors, the ML models learn the wrong things [103].

(iii) **You do not always know *why* it works:** Some ML models, especially the fancy ones like NN or RF, are like black boxes. They can give accurate predictions, but they do not explain how they got there – which makes it hard for scientists to trust or learn from them [103].

(iv) **The way you describe the molecule really matters:** ML models rely on digital representations of molecules (like SMILES strings or molecular fingerprints). If these are chosen poorly or inconsistently, the predictions would not be reliable [119].

(v) **It still needs human help:** Despite the hype, ML is not fully automatic. You often need experts to carefully choose the right features (chemical descriptors) and tune the model. That takes time, and it still depends on human chemistry knowledge [103].

1.6 Summary

Different empirical methods have been introduced in this chapter to predict the crystal densities of important classes of neutral and ionic molecular energetic compounds. Among different group additivity methods, more reliable approaches are introduced in Section 1.1 for estimation of densities of neutral and ionic molecular energetic compounds at room temperature. For neutral energetic compounds, eqs. (1.11)–(1.14) are very simple approaches in comparison with other correlations, but their reliabilities are lower. Due to the high complexity of eq. (1.17), it is recommended to use eqs. (1.15) and (1.16) to assess the crystal densities of various classes of energetic compounds, including polynitroarenes, polynitroheteroarenes, nitroaliphatics, nitrate esters, and nitramines with the general formula $C_aH_bN_cO_d$. For energetic azido compounds with general formula $C_aH_bN_cO_d$, eq. (1.18) is a suitable relationship which can be used to predict the crystal density.

HEDMs are crucial for advanced research due to their high performance and low sensitivity. Nitrogen-rich compounds (exceeding 50% nitrogen by weight) are particularly valuable for explosives, propellants, and gas generators, as they offer high energy output and stability. These materials often feature nitrogen-rich rings or functional groups like $-NO_2$, $-N_3$, and $-NH_2$. Two methods were proposed to predict their densities:

1. **Molecular fragment approach**: Equation (1.19) used functional groups and correction factors to adjust density estimates.
2. **PM6 quantum mechanical method**: Equation (1.20) combined semiempirical calculations with structural parameters and correction terms.

Both methods balance computational efficiency and accuracy, offering practical tools for designing novel energetic materials.

For ionic molecular energetic compounds with general formula $C_aH_bN_cO_d$, eq. (1.22) was used to calculate their crystal densities. Equations (1.23) and (1.24) provided good correlations for predicting the crystal densities of tetrazolium nitrate and tetrazole-N-oxide salts, respectively. HNCSILs are key energetic materials due to their high heat of formation, density, and eco-friendly nitrogen-rich decomposition products. Their density is tunable via functional groups: $-NH_x$ groups enhance it through hydrogen bonding, while $-CN$ or $-N_3$ reduce it by weakening intermolecular interactions, as quantified by predictive equation (1.25).

Traditional computational methods (e.g., DFT) for predicting energetic material properties are accurate but computationally expensive, limiting high-throughput applications. ML has emerged as a faster, data-driven alternative, leveraging large datasets and algorithms (e.g., NN and GNN) to predict properties like density with high accuracy – often surpassing quantum chemical methods. Key advances include graph-based models (e.g., GNNs) that bypass manual descriptor selection and directly map molecular structures to properties, achieving R^2 values. However, ML faces challenges such as limited data availability, poor generalizability to novel compounds, "black box" interpretability issues, and dependence on expert-guided feature selection. The success of ML hinges on high-quality data, optimal molecular representations (e.g., avoiding SMILES for density prediction), and integration with domain expertise to ensure reliability and scalability.

2 Heat of formation

The presence of certain energetic groups in organic compounds results in unstable compounds since the heats of detonation/combustion depend upon the presence of these functional groups [1, 10, 120]. Organic compounds containing energetic groups can decompose, ignite, or explode on exposure to external stimuli such as heat, impact, shock or an electric spark. Thus, the search for new organic energetic compounds with superior performance and lower sensitivity to undesired stimuli is important in both modern civil and military applications.

The condensed (solid or liquid) phase heat (or enthalpy) of formation of an energetic substance at 298.15 K is a measure of its energy content. It is an important factor to consider in the design, assessment, and thermochemical stability of energetic compounds because knowledge of this value is essential in order to allow the evaluation of the performance properties of explosives and propellants such as the detonation pressure, heat of detonation, and specific impulse using theoretical methods [121–123]. Calorimetry is usually used to measure the condensed phase heat of formation of energetic compounds, but it is a time-consuming and destructive technique. Since it requires extremely pure samples [124], calorimetric measurements also have problems that can increase the uncertainty, for example, completeness of the reaction or the production of undesired products [125]. Various computer codes which are used to calculate the detonation and combustion properties of energetic compounds such as EXPLO5 [126], EDPHT [127–129], and NASA-CEC-71 [130] require the value for the condensed phase heat of formation. Appendix A shows the measured values of the condensed phase heats of formation for well-known pure and composite explosives.

2.1 Gas-phase heats of formation of energetic compounds

The molecular architecture of high explosives reveals an interesting design principle. These materials typically combine carbon-rich fuel backbones with specialized energetic groups called explosophores. Particularly promising candidates for next-generation HEDMs often feature nitrogen-rich four-membered heterocycles decorated with functional groups like:
- Nitro ($-NO_2$)
- Nitroso ($-RNO$)
- Nitramino ($-NHNO_2$)
- Azides ($-N_3$)
- Azo bridges ($-N=N-$)

These structural features contribute significantly to explosive performance [131]. Another effective strategy involves incorporating strained ring or cage structures. When

https://doi.org/10.1515/9783112206768-002

these compounds decompose, the release of built-in strain energy boosts detonation power substantially [131].

For maximum performance, many top-performing HEDMs utilize five-membered heterocyclic building blocks like:

- Furazan
- Furoxan
- Isofurazan
- Tetrazole

These compact, dense frameworks not only pack more energy into smaller volumes but also exhibit highly positive formation enthalpies – both critical factors for superior detonation characteristics [132].

Elton et al. [119] conducted a thorough comparison of various molecular featurization techniques, such as the sum over bonds, custom descriptors, Coulomb matrices, bag of bonds, and fingerprints. They found the nitrogen-to-carbon ratio – a well-established indicator of energetic performance. This choice was motivated by the fact that replacing carbon with nitrogen tends to enhance performance, as N=N bonds contribute more to the heat of formation and enthalpy change during detonation compared to C-N or C=N bonds. The authors also emphasized that, when working with limited datasets, manually selecting features based on chemical intuition and domain knowledge can significantly improve model accuracy [119]. For instance, the presence of azide groups in a molecule is known to boost energetic performance while simultaneously increasing shock sensitivity [119].

2.1.1 The evolution of computational thermochemistry from semiempirical to composite quantum methods

Density functional theory (DFT) can provide highly accurate values for quantities such as bond strengths and heats of formation of energetic compounds. Sana et al. [133] introduced the term stabilization energy, which measures special effects such as bond interaction and electron dislocation. It was shown that $-NO_2$ groups result in high destabilization energy. Different quantum mechanical (QM) methods can predict the gas-phase heats of formation of energetic compounds [121]. The determination of the geometries and enthalpies of formation of molecules using molecular mechanics (MM) is widely used for large chemical systems. Thus, MM2 [134], MM3 [135], and MM4 [136] have been parametrized [137]. The MM2 method has been used for the prediction of the gas-phase heats of formation of nitro compounds [138]. MM methods are fast and inexpensive, however, due to the lack of reliable parameters for some compounds, their application is limited [139]. Several semiempirical QM methods have been parametrized to enable the prediction of the energies and enthalpies of formation. In contrast to MM methods, parametrization of the semiempirical methods has been performed over a wide range of atoms and compounds. A more general parametrization leads to an increase in the uncertainties of the semiempirical methods [124]. Thus, several modifications such as AM1 [140], PM3 [141], PM6 [142] and PM7 [143] have been introduced to extend the domain, as well as to increase the accuracy. AM1 and PM3 methods have been used to predict the gas-phase heats of formation of aromatic and aliphatic nitro compounds [144]. The PM3 method has also been used for the indirect prediction of the condensed phase heats of formation of nitroaromatic compounds [145].

For decades, thermochemical calculations have relied on QM methods to predict enthalpic properties. However, a persistent challenge arises: high-accuracy QM methods often come with steep computational costs. Over the years, researchers have developed various approximated methods to strike a balance between accuracy and efficiency. Among these, semiempirical QM methods stand out, as many were explicitly parametrized to predict enthalpies of formation and reaction.

Early semiempirical QM methods, such as MNDO, had notable limitations – for example, a mean absolute error (MAE) of 6.3 kcal/mol for enthalpies of formation on the well-curated CHNO dataset. However, advancements like the ODM2 method reduced this error to 2.63 kcal/mol, rivaling or even surpassing the accuracy of many DFT approaches [146].

For even higher precision, composite QM methods (e.g., G4, G4MP2, W1, CBS-QB3) were designed to achieve chemical accuracy (errors <1 kcal/mol). These combine multiple QM techniques, including high-level ab initio methods, but their computational expense restricts them to small molecules [146].

2.1.2 ML advances in predicting formation enthalpies outperform traditional quantum methods

The rise of ML has introduced a new, data-driven approach to enthalpy prediction. Unlike traditional QM methods, ML models offer near-instantaneous calculations at a fraction of the cost [147]. Some ML models directly predict enthalpies of formation [148], while others refine predictions from baseline methods like DFT, Hartree-Fock, or semiempirical QM methods by correcting errors against experimental or high-level QM data (e.g., G4, G4MP2) [149]. Impressively, several of these models approach chemical accuracy, though most still depend on DFT corrections, limiting their efficiency [150].

Unlike traditional modeling approaches that depend on physical principles like conservation laws and thermodynamics [151], ML uncovers hidden patterns and statistical relationships directly from data to generate reliable, repeatable predictions [152]. Rather than relying on theoretical assumptions, ML leverages algorithms to learn from data itself – a shift that has become increasingly powerful with the rise of big data.

This data-driven approach is transforming materials science, enabling faster discovery and more efficient design of new materials [153]. Studies show that ML can significantly cut R&D costs while accelerating the development of advanced materials [154]. Unsurprisingly, researchers are now applying these techniques to energetic materials, where predictive modeling offers major advantages [155].

The enthalpy of formation is absolutely crucial for predicting how energetic materials will perform during detonation. Mathieu [156] took this research further by testing two advanced deep learning models (ANI-1X and ANI-1ccx). These models could predict electronic energy and vibration frequency well enough to calculate formation enthalpies for regular organic compounds with accuracy rivaling expensive DFT calculations. But here's the catch – when it came to energetic compounds specifically, the models struggled [156]. Mathieu suspects this is because the training database (GDB-11) did not include enough explosive-containing molecules.

A key limitation of these special-purpose ML models is their narrow scope – many were designed only for specific compound classes (e.g., acyclic hydrocarbons, cyclic hydrocarbons, and energetic materials [119]). Additionally, since they rely on precomputed molecular descriptors, they cannot generate or optimize molecular geometries on their own.

Beyond QM and specialized ML models, general-purpose ML methods have emerged, capable of predicting QM potential energies across diverse chemical spaces. These models, such as the ANI family, AIQM1, OrbNet Denali, and DM21, serve as drop-in replacements for QM or force fields in simulations like molecular dynamics and geometry optimizations.

Zheng et al. [146] developed an enthalpy of formation prediction scheme using ANI-1ccx (a neural network potential trained to DLPNO-CCSD(T)/CBS accuracy for C, H, N, and O systems [157]). They also evaluated AIQM1, an ML-enhanced semiempirical QM methods combining ODM2 with ANI-1ccx corrections and dispersion terms. Their results showed that both methods achieve chemical accuracy for many organic molecules – without additional tuning.

Pandey and Roy [118] applied the q-RASPR method to predict the gas-phase heat of formation for energetic compounds. This approach combines QSPR modeling with read-across techniques, improving predictive accuracy compared to traditional QSPR methods. To ensure reliable results, the researchers first curated datasets from multiple sources, refining them for quality. They then represented the molecular structures, calculated key descriptors, and split the data into training and test sets. The models were trained using different methodologies and rigorously validated following OECD guidelines. Additionally, they tested the models' predictive power using multiple machine learning algorithms. They employed PLS regression, deriving the following equation for gas-phase heat of formation:

$$\Delta_f H^\theta (g) = 28.972 + 1.020 RA \; function \, (LK) - 0.298 SD \; activity \, (LK) - 1.884 nCsp3, \quad (2.1)$$

The *RA function* (*LK*) is a composite descriptor derived from read-across analysis, capturing structural and property similarities between compounds – in this model, it positively influences the predicted heat of formation. The *SD activity* (*LK*) represents the weighted standard deviation of observed property values from closely related compounds, serving as a measure of data variability that negatively impacts predictions. Finally, *nCsp3* simply counts the number of sp³-hybridized carbon atoms in a molecule, with its negative coefficient suggesting that more aliphatic carbons tend to decrease the heat of formation, likely due to their stabilizing effect. Together, these descriptors create a robust predictive framework where molecular similarity, data consistency, and structural features collectively determine the thermodynamic property.

2.2 Condensed phase heats of formation of energetic compounds

When estimating an explosive material's detonation heat, researchers typically analyze two key factors: the condensed phase heat of formation of the explosive itself and the standard heat of formation of its detonation products [158, 159]. Generally, compounds with higher positive heat of formation values (per unit weight) tend to release more energy during detonation, leading to better explosive performance [158, 159].

To efficiently extract key physicochemical properties from a small dataset, Chen et al. [160] introduced spatial matrix descriptors. These included the volume occupation spatial matrix and heat contribution spatial matrix, which helped machine learning models capture the spatial distribution of mass and energy at the atomic level. These descriptors were specifically designed to predict crystalline density and solid-phase heat of formation [160]. The innovation behind spatial matrices was to minimize redundant information in Coulomb matrices by incorporating relevant physicochemical causality relationships [160].

Different methods have been used to predict the condensed phase heat of formation of compounds containing energetic groups. Prediction of the condensed phase heat of formation can be achieved using different computational methods, in particular: group additivity, MM, QM, and empirical methods, or QSPR based on the molecular structure [161, 162]. These methods can be classified in two different approaches for estimation of the condensed phase heat of formation. The first approach uses indirectly the heat of phase change property, that is, the heats of sublimation and vaporization. The second approach utilizes directly the molecular structure of energetic materials wherein the methods of QSPRs were used. Many energetic compounds exist in the solid phase at room temperature. Thus, different methods were introduced to calculate the heat of sublimation in the first approach, that is, group additivity [163, 164], QSPR based on complex descriptors [165], molecular surface electrostatic potential [166], and other QSPR schemes based on simple structural descriptors [167–170]. Many of the mentioned methods require complex computer codes but some models require the molecular structure of energetic materials, which are illustrated here.

2.2.1 Thermochemical prediction of condensed phase heats of formation in energetic materials

Hess's law can be used to calculate the solid- and liquid-phase heats of formation using the predicted heats of vaporization and sublimation. The heat of sublimation or sublimation enthalpy has an important role in the assessment of the strength of inter-molecular cohesion. It provides a valuable indicator of crystal stability and vapor pressure of organic compounds containing energetic groups such as $-O-O-$, $-N_3$, $-ON=O$, $-NO_2$, $-ONO_2$, and $-NNO_2$ in the solid state [171]. Energetic compounds may exist in the solid or liquid phase at room temperature. Solid- and liquid-phase heats of formation of energetic compounds can be obtained by subtraction of the heat of sublimation from the gas-phase value as [172]:

$$\Delta_f H^\theta(s) = \Delta_f H^\theta(g) - \Delta_{sub} H^\theta \tag{2.2}$$

$$\Delta_f H^\theta(l) = \Delta_f H^\theta(g) - \Delta_{vap} H^\theta, \tag{2.3}$$

where $\Delta_f H^\theta(s)$ is the standard solid-phase heat of formation; $\Delta_f H^\theta(l)$ is the standard liquid-phase heat of formation; $\Delta_f H^\theta(g)$ is the standard gas-phase heat of formation; $\Delta_{sub} H^\theta$ is the standard heat of sublimation; $\Delta_{vap} H^\theta$ is the standard heat of vaporization. Some accurate methods exist for the calculation of $\Delta_f H^\theta(g)$ for neutral energetic compounds. Byrd and Rice [173] used DFT coupled with an atom and group contribution method for estimation of $\Delta_f H^\theta(g)$. For a test set of 45 energetic compounds, Oh-linger et al. [174] introduced T1 as a novel multilevel computational method to compare the ability of six different methods for accurate calculation of $\Delta_f H^\theta(g)$. Akutsu et al. [138] combined the heats of vaporization and sublimation using the additivity rule with the calculated gas-phase heats of formation data from the PM3 and MM2 methods to calculate the condensed phase heats of formation. It was shown that there is a correlation between the statistically-based quantities of electrostatic potentials mapped onto isodensity surfaces of isolated molecules and their heats of sublimation and vaporization [175]. Rice and coworkers [176, 177] used the 6–31G* basis set and the hybrid B3LYP density functional to convert QM energies of molecules into gas-phase heats of formation. They used the surface electrostatic potentials of individual molecules for computation of the heats of sublimation and vaporization.

Chapter 5 provides an explanation of different methods for calculating $\Delta_{sub} H^\theta$, and several standard QM techniques are illustrated here. Politzer et al. [178] introduced the following equation for calculation of the enthalpy of sublimation:

$$\Delta_{sub} H^\theta = w_1 (SA)^2 + w_2 \left(\sigma_{tot}^2 \nu\right)^{0.5} + w_3, \tag{2.4}$$

where SA is molecular surface area; σ_{tot}^2 indicates the variability of the potential on the molecular surface; ν shows a degree of the balance between positive and negative regions. The mentioned parameters w_1, w_2, and w_3 in eq. (2.4) can be determined from least-squares fitting to reliable values of the enthalpies of sublimation of organic compounds containing energetic groups. Byrd and Rice [173] used experimental data $\Delta_{sub}H^{\theta}$ for 23 different energetic compounds to obtain these parameters. Equation (2.4) was widely used in recent years for the prediction of the $\Delta_f H^{\theta}(s)$ values of newly designed energetic compounds [179–185]. Some efforts have been made to improve the prediction accuracy of eq. (2.4) by optimizing the parameters w_1, w_2, and w_3 for a very narrow class of compounds such as energetic tetrazine derivatives [186].

Politzer et al. [178] inserted an average deviation of electrostatic potential (Π) in eq. (2.4) to measure local polarity as an additional parameter:

$$\Delta_{sub}H^{\theta} = w_1(SA)^2 + w_2\left(\sigma_{tot}^2 v\right)^{0.5} + w_3 + w_4\Pi. \tag{2.5}$$

Suntsova and Orofeeva [166] adjusted the parameters w_1, w_2, w_3 and w_4 in eqs. (2.4) and (2.5) using the reported enthalpies of sublimation for 185 compounds, including 148 and 37 compounds for training and test sets, respectively. The overall performances of eqs. (2.4) and (2.5) are close to each other but the maximum absolute deviation of eq. (2.5) is lower [166]. Equation (2.5) was used to predict $\Delta_f H^{\theta}(s)$ some tetrazole-, tetrazine-, furazan-, and furoxan-based energetic compounds [166].

Group additivity methods can be used to estimate the ideal gas-phase heats of formation, using, for example, the methods of Benson, Yoneda, and Joback [187]. For these methods, assemblies of adjacent atoms are defined as groups, so that the enthalpy of the molecule is calculated by the summation of the contributions of the groups. Further parameters may be used to consider the effects of strain, resonance and conjugation. Group additivity methods can also be used to calculate the condensed phase heat of formation for some specific classes of energetic compounds. Due to the presence of different molecular interactions, molecular packing, and polymorphism in some solid compounds, prediction of the condensed phase heat of formation is more difficult than of the gas-phase heat of formation [188]. Several group additivity methods have been used to predict the condensed phase heat of formation of common CHNO energetic materials [189, 190]. For example, Bourasseau [190] applied the group additivity method to predict the standard heats of formation at 298 K of aliphatic and alicyclic polynitrocompounds. Salmon and Dalmazzone [189] also introduced a group contribution method that can be applied to large classes of CHNO energetic compounds to predict enthalpies of formation in the solid state (at 298.15 K). Argoub et al. [188] introduced a suitable method which provides significant improvements in accuracy and applicability as compared to the various group contribution methods for estimating $\Delta_f H^{\theta}(s)$. Their model has a simple linear form as [188]:

$$\Delta_f H^{\theta}(s) = \sum_i^N n_i \times A_i, \tag{2.6}$$

where A_i shows the contribution of the first-, second-, or third-order group of type i, which occurs n_i times [188] in a desired organic compound. Reliability of group contribution methods is lower than that of QM methods as well as of QSPR models based on molecular moieties for estimation of $\Delta_f H^\theta(s)$ in high-energy content organic compounds [128, 191–195]. It was shown that the reliability of the method of Argoub et al. [188] is lower than that of QSPR models based on molecular fragments [191]. However, group additivity methods have some restrictions, that is:

(1) they cannot be used for energetic materials which exist in the liquid state at 298.15 K,
(2) the group contributions of some functional groups have not yet been defined in the models, and
(3) for compounds which exhibit unusual chemical structures, additivity methods cannot be used.

2.2.2 Empirical approaches or QSPR methods on the basis of structural parameters

There are several methods which can be used to predict the condensed phase heats of formation of some classes of energetic compounds. Among these methods, there is a complex method in which the solid-phase heat of formation of a desired CHNO explosive in the range $Q_{corr} > 4{,}602$ kJ/g [196] (Q_{corr} is the corrected heat of detonation on the basis of Kamlet's method [197]), can be predicted based on its approximate detonation temperature. The other methods use the molecular structures of energetic compounds, and are reviewed here.

2.2.2.1 Simple procedure for nitroaromatic energetic materials

It was shown that the following general equation is suitable for calculating the condensed phase heat of formation for most nitroaromatic and benzofuroxan-based energetic compounds with general formula $C_a H_b N_c O_d$ [198]:

$$\Delta_f H^\theta(c) = 32.76a - 33.96b + 69.12c - 116.32d$$
$$+ 124.8 n_{NO_2} - 65.10 n_{Ar-NH} - 93.64 n_{OH} - 202.3 n_{COOH} \qquad (2.7)$$
$$+ 13.56(n_{Ar} - 1) + 121.4 n_{-N=N-} + 223.2 n_{cyclo-N-O-N},$$

where $\Delta_f H^\theta(c)$ is the standard heat of formation of a specific compound in the condensed phase (solid or liquid) in kJ/mol; n_{NO_2}, n_{OH}, and n_{COOH} are the number of nitro, hydroxyl, and carboxyl functional groups, respectively; n_{Ar-NH} is the number of NH (or NH$_2$) functional groups attached to aromatic rings, n_{Ar} is the number of aromatic rings, $n_{-N=N-}$ is the number of noncyclic –N=N– groups and $n_{cyclo-N-O-N}$ is the number of O (in benzofuroxan compounds) groups.

Example 2.1: The condensed phase heat of formation of

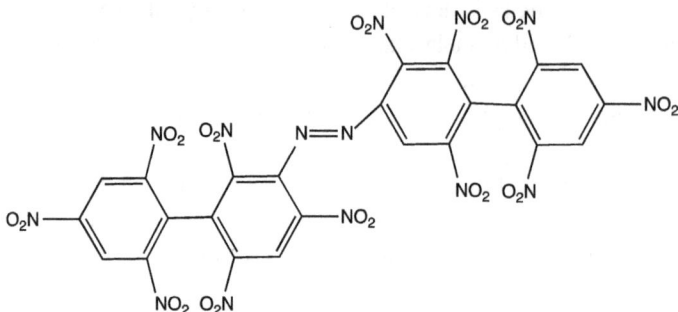

is calculated as follows:

$$\Delta_f H^\theta (c) = 32.76a - 33.96b + 69.12c - 116.32d$$

$$+ 124.8 n_{NO_2} - 65.10 n_{Ar-NH} - 93.64 n_{OH} - 202.3 n_{COOH}$$

$$+ 13.56(n_{Ar} - 1) + 121.4 n_{-N=N-} + 223.2 n_{cyclo-N-O-N}$$

$$= 32.76(24) - 33.96(6) + 69.12(14) - 116.32(24)$$

$$+ 124.8(12) - 65.10(0) - 93.64(0) - 202.3(0)$$

$$+ 13.56(4 - 1) + 121.4(1) + 223.2(0)$$

$$= 417.8 \text{ kJ/mol.}$$

The measured $\Delta_f H^\theta (c)$ of this compound is 480.3 kJ/mol [199].

2.2.2.2 More reliable approach for nitroaromatic energetic materials

It was shown that a more reliable approach can be used to calculate the condensed phase heat of formation for nitroaromatic compounds that have complex and different molecular structures according to [194]:

$$\Delta_f H^\theta (c) = \frac{\begin{array}{c} 2.690a - 2.896b + 2.876c - 2.784d \\ -1.701(n'_{Ar} - 1) - 1.607\left(\dfrac{n_{NO_2}}{n_{DFG/SP}} \times E\right) + 3.246\left(\dfrac{n_{IFG/SP}}{n_{NO_2}} \times F\right) \end{array}}{Mw \times 10^{-4}}, \quad (2.8)$$

where Mw is the molecular weight of nitroaromatic compound. Three variables n'_{Ar}, E, and F can be predicted based on the following situations:

(1) n'_{Ar}: The value of n'_{Ar} is equal to n_{Ar}. Therefore, for the presence of one and two aromatic rings, n'_{Ar} is equal to one and two, respectively. In the case of $n_{Ar} \geq 3$, n'_{Ar} equals n_{Ar} except if there is one nitro group (e.g., in 2,2',2'',4,4',4'',6,6',6''-nonanitro-1,1':3',1''-terphenyl) or nitrogen atom (e.g., in 2,4,6-tripicryl-1,3,5-triazine)

between two aromatic rings in which $n'_{Ar} = 0$. If the $-N=N-$ group is attached to an aromatic ring, the value of n'_{Ar} is also equal to zero.

(2) $n_{DFG/SP}$ and E: The ratio $n_{NO_2}/n_{DFG/SP}$, and E are defined for different functional groups or structural parameters as follows:

(a) $n_{DFG/SP} = n_{OH}$: For $n_{NO_2} = 1$, E is equal to 1.0. However, if $n_{NO_2} > 1$, then $E = 0.5$ and 1.75 for $n_{OH} = 1$ and 2, respectively.

(b) $n_{DFG/SP} = n_{NH_x}$: For $n_{NO_2} = 1$, E is equal to 0.0. However, if $n_{NO_2} > 1$, $E = 0.75$ and 0.667 for $n_{NH_x} = 1$ and >1, respectively. For the presence of a ring containing >NH group attached to two aromatic rings (e.g., tetranitrocarbazole), $E = 1.25$.

(c) $n_{DFG/SP} = n_{-C(=O)-}$: If the $-C(=O)-OH$ group is present, $E = 1.75$. The value of $E = 0.875$ for the attachment one $-C(=O)-$ to two aromatic rings (e.g., (2,4-dinitrophenyl)(2,4,6-trinitrophenyl)methanone).

(d) Polynitronaphthalene: $n_{NO_2}/n_{DFG/SP} \times E = 2.0$.

The higher value of $n_{NO_2}/n_{DFG/SP} \times E$ can be used for the presence of multiple types of functional groups, for example, $-OH$ and $-NH_2$. Thus, $n_{NO_2}/n_{DFG/SP} \times E = 0.75$ in 2-amino-4,6-dinitro-phenol.

(3) $n_{IFG/SP}$ and F: The ratio $n_{IFG/SP}/n_{NO_2}$ can be used to predict the values of F as follows:

(a) $n_{IFG/SP} = n_{-R/-OR}$: on attachment of $-R$ or $-OR$ to an aromatic ring, $F = 2.0$.

(b) $n_{IFG/SP} = n_{-NH-NH_2}$: on attachment of hydrazine to an aromatic ring, $F = 1.0$.

(c) $n_{IFG/SP} = n_{-N=N-}$: If the ratio $n_{-N=N-}/n_{NO_2} \geq 0.167$, $n_{NO_2}/n_{DFG/SP} \times F = 12.0$. For other nitro compounds containing $-N=N-$ groups, $n_{NO_2}/n_{DFG/SP} \times F = 6.0$.

Example 2.2: The condensed phase heat of formation of the compound given in Example 2.1 is calculated as

$$\Delta_f H^\theta(c) = \frac{\begin{matrix} 2.690a - 2.896b + 2.876c - 2.784d \\ -1.701(n'_{Ar} - 1) - 1.607\left(\dfrac{n_{NO_2}}{n_{DFG/SP}} \times E\right) + 3.246\left(\dfrac{n_{IFG/SP}}{n_{NO_2}} \times F\right) \end{matrix}}{Mw \times 10^{-4}}$$

$$= \frac{\begin{matrix} 2.690(24) - 2.896(6) + 2.876(14) - 2.784(24) \\ -1.701(0) - 1.607(0) + 3.246(6) \end{matrix}}{874.4 \times 10^{-4}}$$

$$= 458.7 \text{ kJ/mol}.$$

Since the measured $\Delta_f H^\theta(c)$ of this compound is 480.3 kJ/mol [199], the predicted $\Delta_f H^\theta(c)$ result obtained using eq. (2.8) is close to the value obtained from the experimental data.

2.2.2.3 Using the estimated gas-phase enthalpies of formation from the PM3 and B3LYP methods

For nitroaromatic energetic compounds, it was found that the estimated gas-phase heat of formation can be used to predict the condensed phase heat of formation. Two suitable correlations on the basis of the B3LYP/6–31G* and PM3 as follows [145]:

$$\Delta_f H^{\theta}(c) = 0.874 \left[\Delta_f H^{\theta}(g)\right]_{B3LYP/6-31G*} + 35.575a - 22.59b - 31.947d$$

$$+ 30.5n_{NO_2} - 141.91n_{Ar}, \tag{2.9}$$

$$\Delta_f H^{\theta}(c) = 0.911 \left[\Delta_f H^{\theta}(g)\right]_{PM3} - 10.8b + 26.968d$$

$$+ 46.17n_{NO_2} - 101.14n_{N_2} - 319.129n_{TRs}, \tag{2.10}$$

where $\left[\Delta_f H^{\theta}(g)\right]_{B3LYP/6-31G*}$ and $\left[\Delta_f H^{\theta}(g)\right]_{PM3}$ are the calculated gas-phase heats of formation in kJ/mol using the B3LYP/6–31G* and PM3 methods, respectively; n_{N_2} is the number of –N=N– or –N≡N groups; the n_{TRs} is the number of the attachment of three aromatic rings. These correlations are more complex than eqs. (2.7) and (2.8) because they require QM computations.

2.2.2.4 Simple method for nitramines, nitrate esters, and nitroaliphatics

From studying various nitramines, nitrate esters, nitroaliphatics, and related energetic compounds, it could be shown that the following equation can provide a suitable pathway for obtaining $\Delta_f H^{\theta}(c)$ [200]:

$$\Delta_f H^{\theta}(c) = 29.68a - 31.85b + 144.2c - 90.71d$$

$$- 88.84n_{OH} - 39.14n_{N-NO_2} - 45.62n_{>C=O} + 256.3n_1^0$$

$$- 380.5n_{=C<\overset{N}{N}} + 30.20n_{O-NO_2}, \tag{2.11}$$

where n_{OH}, n_{N-NO_2}, $n_{>C=O}$, and n_{O-NO_2} are the number of specified functional groups, $n_1^0 = 0$ if hydrogen is present in the molecule, and $n_1^0 = 1$ for hydrogen-free compounds; $n_{=C<\overset{N}{N}}$ is the number of $=C<\overset{N}{N}$ structural moieties present in the energetic compound.

Example 2.3: The condensed phase heat of formation of the following compound:

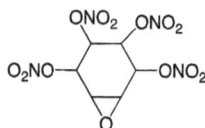

is calculated as

$$\Delta_f H^\theta(\text{c}) = 29.68a - 31.85b + 144.2c - 90.71d$$

$$- 88.84n_{\text{OH}} - 39.14n_{\text{N-NO}_2} - 45.62n_{>\text{C=O}} + 256.3n_1^0$$

$$- 380.5n_{=\text{C}<\overset{\text{N}}{\underset{\text{N}}{}}} + 30.20n_{\text{O-NO}_2}$$

$$= 29.68(6) - 31.85(6) + 144.2(4) - 90.71(13)$$

$$- 88.84(0) - 39.14(0) - 45.62(0) + 256.3(0)$$

$$- 380.5(0) + 30.20(4)$$

$$= -494.6 \text{ kJ/mol}.$$

The measured $\Delta_f H^\theta(\text{c})$ of this compound is −444.3 kJ/mol [199].

2.2.2.5 More reliable method for acyclic and cyclic nitramines, nitrate esters and nitroaliphatic energetic compounds

It was shown that a more reliable correlation than that of eq. (2.11) can be established, which is based on the molecular structure of energetic compounds. This correlation can be formulated as follows [193]:

$$\Delta_f H^\theta(\text{c}) = 39.10a - 37.89b + 96.73c - 66.07d$$

$$+ 215.4 \sum_i n_{i,\text{DE}} - 217.6 \sum_j n_{j,\text{IE}}. \tag{2.12}$$

The factors n_{DE} and n_{IE} are the number of some specific functional groups or structural parameters, which may decrease or increase the value of $\Delta_f H^\theta(\text{c})$, respectively. The values of n_{DE} and n_{IE} can be determined as follows:

(1) Prediction of n_{DE}

    (a) –OH group: If the ratio of the number of hydroxyl groups to the number of nonaromatic nitro or nitrate groups $\left(n_{\text{OH}}/n_{\text{NO}_2 \text{ or ONO}_2}\right) \geq 1$, then $n_{\text{DE}} = n_{\text{OH}}/n_{\text{NO}_2 \text{ or ONO}_2} \times 0.5$ (e.g., for 2-hydroxymethyl-2-nitro-propane-1,3-diol, $n_{\text{OH}}/n_{\text{NO}_2 \text{ or ONO}_2} = 3$ and $n_{\text{DE}} = 1.5$). For $n_{\text{OH}}/n_{\text{NO}_2 \text{ or ONO}_2} < 1$, $n_{\text{DE}} = 0.25$ (e. g., for 2,2,2-trinitro-ethanol, $n_{\text{OH}}/n_{\text{NO}_2 \text{ or ONO}_2} = 0.33$).

    (b) >C=O group: For compounds containing the >C=O group, $n_{\text{DE}} = n_{>\text{C=O}} \times 0.6$ (e.g., for 4,4,4-trinitro-butyric acid 2,2,2-trinitro-ethyl ester, $n_{\text{DE}} = 0.6$).

    (c) Cyclic and acyclic ether functional groups: For six-membered cyclic ether rings only, $n_{\text{DE}} = 0.25$. If there is an ether functional group of the type ROCH$_2$CH$_2$OR', $n_{\text{DE}} = 0.5$ (e.g., 1-nitrooxy-2-[2-(2-nitrooxy-ethoxy)-ethoxy]-ethane). The value of n_{DE} is found to be 0.25 for other types of acyclic ethers (e.g., 1-nitrooxy-2-(2-nitrooxy-ethoxy)-ethane).

    (d) Some specific molecular structures: if the $\overset{\text{O}}{\underset{\text{N}\quad\text{N}}{\|}}$, $\overset{\text{NO}_2}{\underset{\text{N}\quad\text{N}}{|}}$, or –NH–NO$_2$ molecular fragments are present, the values of n_{DE} correspond to 1.0, 0.5, and 0.2, respectively. For example, there are two ⟨structure⟩ and one –NH–NO$_2$ molecu-

lar fragments in 1-nitro-3-guanidinourea which gives $\sum_i n_{DE} = 2 + 0.2 = 2.2$.

(2) Prediction of n_{IE}

(a) Acyclic and cyclic nitramines containing only one C–N(NO$_2$)–C fragment: The value of $n_{IE} = 0.3$ (e.g., N-ethyl-N-nitro-ethanamine).

(b) The number of nitro groups attached to cubane (C$_8$H$_8$): The value of n_{IE} is 0.2. For example, $\sum_i n_{IE} = 8 \times 0.2 = 1.6$ for octanitrocubane.

(c) Hydrogen-free nitroalkanes: The value of $n_{IE} = 1.0$ (e.g., tetranitomethane).

Example 2.4: The condensed phase heat of formation of the compound given in Example 2.3 is calculated as

$$\Delta_f H^\theta(c) = 39.10a - 37.89b + 96.73c - 66.07d$$
$$+ 215.4 \sum_i n_{i, DE} - 217.6 \sum_j n_{j, IE}$$
$$= 39.10(6) - 37.89(6) + 96.73(4) - 66.07(13)$$
$$+ 215.4(0) - 217.6(0)$$
$$= -464.7 \text{ kJ/mol.}$$

Since the measured $\Delta_f H^\theta(c)$ of this compound is -444.3 kJ/mol [199], the predicted $\Delta_f H^\theta(c)$ obtained using eq. (2.12) is close to the value obtained from experimental data.

2.2.2.6 Prediction of the condensed phase heats of formation of polynitroarenes, polynitroheteroarenes, acyclic and cyclic nitramines, nitrate esters, and nitroaliphatic compounds

For polynitroarenes, polynitroheteroarenes, acyclic, and cyclic nitramines, nitrate esters, and nitroaliphatic compounds, the following correlation can be used to predict the condensed phase heat of formation [195]:

$$\Delta_f H^\theta(c) = 32.33a - 39.49b + 92.41c - 63.85d$$
$$+ 105.0\Delta_f H^\theta_{IEC} - 106.6\Delta_f H^\theta_{DEC}, \tag{2.13}$$

where $\Delta_f H^\theta_{IEC}$ and $\Delta_f H^\theta_{DEC}$ are two correcting functions which can be specified based on the presence or absence of some groups, which are described in the following situations.

(1) Prediction of $\Delta_f H^\theta_{DEC}$:

(a) –OH group: The values of $\Delta_f H^\theta_{DEC}$ are 1.4 and 1.0 for energetic compounds containing the –OH group as Ar–OH and R–OH, respectively (e.g., $\Delta_f H^\theta_{DEC} = 2 \times 1.4 = 2.8$ for 2,4,6-trinitrobenzene-1,3-diol).

(b) $-NH_x$ groups: If the $-NH_2$ or $>NH$ (or $-NH-NH_2$) group is present in the energetic compound, the value of $\Delta_f H^\theta_{DEC}$ is equal to 0.7 (e.g., $\Delta_f H^\theta_{DEC} = 3 \times 0.7 = 2.1$ for 2,4,6-trinitrobenzene-1,3,5-triamine).

(c) Acyclic and cyclic ether functional groups: The values of $\Delta_f H^\theta_{DEC}$ are 0.5 and 0.9 (except cyclic ethers with three-membered rings), respectively (e.g., $\Delta_f H^\theta_{DEC} = 1 \times 0.5 = 0.5$ for 2-methoxy-1,3,5-trinitrobenzene).

(d) Other specific polar groups: If the $-COOH$ (or $-O^-NH_4^+$, $-COCO-$, and $-NH-CO-$), $-N-CO-N-$, $-COO-$ (or acyclic $-CO-$), $-CO-H$, cyclic-$CO-$, or $-NH-NO_2$ functional groups are present, the contributions of these groups to $\Delta_f H^\theta_{DEC}$ correspond to 2.8, 2.4, 1.4, 1.0, 0.5, and 0.3, respectively (e.g., $\Delta_f H^\theta_{DEC} = 1 \times 2.8 = 2.8$ for 3,5-dinitrobenzoic acid).

(e) The number of nitrogen heteroatoms in six-membered rings: For nitroaromatics containing more than one six-membered aromatic ring, the contribution of each nitrogen heteroatom in the six-membered ring is 0.33 (e.g., $\Delta_f H^\theta_{DEC} = 3 \times 0.33 = 0.99$ for 2,4,6-tris(2,4,6-trinitrophenyl)-1,3,5-triazine).

(2) Prediction of $\Delta_f H^\theta_{IEC}$:

(a) Cyclic and acyclic nitramines: The values of $\Delta_f H^\theta_{IEC}$ are 0.5 and 0.8 for the acyclic and cyclic functional groups, respectively (e. g., $\Delta_f H^\theta_{IEC} = 0.5$ for 1,3,5-trinitro-1,3,5-triazinane).

(b) $R-NO_2$ (or $R-ONO_2$) and $Ar-N_3$: The value of is 0.3 for the attachment of an $-NO_2$ or $-ONO_2$ group to a nonaromatic carbon atom, and for the attachment of the $-N_3$ group to an aromatic ring, (e. g. $\Delta_f H^\theta_{IEC} = 0.3$ for 1,1-dinitropropane). There are some exceptions:

(i) if the $-C(NO_2)_3$ group is present, $\Delta_f H^\theta_{IEC} = 0.8$ (e.g., $\Delta_f H^\theta_{DEC} = 1.0$ and $\Delta_f H^\theta_{IEC} = 0.8$ for 2,2,2-trinitroethanol);

(ii) for linear mono-nitroalkanes ($a \geq 4$), $\Delta_f H^\theta_{IEC} = 0.6$ (e.g., $\Delta_f H^\theta_{IEC} = 0.6$ for 1-nitrobutane);

(iii) for hydrogen-free nitroalkanes, $\Delta_f H^\theta_{DEC} = 2.2$ (e. g., $\Delta_f H^\theta_{IEC} = 2.2$ for tetranitromethane);

(iv) for symmetric linear di-nitroalkanes, $\Delta_f H^\theta_{IEC} = 0.0$ (e. g., $\Delta_f H^\theta_{DEC} = 0.0$ for 1,2-dinitroethane).

(c) $-N=N-$ and $\overset{\diagdown}{N}-O$: For the molecular fragments $\overset{\diagdown}{N}-O$ and $-N=N-$, the contributions to $\Delta_f H^\theta_{IEC}$ are 0.7 and 0.8, respectively ($\Delta_f H^\theta_{IEC} = 3 \times 0.7 = 2.1$ for benzenetrifuroxan).

(d) Number of carbocyclic aromatic rings ($n_{Ar,car}$): For energetic compounds which only contain carbocyclic aromatic rings, the contribution to $\Delta_f H^\theta_{IEC}$ is $(n_{Ar,car}-1) \times 0.3$ (e. g. $\Delta_f H^\theta_{IEC}$ is $(2-1) \times 0.3 = 0.3$ for 2,2',4,4',6,6'-hexanitrobiphenyl).

(e) Attachment of alkyl groups to an aromatic ring: The values of $\Delta_f H^\theta_{IEC}$ are 0.2 and 0.8 for the attachment of methyl and longer carbon chain alkyl groups (or –CH=CH–) to aromatic rings, respectively (e.g., $\Delta_f H_{IEC}^{\theta} = 0.2$ for 1-methyl -2,4-dinitrobenzene). If $n_{Ar, car} > 1$, it is not necessary to include condition (d).

(f)　Nitro groups attached to nonaromatic four-membered rings: $\Delta_f H_{IEC}^{\theta} = 0.45 \times$ the number of –NO$_2$ groups attached to a nonaromatic four-membered ring (e.g., $\Delta_f H_{IEC}^{\theta} = 8 \times 0.45$ for octanitrocubane).

Example 2.5: The value of $\Delta_f H^{\theta}(c)$ for the following molecular structure is calculated as

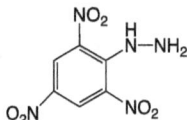

$$\Delta_f H^{\theta}(c) = 32.33a - 39.49b + 92.41c - 63.85d + 105.0\Delta_f H_{IEC}^{\theta} - 106.6\Delta_f H_{DEC}^{\theta}$$

$$= 32.33(6) - 39.49(5) + 92.41(5) - 63.85(6) + 105.0(0) - 106.6(0.7)$$

$$= 0.8 \text{ kJ/mol.}$$

The measured $\Delta_f H^{\theta}(c)$ of this compound is 36.5 kJ/mol [199].

2.2.2.7 General correlation for organic energetic materials

It has been established that the following general correlation can be used for organic energetic compounds containing different types of energetic groups such as –NO$_2$, –ONO$_2$, –NNO$_2$, –ON=O, –O–O–, and –N$_3$ [201]:

$$\Delta_f H^{\theta}(c) = -111.4 + 33.11a - 28.84b + 86.80c - 62.03d$$

$$-79.25\left(\Delta_f H_{add,DHC}^{\theta} + \Delta_f H_{nonadd,DHC}^{\theta}\right)$$

$$+ 153.3\left(\Delta_f H_{add,IHC}^{\theta} + \Delta_f H_{nonadd,IHC}^{\theta}\right), \tag{2.14}$$

whereby the subscripts DHC and IHC in the $\Delta_f H_{add,DHC}^{\theta}$, $\Delta_f H_{nonadd, DHC}^{\theta}$, $\Delta_f H_{add,IHC}^{\theta}$, and $\Delta_f H_{nonadd, IHC}^{\theta}$ show decreasing and increasing heat contents in the compounds, respectively. Tables 2.1–2.4 summarize the values of $\Delta_f H_{add,DHC}^{\theta}$, $\Delta_f H_{nonadd, DHC}^{\theta}$, $\Delta_f H_{add,IHC}^{\theta}$, and $\Delta_f H_{nonadd, IHC}^{\theta}$ for various different functional groups and molecular fragments.

Table 2.1: Summary of the correcting function $\Delta_f H^\theta_{add,DHC}$.

Molecular moieties	Compound	$\Delta_f H^\theta_{add,DHC}$	Example	Exception
Hydroxyl group	Ar–OH	1.2	2 × 1.2 = 2.4 for 4,6-dinitroresorcinol	
	R–OH	1.0	3 × 1.0 = 3.0 for 2-(hydroxymethyl)-2-nitro-1,3-propanediol	
–NH₂ or >NH group	More than one –NH₂ or C–NH– group	0.6	3 × 0.6 = 1.8 for 2,4,6-trinitrobenzene-1,3,5-triamine	The value of $\Delta_f H^\theta_{add,DHC}$ = 0.7 and 1.4 for the presence of Ar–NH–Ar and 1,2,4-triazole molecular fragments, respectively.
Specific polar groups	–COOH (or –ONH₄, –NHCOO–, NHCONH–, –NHCOC–, and –COCO–)	2.8	2.8 for 3,5-dinitrobenzoic acid	
	–COOCO–	2.5	2.5 for bis(1-oxobutyl) peroxide	
	–NHC(=NH)NH–	1.7	1.7 for nitroguanidine (NQ)	
	–COO–C	1.4	1.4 for butanoic acid, 4,4,4-trinitro-, 2,2,2-trinitroethyl ester	
	Cyclic or acyclic ketone	1.2	1.2 for tetramethylolcyclopentanone tetranitrate	
Alkyl or aryl halide	Fluorine atoms attached to nonaromatic carbon atoms	$6n_{CF_2} - 2$, where n_{CF_2} is the number of CF₂ groups	6 × 2 – 2 = 10 for 1,1,2,2-tetrafluoro-1,2-dinitroethane	
Nitrate salts	–CH$_n$–NH₂ (or –(CH$_n$)₂–NH, cyclic –(CH$_n$)₃N, –NH$_n$–NH₂, and NH₄⁺)	1.5	2 × 1.5 = 3.0 for ethylenediamine dinitrate (EDD)	
	–C(=O or NH)–NH₂	4.2	4.2 for guanidine nitrate	

Table 2.2: Summary of the correcting function $\Delta_f H^\theta_{nonadd, DHC}$.

Molecular moieties	Compound	$\Delta_f H^\theta_{nonadd, DHC}$	Example	Exception
Ether	Acyclic ether	0.5	2,2'-[Oxybis(methylene)]bis [2-[(nitrooxy)methyl] propane-1,3-diyl] tetranitrate	Ar–O–Ar and Ar–O–R, where R is an alkyl group with more than one carbon atom
	Six member cyclic ether	1.7	D-Glucopyranose pentanitrate	
	–O–C–O–C– O–	3.4	Saccharose octanitrate	
Alkyl fluoride	Monofluoro derivative of R–F	2	Fluorotrinitromethane	

Table 2.3: Summary of the correcting function $\Delta_f H^\theta_{add, IHC}$.

Molecular moieties	Compound	$\Delta_{bff} H^\theta_{add, IHC}$	Example	Exception
Molecular fragments –N=N– (or –N–N–N– and –N–N–N–N–), 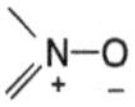 $\overset{\backslash}{\underset{/}{N}}\!\!-\!\!O$ (or azido group)	–N=N– (or –N–N–N– and –N–N–N–N–)	0.6	2 × 0.6 = 1.2 for 2,6-bis (picrylazo)-3,5-dinitropyridine	
	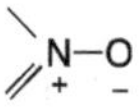 $\overset{\backslash}{\underset{/}{N}}\!\!-\!\!O$	0.7	3 × 0.7 = 2.1 for benzo [1,2-c:3,4-c':5,6-c"]tris [1,2,5]oxadiazole, 1,4,7-trioxide (benzenetrifuroxan)	
Nitro groups attached to a nonaromatic four-membered ring	Attachment of –NO$_2$ groups to a nonaromatic four-membered ring	0.45	8 × 0.45 = 3.6 for octanitrocubane	
Number of carbocyclic aromatic rings (n_{Ar})	For energetic compounds containing only carbocyclic aromatic rings	$(n_{Ar} - 1) \times 0.3$	(2 – 1) × 0.3 = 0.3 for 2,2',4,4',6,6'-hexanitrobiphenyl	

Table 2.4: Summary of the correcting function $\Delta_f H^\theta_{nonadd,IHC}$.

Molecular moieties	Compound	$\Delta_f H^\theta_{nonadd,IHC}$	Example	Exception
Cyclic and acyclic nitramine (or nitroso amine) functional groups	Acyclic nitramine	0.8	N-Methyl-N-nitro-methanamine	
	Cyclic nitramine (or nitroso amine)	0.5	1,4-Dinitropiperazine	
R-NO$_2$, R-ONO$_2$, R-CN, mono and polynitrobenzene, -NH-NO$_2$ (or -NF-NO$_2$), -ONO (nitrite), -O-OH and azide derivatives of carbocyclic aromatic compounds or R-N$_3$	Attachment of more than one -NO$_2$ group to nonaromatic carbon atom(s)	0.6	1,1-Dinitropropane	(a) For the presence -C(NO$_2$)$_3$, $\Delta_f H^\theta_{nonadd, IHC} = 0.8$ (e.g., 2,2,2-trinitroethanol);
	Attachment of the -N$_3$ group to a carbocyclic aromatic ring (or R-N$_3$)	0.7	(Azidomethyl)-benzene	
	R-ONO$_2$ (other functional groups except the -NO$_2$ group are absent)	0.5	Ethylene glycol, dinitrate (nitroglycol)	(b) For linear mono-nitroalkanes, $\Delta_f H^\theta_{nonadd, IHC} = 0.6$ (e.g., 1-nitrobutane);
	R-ONO$_2$ (in the presence of the other functional groups)	0.1	N-Butyl-N-(2-nitroxyethyl) nitramine (BuNENA)	(c) For hydrogen-free nitroalkanes, $\Delta_f H^\theta_{nonadd, IHC} = 2.2$ (e.g., for tetranitromethane)
	Mono and polynitrobenzene	0.7	1,2-Dinitrobenzene	
	-ONO (nitrite)	0.5	Propyl nitrite	
	-O-OH	0.5	tert-Butyl hydroperoxide	
	-CN	0.4	Trinitroacetonitrile	
	-NH-NO$_2$ (or -NF-NO$_2$)	0.2	Ethylenedinitramine (EDNA)	

(continued)

Table 2.4 (continued)

Molecular moieties	Compound	$\Delta_f H^\theta_{\text{nonadd, IHC}}$	Example	Exception
Furan derivatives and cyclic ethers smaller than a six-membered ring	Furan derivatives	0.25	2-Nitrofuran	
	Cyclic ethers amller than a six membered ring	0.5	α-Epoxyconduritole tetranitrate	
Attachment of alkyl groups to aromatic ring	Attachment of methyl groups to an aromatic ring	0.5	2,4,6-Trinitrotoluene (TNT)	
	Attachment of longer carbon-chain alkyl groups (or –CH=CH–) to an aromatic ring	0.6	1-Ethyl-2-nitrobenzene	
Cyclic and acyclic peroxide	Cyclic and acyclic peroxide with general formula R_1–O–O–R_2 in which R_1 or R_2 has four or less carbon atoms	0.7	Hexamethylenetriperoxide Diamine (HMTD)	
Alkyl or aryl halide	Monochloro derivative of Ar–Cl	0.2	1-Chloro-2-nitrobenzene	
	Monochloro derivative of R–Cl	0.5	1-Chloro-1,1-dinitroethane	
	Monobromo derivative of R–Br	0.8	Bromotrinitromethane	
	Monoiodo derivative of R–I	1.2	Iodotrinitromethane	

Example 2.6: Glycerol-1,2-dinitrate has the following molecular structure:

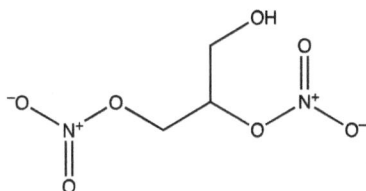

Applying eq. (2.14) gives the condensed phase heat of formation as

$$\Delta_f H^\theta(c) = -111.4 + 33.11a - 28.84b + 86.80c - 62.03d$$

$$-79.25\left(\Delta_f H^\theta_{add,DHC} + \Delta_f H^\theta_{nonadd,DHC}\right)$$

$$+153.3\left(\Delta_f H^\theta_{add,IHC} + \Delta_f H^\theta_{nonadd,IHC}\right)$$

$$= -111.4 + 33.11(3) - 28.84(6) + 86.80(2) - 62.03(7)$$

$$-79.25(1+0) + 153.3(0+0.1)$$

$$= -509.6 \text{ kJ/mol}.$$

The measured $\Delta_f H^\theta(c)$ of this compound is −472.4 kJ/mol [199].

2.3 Energetic compounds with high nitrogen contents

Significant amounts of solid carbon and nonoxidized organic species are produced during the detonation of common high explosives because of the negative oxygen balance of these compounds. However, the energy that nitrogen-rich explosives liberate is due to their high positive heats of formation rather than as a result of the oxidation of the carbon backbone [202]. The main detonation product of materials with high nitrogen contents is N_2 gas, which means that the detonation process is clean [203]. High-nitrogen materials have high densities and good oxygen balances because they have a low carbon and hydrogen content [204]. Common nitramine high explosives have nitrogen contents of up to 38 % (in RDX, HMX, and Cl-20), while some derivatives of triazole, tetrazole, triazine, tetrazine, furazan, and organic nitrogen-containing chains have up to 88 % nitrogen-contents. High-nitogen materials have been defined as those containing more than 50 % w/w nitrogen in their molecular structure [205]. High-nitrogen compounds are suitable candidates for insensitive high explosives [206], gun propellants [207, 208], and clean gas generators in vehicle airbags [209]. Two different empirical methods are introduced here for the calculation of the condensed phase heat of formation of these compounds.

2.3.1 Using the molecular structure

It has been shown that the molecular structure of high-nitrogen materials can be used to predict their condensed phase heat of formation as follows [210]:

$$\Delta_f H^\theta(c) = 39.24a - 40.01b + 83.63c - 49.61d + 115.5 \sum_i IF_i - 177.4 \sum_j DF_j, \qquad (2.15)$$

where IF and DF are correcting factors, which can be estimated as follows:
(1) Azido group: For the attachment of the $-N_3$ group to a tetrazole or tetrazine ring, as well as to aliphatic N-containing compounds, IF is the number of azido groups. Meanwhile, the contribution of $-N_3$ for high-N compounds containing only triazole and triazine rings is zero.
(2) Azo and azoxy groups: In tetrazine, triazine, triazole, and furazan derivatives which contain $-N{=}N-$ or $-N{=}N^+O^--$ bridges, $IF = 1.5$. For tetrazole derivatives, $IF = 0.2$.
(3) Guanidino, carbonyl and hydroxide groups: if any of these functional groups are present, $DF = 1.2 \times$ the number of these groups in the molecule.
(4) Amino groups: $DF(-NH_2) = 0.6 \times$ (the number of $-NH_2$ groups) and $DF(-NH-) = 0.3 \times$ (the number of $-NH-$ groups).

Example 2.7: The condensed phase heat of formation of 3,6-diazido-1,2,4,5-tetrazine with the following molecular structure is calculated as

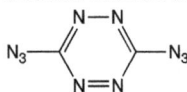

$$\Delta_f H^\theta(c) = 39.24a - 40.01b + 83.63c - 49.61d + 115.5 \sum_i IF_i - 177.4 \sum_j DF_j$$

$$= 39.24(2) - 40.01(0) + 83.63(10) - 49.61(0) + 115.5(2) - 177.4(0)$$

$$= 1146 \text{ kJ/mol}.$$

The experimental $\Delta_f H^\theta(c)$ was reported to be 1101 kJ/mol [211].

2.3.2 Gas-phase information

It was shown that the value of $\Delta_f H^\theta(c)$ can be calculated using the $\Delta_f H^\theta(g)$ value obtained from either the B3LYP/6–31G* or PM6 method as follows [191]:

$$\Delta_f H^\theta(c) = -57.28 + 0.9350 \left[\Delta_f H^\theta(g)\right]_{B3LYP/6-31G*}$$

$$+ 27.86 \sum_i ICF_i - 30.06 \sum_j DCF_j \qquad (2.16)$$

$$\Delta_f H^\theta(c) = -34.99 + 0.8692 \left[\Delta_f H^\theta(g)\right]_{PM6}$$

$$+ 30.64 \sum_i ICF_i - 25.95 \sum_j DCF_j, \qquad (2.17)$$

where ICF and DCF are increasing and decreasing correcting factors, respectively, which are estimated according to the following:

(1) Azido group: The value of ICF equals two times the number of azido groups in the molecule.
(2) Azo and azoxy groups: The value of ICF equals the total number of –N=N– or –N=N⁺O⁻ bridges.
(3) Hydrogen bonding effect: The value of DCF is the number of hydrogen bonds in each molecule.
(4) Ethylene groups: The value of DCF is two times the total number of ethylene groups in the molecule.

Appendix B shows the calculated $\left[\Delta_f H^\theta(g)\right]_{B3LYP/6-31G*}$ and $\left[\Delta_f H^\theta(g)\right]_{PM6}$. Table B.1 provides the calculated total energies and formation enthalpies (at 0 K) for 100 energetic materials with high nitrogen contents by the B3LYP/6–31G*. Table B.2 also gives the predicted gas-phase standard enthalpies of formation for 100 energetic materials with high nitrogen content.

Example 2.8: The calculated gas-phase heat of formation of 1,2-di(1H-tetrazol-5-yl)ethane with the following molecular structure is $[\Delta_f H^\theta(g)]_{B3LYP/6-31G*} = 639.5$ kJ/mol and $[\Delta_f H^\theta(g)]_{PM6} = 613.9$ kJ/mol:

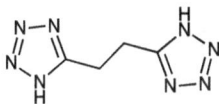

The use of eqs. (2.16) and (2.17) gives the values of $\Delta_f H^\theta(c)$ as

$$\Delta_f H^\theta(c) = -57.28 + 0.9350 \left[\Delta_f H^\theta(g)\right]_{B3LYP/6-31G*} + 27.86 \sum_i ICF_i - 30.06 \sum_j DCF_j$$

$$= -57.28 + 0.9350(639.5)_{B3LYP} + 27.86(0) - 30.06(2+2)$$

$$= 420.4 \text{ kJ/mol},$$

$$\Delta_f H^\theta(c) = -34.99 + 0.8692 \left[\Delta_f H^\theta(g)\right]_{PM6} + 30.64 \sum_i ICF_i - 25.95 \sum_j DCF_j$$

$$= -34.99 + 0.8692(613.9)_{PM6} + 30.64(0) - 25.95(2+2)$$

$$= 398.4 \text{ kJ/mol}.$$

The experimental $\Delta_f H^\theta(c)$ value was reported to be 444.4 kJ/mol [212].

2.4 The condensed phase heat of formation of energetic ionic liquids and salts

2.4.1 Complex approach

The presence of energetic cations or anions in some classes of ionic liquids or salts provides energetic ionic liquids or salts. Energetic ionic liquids or salts (EILoS) can contain high nitrogen (high-N) organic cations and bulky anions. Anions include one or more energetic groups, that is, $-NO_2$, $-N_3$, and $-CN$. Since EILoS can give suitable thermally stability, they may be used as explosives, pyrotechnics, and propellants [213–217]. The value of the condensed phase heat of formation of an EILoS is very important for the assessment of its detonation velocity and detonation pressure [219]. For an EILoS with the general formula $(cation)_i(anion)_j$, computation of its $\Delta_f H^\theta(c)$ can be done using gas-phase heats of formation of anion and cation as well as the heat of phase transition according to Hess's law of constant summation (Born–Haber energy cycle) as [220]:

$$\Delta_f H^\theta(c)[\text{EILoS}] = \sum_i \Delta_f H_i^\theta(g)[\text{cation}] + \sum_j \Delta_f H_j^\theta(g)[\text{anion}] - \Delta H_{PT} \tag{2.18}$$

where $\Delta_f H^\theta(c)[\text{EILoS}]$ is the condensed phase heat of formation of EILoS; and $\sum_j \Delta_f H_j^\theta(g)[\text{anion}]$ are the sum of gas-phase heats of formation of cations and anions, respectively; ΔH_{PT} is the heat of phase transition. Gao et al. [221] used eq. (2.18) to compute $\Delta_f H^\theta(c)[\text{EILoS}]$ for 119 energetic salts including imidazolium, triazolium, and tetrazolium-based cations. They also used isodesmic reactions to compute $\sum_i \Delta_f H_i^\theta(g)[\text{cation}]$ and $\sum_j \Delta_f H_j^\theta(g)[\text{anion}]$. For computation of ΔH_{PT}, they used the method of Jenkins et al. [222]. They considered all species as closed shells in nature. They carried out various geometric optimization and frequency analyses for molecules with more than 10 heavy atoms using B3-LYP functional with $6–31 + G^{**}$ basis set. They calculated single energy points at the MP2-(full)/$6–311++G^{**}$ level. They characterized all of the optimized structures to obtain true local energy minima on the potential energy surface without imaginary frequencies. They computed the molecular energies (proton affinities and ionization energies) at the G215 or G2MP2 level. Zhang et al. [223] studied the effects of different energetic substituents containing $-NO_2$, $-NF_2$, $-CN$, $-N_3$, and $-NH_2$ $\Delta_f H^\theta(c)[\text{EILoS}]$ for some tetrazole salts using a similar complex approach as that used by Gao et al. [221]. They indicated that the presence of energetic groups in most cases can increase $\Delta_f H^\theta(c)[\text{EILoS}]$. The use of this approach has several shortcomings because it requires a high-speed computer, specific computer codes, and expert users. Thus, two simple QSPR methods based on fragments of EILoS for prediction of $\Delta_f H^\theta(c)[\text{EILoS}]$ are discussed here [224, 225].

2.4.2 QSPR methods

2.4.2.1 Imidazolium-based ionic liquids or salts

A simple approach was introduced to estimate $\Delta_f H^\theta(c)$ of imidazolium-based ionic liquids or salts [224]. For some imidazolium-based ionic liquids or salts, the reliability of the model was higher than the available outputs of complex QM methods [224]. This approach is based on the following form [224]:

$$\Delta_f H^\theta(c)[\text{IMILoS}] = 1594 - 76.32 b_{\text{cat}} - 433.0 e_{\text{ani}} + 289.3 \Delta_f H^\theta_{\text{Inc}}[\text{IMILoS}]$$

$$- 304.3 \Delta_f H^\theta_{\text{Dec}}[\text{IMILoS}] \tag{2.19}$$

where $\Delta_f H^\theta(c)[\text{IMILoS}]$ is the condensed phase heat of formation of imidazolium-based ionic liquids or salts; b_{cat} and e_{ani} denote the number of hydrogen and fluorine atoms in cation and anion, respectively; $\Delta_f H^\theta_{\text{Inc}}[\text{IMILoS}]$ and $\Delta_f H^\theta_{\text{Dec}}[\text{IMILoS}]$ contribute to increasing and decreasing heat contents in imidazolium-based ionic liquids or salts, respectively [224]. Table 2.5 summarizes the values of $\Delta_f H^\theta_{\text{Inc}}[\text{IMILoS}]$ and $\Delta_f H^\theta_{\text{Dec}}[\text{IMILoS}]$.

Since the coefficients of b_{cat} and e_{ani} have negative signs, decreasing their values gives a more positive value of $\Delta_f H^\theta(c)[\text{IMILoS}]$. Thus, it is desirable to design imidazolium-based ionic liquids or salts with low values of b_{cat} and e_{ani} because they provide high detonation or combustion performance. Moreover, the existence and absence of $\Delta_f H^\theta_{\text{Inc}}[\text{IMILoS}]$ and $\Delta_f H^\theta_{\text{Dec}}[\text{IMILoS}]$, respectively, can help to improve the stored chemical energy in a designed sample.

Example 2.9: Calculate the value of $\Delta_f H^\theta(c)[\text{IMILoS}]$ for the following ionic liquid:

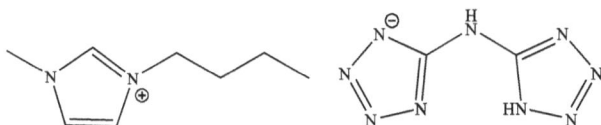

The use of eq. (2.19) and Table 2.5 provides the value of $\Delta_f H^\theta(c)[\text{IMILoS}]$ as

$$\Delta_f H^\theta(c)[\text{IMILoS}] = 1594 - 76.32 b_{\text{cat}} - 433.0 e_{\text{ani}} + 289.3 \Delta_f H^\theta_{\text{Inc}}[\text{IMILoS}]$$

$$- 304.3 \Delta_f H^\theta_{\text{Dec}}[\text{IMILoS}]$$

$$= 1594 - 76.32 \times 30 - 433.0 \times 0 + 289.3 \times 0.4 (3-1) - 304.3 \times 0$$

$$= 680 \text{ kJ/mol} \tag{2.20}$$

The reported value of $\Delta_f H^\theta(c)[\text{IMILoS}]$ is 682.3 [225].

Table 2.5: Different values of $\Delta_f H^\theta_{Inc}$[IMILoS] and $\Delta_f H^\theta_{Dec}$[IMILoS].

Cation	Anion	$\Delta_f H^\theta_{Inc}$ [IMILoS]	$\Delta_f H^\theta_{Dec}$ [IMILoS]	Condition
		$0.4(n-1)$	0	$1 \le n \le 7$
		0	$1.75 - 0.33n$	$1 \le n \le 3$
		0	4.4	$n = 1$
		$3.5 - \dfrac{9-n}{2}$	0	$n \le 9$
		0	4.0	$n \ge 13$
		$3.0 - \dfrac{9-n}{2}$	0	$3 \le n \le 9$
		0	4.3	$n \ge 11$
		0	1.2	$n = 3$
		0	5.7	$n \ge 11$
		0	0.3	–
		3.0	0	–

Table 2.5 (continued)

Cation	Anion	$\Delta_f H^\theta_{Inc}$ [IMILoS]	$\Delta_f H^\theta_{Dec}$ [IMILoS]	Condition
(imidazolium structure, O$_2$N-substituted)	(tetrazole/azo anion structure)			
(imidazolium structure with (CH$_2$)n)	Br$^\ominus$	0	$-0.4n + 2.9$	$3 \leq n \leq 4$
(imidazolium structure with (CH$_2$)n groups)		0	0.8	$n \leq 3$
		3.2	0	$n \geq 7$

2.4.2.2 Triazolium-based energetic ionic liquids or salts

A simple method has been introduced to estimate $\Delta_f H^\theta(c)$ of triazolium-based energetic ionic liquids or salts [226]. It requires some specific elemental composition of cations and anions as well as two correcting functions. It is given as follows [226]:

$$\Delta_f H^\theta(c)[\text{TAILoS}] = -27.31 b_{cat} + 102.7 c_{cat} + 259.8 a_{ani} - 319.2 b_{ani} + 45.32 c_{ani} - 125.9 d_{ani}$$

$$+ 632.6 f_{ani} + 79.78 \Delta_f H^\theta_{Inc}[\text{TAILoS}] - 74.50 \Delta_f H^\theta_{Dec}[\text{TAILoS}]$$

$$(2.20)$$

where $\Delta_f H^\theta(c)[\text{TAILoS}]$ is the value of $\Delta_f H^\theta(c)$ for triazolium-based energetic ionic liquids or salts; c_{cat} is the number of nitrogen atoms in cation; a_{ani}, b_{ani}, c_{ani}, d_{ani}, and f_{ani} give the number of carbon, hydrogen, nitrogen, oxygen, and chlorine atoms in the anion, respectively; $\Delta_f H^\theta_{Inc}[\text{TAILoS}]$ and $\Delta_f H^\theta_{Dec}[\text{TAILoS}]$ give increasing and decreasing functions in the triazolium-based EILoS, respectively. Table 2.6 gives the summary of the values of $\Delta_f H^\theta_{Inc}[\text{TAILoS}]$ and $\Delta_f H^\theta_{Dec}[\text{TAILoS}]$.

Table 2.6: Contribution of structural parameters in predicting $\Delta_f H^\theta_{Inc}$ [TAILoS] and $\Delta_f H^\theta_{Dec}$ [TAILoS].

Cation	Anion	$\Delta_f H^\theta_{Inc}$ [TAILoS]	$\Delta_f H^\theta_{Dec}$ [TAILoS]
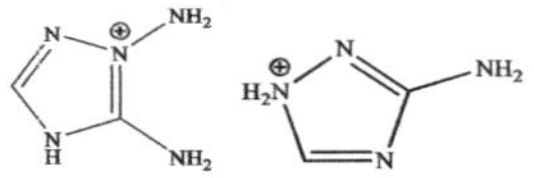	NO_3^-	0	1.5
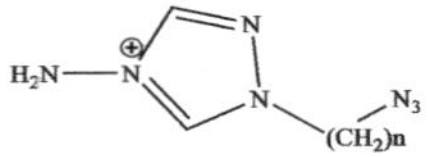		2.0	0
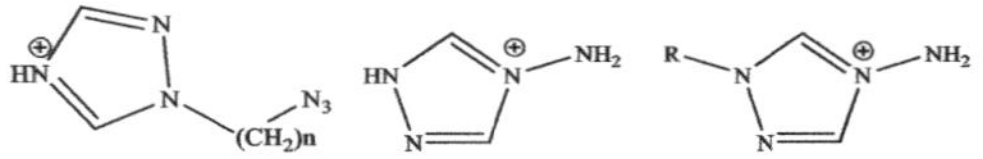	ClO_4^-	0	1.5
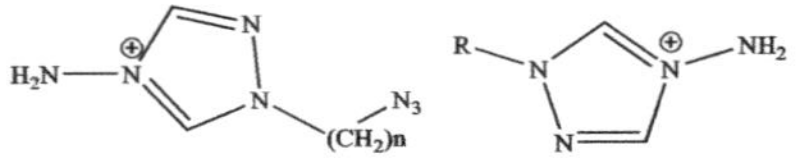		2.5	0
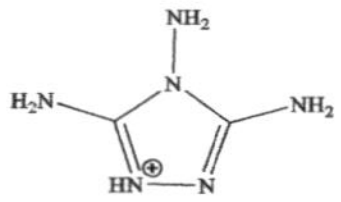	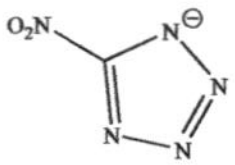	0	2.0
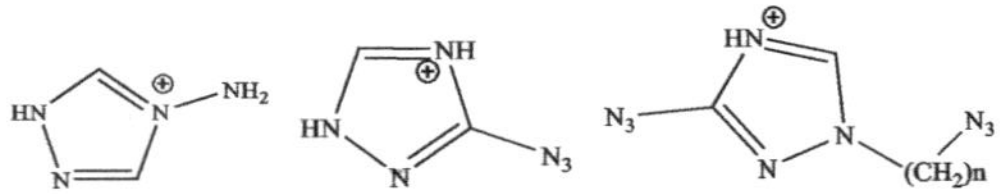		2.5	0
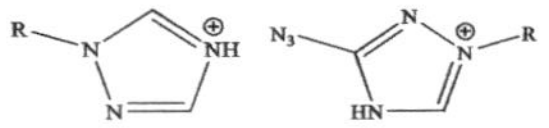		5.0	0

2.5	0
3.5	0
0	1.5
0	1.5
1.5	0
3.0	0
5.5	0

(continued)

Table 2.6 (continued)

Cation	Anion	$\Delta_f H^\theta_{Inc}$ [TAILoS]	$\Delta_f H^\theta_{Dec}$ [TAILoS]
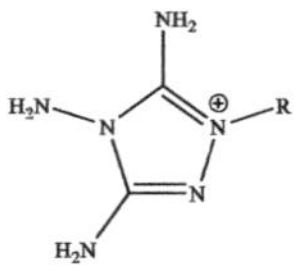	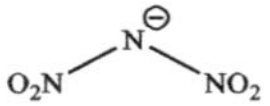	3.0	0
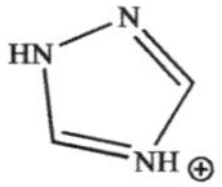 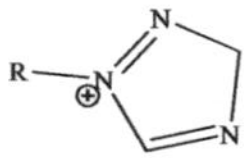 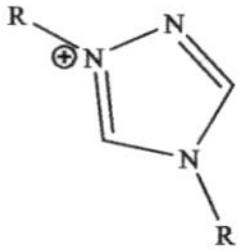	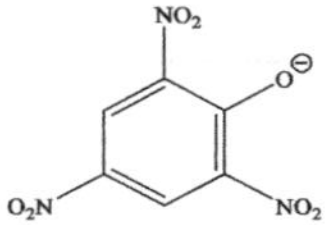	1.5	0
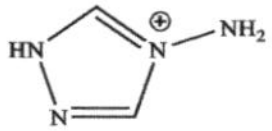	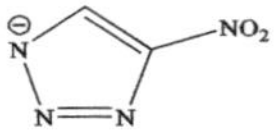	0	3.5

Example 2.10: Predict the value of $\Delta_f H^\theta(c)$[TAILoS] for the following ionic liquid:

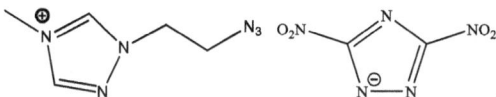

The use of eq. (2.20) and Table 2.6 gives the value of $\Delta_f H^\theta(c)$[TAILoS] as follows:

$$\Delta_f H^\theta(c)[\text{TAILoS}] = -27.31b_{cat} + 102.7c_{cat} + 259.8a_{ani} - 319.2b_{ani} + 45.32c_{ani} - 125.9d_{ani}$$

$$+ 632.6f_{ani} + 79.78\Delta_f H^\theta_{Inc}[\text{TAILoS}] - 74.50\Delta_f H^\theta_{Dec}[\text{TAILoS}]$$

$$= -27.31 \times 9 + 102.7 \times 6 + 259.8 \times 2 + 45.23 \times 5 - 125.9 \times 4 + 79.78 \times 1.5$$

$$= 732.2 \text{ kJ/mol.}$$

The measured value of $\Delta_f H^\theta(c)$[TAILoS] is 771.3 [227]. The calculated value of $\Delta_f H^\theta(c)$[TAILoS] by complex QM method using eq. (2.19) is 576.0 kJ/mol [221], which has a larger deviation (−195.3 kJ/mol).

Appendix C, including Tables C.1 and C.2, shows the measured and calculated values of $\Delta_f H^\theta(c)$[IMILoS] and $\Delta_f H^\theta(c)$[TAILoS] for some energetic ionic liquids and salts.

2.5 Summary

This chapter has introduced different empirical methods for the prediction of the condensed phase heat of formation of important classes of energetic compounds, as well as of high-nitrogen compounds (i.e., compounds with more than 50 % w/w nitrogen in their molecular structure). Equations (2.7) and (2.10) provide the simplest approaches for the prediction of the $\Delta_f H^\theta(c)$ values of nitroaromatics, and also for nitramines, nitrate esters, nitroaliphatics, and related energetic compounds. The other correlations outlined in Section 2.2.2 give more reliable predictions, but they are considerably more complex. Among the correlations that were introduced, eq. (2.14) provides the best method for the reliable calculation of $\Delta_f H^\theta(c)$ for different types of energetic compounds containing different types of energetic groups, that is, $-NO_2$, $-ONO_2$, $-NNO_2$, $-ON{=}O$, $-O{-}O{-}$, and $-N_3$. For high-nitrogen compounds, eq. (2.16) is simpler than eq. (2.17). The reliability of both correlations is good, but eq. (2.17) should be used preferably for high-nitrogen compounds containing unfamiliar molecular fragments.

Section 2.4 introduces eq. (2.18) as a complex QM approach for the calculation of $\Delta_f H^\theta(c)$[EILoS]. Equations (2.19) and (2.20) give two simple and reliable correlations for predicting the values of $\Delta_f H^\theta(c)$[IMILoS] and $\Delta_f H^\theta(c)$[TAILoS], respectively.

3 Melting point

The knowledge of the melting point of the desired compound can be used for its identification and purification. It may be used to calculate other properties, for example,
aqueous solubility, liquid viscosity, and vapor pressure [228]. Since there is a close relationship of melting point to solubility, it is important in environmental studies and
the assessment of the purity of a chemical substance [229]. In the measurement of
melting points some difficulties may occur such as the relationship of the melting
point to the phase transition from the solid to the liquid state and the possible existence of different crystalline structures below the melting point [230]. The melting
point of organic compounds containing energetic groups is a significant property especially for melting cast explosives [231]. It is important to have reliable methods for
the prediction of melting points of different classes of organic energetic compounds
because they are significant in the design of low-melting-point candidates for casting
medium.

Energetic melt-castable materials play a crucial role in the world of explosives
and propellants. Unlike typical energetic materials, these compounds can reversibly
transition between solid and liquid states within a relatively narrow melting range
(70–120 °C) [232, 233]. This unique property makes them ideal carriers for embedding
other crystalline energetic (or nonenergetic) components into various shapes and dimensions needed for practical applications [234]. Despite their importance, the most
commonly used melt-castable material today is still 2,4,6-trinitrotoluene (TNT) – a
high-density aromatic compound discovered over 150 years ago. However, TNT has
significant drawbacks, including low energy output, toxic byproducts, and environmental concerns [235]. Over the past few decades, researchers have explored potential
replacements, such as:

- 1-Methyl-3,4,5-trinitropyrazole (MTNP) [236]
- 1,3,3-Trinitroazetidine (TNAZ) [237]
- Bis(1,2,4-oxadiazole)bis(methylene) dinitrate (BOM) [238]

While these alternatives show promise, their real-world performance still falls short,
keeping them confined to laboratory testing rather than industrial use.

Designing new energetic melt-castable materials is no easy task. Scientists must
balance four key properties [239, 240]:

1. Energy output
2. Sensitivity (safety)
3. Melting point
4. Decomposition temperature

This is far more complex than designing conventional energetic materials, where only
energy and sensitivity are major concerns.

https://doi.org/10.1515/9783112206768-003

The search for new energetic compounds with "ideal" physical properties is a problem of the utmost importance to both research chemists and chemical industry. Since the melting point is one of the fundamental physical properties which is used in the identification and purification of a chemical, and is also highly valuable for the calculation of other important physicochemical properties such as the vapor pressure and aqueous solubility, it is important to have reliable methods which can be used to predict the melting point of an energetic compound. Group additivity, QSPR, quantum mechanical and empirical methods are all different approaches which have been developed to enable prediction of the melting points of different classes of organic compounds and ionic liquids. Some simple and reliable empirical methods are discussed in detail in this chapter, and the other methods are also briefly described.

3.1 Group additivity, QSPR, ML, and quantum mechanical methods

The melting point marks the transition where a solid's orderly molecular structure gives way to a disordered liquid state – a critical property for pharmaceuticals since it directly impacts drug solubility, which determines how effectively medicines work in the body [241]. While experimental melting point measurements are straightforward, drug discovery urgently needs predictive methods to screen vast libraries of unsynthesized compounds and avoid costly late-stage development issues [241].

Among the different approaches which can be used to predict the melting points of organic compounds, group additivity methods are widely used [187]. The group additivity methods can predict the melting point of a desired organic compound by summing the number of each group multiplied by its contribution [187]. Although they are simple and provide quick estimations, many have questionable accuracy and their reliability with respect to the prediction of melting points of organic energetic compounds is unknown. The Joback–Reid [242] approach for the estimation of the melting points of pure compounds is the simplest group additivity method. The estimated melting points (T_m) from the Joback–Reid group additivity method [242] can be estimated as follows:

$$T_m = 122.5 + \sum n_i GAV_i, \tag{3.1}$$

where T_m is in K; n_i is the number of groups of type i and GAV_i is the group contribution of the melting points of group i in the molecule. It was found that the results predicted using the group-contribution method of Joback–Reid [242] show an average deviation of 37.6% for 60 carbocyclic nitroaromatic compounds which were considered [243]. Thus, the group contribution method can only be used to obtain a very approximate guess. Some of the other group additivity methods include those of Lydersen [244], Ambrose [245], Klincewicz and Reid [246], Lyman et al. [247], Constantinou et al. [248–251], Marrero-Morejón and Pardillo-Fontdevilla [252], and Marrero and Gani

[253]. In contrast to other physicochemical properties, melting points are not well estimated by the group additivity methods [244, 254–256].

Quantum mechanical calculations are another approach that has been used for simulating solid to liquid phase transitions in energetic materials, in order to predict their melting points [257–260]. MD is a complex method which can be used to simulate solid-to-liquid phase transitions in energetic materials to predict their melting points [257–260]. Due to the free energy barrier for the formation of a liquid-solid interface, estimation of the melting point can be considered difficult problems through MD. This situation can cause superheating in a perfect crystal, which results in an overestimation of the melting point. Various ways of performing the simulations have been investigated in order to determine the most practical one for use for energetic materials [257–260]. Methods of quantitative structure-property relationships (QSPR) have been developed for organic molecules – including a number of drugs and/or homologous series [261–263].

SPARC performs automated reasoning in chemistry (PARC) is a complex physicochemical calculator, which can be used to estimate chemical reactivity parameters and physical properties of organic molecules [197]. It uses mechanisms to calculate these properties by incorporating trained parameters derived from experimental measurements and best fit using linear regression [264]. Whiteside et al. [265] used the SPARC platform to estimate the entropy of fusion, enthalpy of fusion, and the melting point of organic compounds through three models. The first model combines interaction terms and physical descriptors for the calculation of the entropy of fusion. The second model is a function of the entropy of fusion, boiling point, and flexibility of the molecule to estimate the enthalpy of fusion. The third model is based on the division of the enthalpy of fusion by the entropy of fusion for the prediction of the melting point. The predicted RMSE (Root Mean Square Error) of 904 organic compounds for the entropy, enthalpy, and melting models are 12.5 J/mol K, 4.87 kJ/mol, and 54.4 K, respectively.

Yalkowsky and coworkers [266–269] have developed Unified Physical Property Estimation Relationships (UPPER) as a system of empirical and theoretical relationships for the estimation of physicochemical properties of organic molecules. The UPPER can calculate the entropy of fusion, enthalpy of fusion, and melting point of organic compounds. It has been used for a wide range of organic compounds (over 2000 compounds) as compared to SPARC (SPARC Performs Automated Reasoning in Chemistry) with a similar pattern for calculation of melting point, that is, the ratio of enthalpy of fusion to the entropy of fusion. It uses the group additivity method for the calculation of the enthalpy of fusion. It also uses molecular geometry for the assessment of the degree of restriction of molecular motion in the crystal to that of the liquid, that is, symmetry, eccentricity, chirality, flexibility, and hydrogen bonding.

Al-Fakih et al. [229] introduced a QSPR model with four complex descriptors for the prediction of melting points of 92 carbocyclic nitroaromatic compounds based on the proposed penalized adaptive bridge (PBridge) as:

$$T_m = 11.034 + 0.542 Eig02\_AEA(dm) - 42.311 Mor29v - 2.547 Mor13u + 4.592 D/Dtr09,$$

$$(3.2)$$

where $Eig02\_AEA(dm)$ is edge adjacency index corresponding to Eigenvalue n. 2 from edge adjacency mat. weighted by dipole moment; $Mor29v$ and $Mor13u$ are two 3D-MoRSE descriptors corresponding to signal 29/weighted by van der Waals volume and signal 13/unweighted, respectively; $D/Dtr09$ is ring descriptors corresponding to distance/detour ring index of order 9.

Current prediction models may use complex chemical descriptors or fragment-based approaches rooted in quantum chemistry [270], but Mi et al. [271] developed a solution using natural language processing (NLP) that only requires SMILES strings – the chemical equivalent of a text message that encodes molecular structure in ASCII characters [272]. Inspired by how NLP powers translation apps and chatbots by detecting patterns in language [273], Mi et al. [271] applied similar logic to chemistry, since molecules with similar structures typically share properties like melting points, and stability [274]. AI model of Mi et al. [271] bypasses traditional chemistry theories entirely, instead treating SMILES strings as a chemical language to predict Tm through pure pattern recognition. Mi et al. [271] validated this approach against existing methods while accounting for how SMILES formatting and molecular complexity affect accuracy, though Mi et al. [271] intentionally excluded rare polymorphic cases with extreme T_m variations to focus on proving NLP's core predictive capability.

Pandy and Roy [118] developed QSPR models to predict key properties of energetic materials, including melting points. Their melting point model used a single descriptor – the RA function (LK) – which helped improve prediction accuracy as follows ($R^2 = 0.746$):

$$T_m = 9.081 + 0.952 RA\,function\,(LK),$$

$$(3.3)$$

Recent studies have begun exploring how different combinations of ML models and molecular descriptors can improve predictions of melting temperatures T_m for organic compounds [275]. These ML-based approaches have shown significantly better performance than traditional tools like the EPI Suite, highlighting the growing potential of advanced computational methods in predicting T_m.

Many of these studies rely on convolutional neural networks (CNNs) to generate molecular descriptors directly from molecular graphs [276]. While these CNN-based embeddings achieve impressive accuracy, they come with major drawbacks – particularly in terms of model deployment and portability. Because the embedding and prediction steps are tightly integrated within the CNN architecture, the resulting descriptors are highly specific to the training dataset. This means they can't easily be transferred to other applications, and retraining the model for new datasets requires substantial computational resources. Essentially, these models are powerful but inflexible, limiting their broader use in chemical research. In contrast, methods like mol2vec offer a more adaptable solution by generating universal molecular embeddings through a pretrained model. These embeddings have achieved state-of-the-art performance in predicting various molecular properties and can be seamlessly applied to diverse chemical datasets without retraining.

Galeazzo and Shiraiwa [277] presented the first ML-driven method for predicting using molecular embeddings. By leveraging different machine learning algorithms, they accounted for molecular structure, functional groups, and atomic connectivity, achieving better accuracy than previous empirical models. Their approach not only reproduces experimental T_m values but also captures how the number and positioning of functional groups influence this property.

Recent advances in big chemical data and deep neural networks (DNNs) have created exciting opportunities to improve melting point predictions using QSPR approaches [278]. Among these, graph neural networks (GNNs) – a type of DNN with "self-learning" capabilities – have shown particular promise for building generalized models when trained on large, structurally diverse datasets [279]. However, in specialized fields like energetic melt-castable materials and energy-harvesting materials, limited training data often prevent GNNs from accurately learning key properties like melting points. To address this, integrating prior chemical knowledge into these self-learning models could offer a solution [280]. To tackle these challenges, Song et al. [281] assembled a melting point dataset of 98,254 organic compounds and trained a GNN model on it. Using a convolutional operation on molecular graphs, they extracted graph neural fingerprints (GNFs), achieving a mean absolute error (MAE) of 25.0 K on the test set – a significant improvement for diverse molecular structures. When they compared different featurization methods (including ECFP, handcrafted descriptors, and GNFs), the GNN consistently outperformed others in capturing melting point trends.

Interestingly, combining handcrafted descriptors with GNFs further boosted accuracy for an independent dataset of 635 energetic melt-castable materials. This hybrid approach leverages both data-driven learning and chemical intuition, helping overcome data scarcity in niche applications. Our findings suggest that merging state-of-the-art GNNs with knowledge-based descriptors could unlock better predictions for challenging molecular properties, especially in areas where experimental data is sparse.

3.2 Simple empirical methods on the basis of molecular structure

Several simple correlations have recently been introduced to predict the melting points of certain classes of energetic compounds. If impurities are present in the energetic compounds, or if the energetic compound exhibits thermal instability, experimental determination of the melting point may be thwarted. It is important to select a reliable predictive model for organic compounds containing energetic functional groups such as $Ar-NO_2$, $C-NO_2$, $C-ONO_2$, or $N-NO_2$. According to Carnelley's rule [282, 283], the more symmetrical the organic isomer is, the higher its melting point. The dipole moment is one of the factors which directly control the melting point. Thus, the sum of all of the local dipole moments has a more pronounced effect than the net dipole moment on the melting point. Due to interactions between local dipole moments of neighboring atoms or groups, molecular interactions can result in close proximity of the molecules in the crystal. It can be expected that the more symmetrical paraisomers and the local dipole moments of some polar groups have a distinct effect in increasing the melting point. The presence of alkyl and alkoxy groups attached to nitroaromatic rings can decrease the planarity of the molecules, which in turn reduces the packing efficiency of molecules in the crystals. This results in a decrease in the interaction between local dipole moments of neighboring nitro groups. Although the attachment of a polar nitramine group may increase the molecular interactions, the presence of an alkyl substituent can change its effect because it can make the molecule less planar. The molecules become further apart if an alkyl nitramine group is introduced. In this section, some of the best available, simple methods for predicting melting points for several selected classes of energetic compounds will be reviewed.

3.2.1 Nitroaromatic compounds

The study of various carbocyclic nitroaromatic organic compounds with general formula $C_aH_bN_cO_d$ has shown that the following equation can be used to predict the melting point [243]:

$$T_m = 282.96 - 2.7543b + 46.570c + 94.318T_{SFG} + 54.752T_{o,p}, \qquad (3.4)$$

where T_{SFG} is the contribution of a specific functional group and $T_{o,p}$ is a parameter that can be applied to disubstituted benzene rings. The values of T_{SFG} and $T_{o,p}$ are predicted based on the following conditions:

(1) T_{SFG}: If –OH and –NH$_2$ are *ortho* to the –NO$_2$ group, the value of T_{SFG} equals zero, whereas if the –NH$_2$ and –OH groups in are *meta* or *para* positions relative to the –NO$_2$ group then T_{SFG} has the value 0.3. $T_{SFG} = 1.0$ if the –COOH, –CON– or –COO– functional groups are present. T_{SFG} also has the value –1.2 if the nitramine (N–NO$_2$) functional group is present.

(2) $T_{o,p}$: The value of $T_{o,p} = 1.0$ for *para*-disubstituted benzene rings. The presence of alkyl (–R) or alkoxy (–OR) groups in *ortho* positions relative to the –NO$_2$ group may result in a decrease in the melting point and therefore $T_{o,p}$ is equal to –0.7.

Example 3.1: 2,6-Bis(picryamino)-3,5-dinitropyridine (PYX) has the following molecular structure:

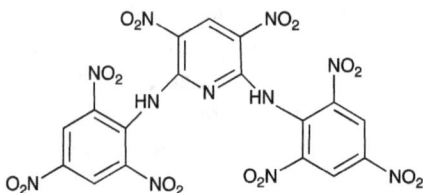

The use of eq. (3.4) gives T_m as follows:

$$T_m = 282.96 - 2.7543b + 46.570c + 94.318T_{SFG} + 54.752T_{o,p}$$

$$= 282.96 - 2.7543(7) + 46.570(11) + 94.318(0) + 54.752(0)$$

$$= 775.95 \text{ K}.$$

The experimental T_m was reported to be 733 K [199].

3.2.2 Polynitroarene and polynitroheteroarene compounds

Another correlation has recently been introduced that can be applied to a wider classes of energetic compounds, including polynitroarenes and polynitroheteroarenes, and is as follows [284]:

$$T_m = 355.0 + 3.33a - 5.63b + 14.57c + 90.83ISSP - 63.75DSSP, \tag{3.5}$$

where ISSP and DSSP are increasing and decreasing effects of some specific structural features, which can be specified according to the following situations.

3.2.2.1 *ISSP*
(1) The presence of $-NH_2$ groups:
   (a) Polynitroarenes: The value of ISSP is 0.5 if $n_{NH_2}/n_{NO_2} < 0.5$ except for mononitro anilines, in which ISSP is equal to 0.5 and 0.0 for *para*-nitroaniline and *ortho*- (or *meta*)-aniline, respectively. If $n_{NH_2}/n_{NO_2} \geq 0.5$, ISSP equals 2.0 for the attachment of $-NH_2$ groups to carbocyclic nitroaromatic compounds.
   (b) Polynitroheteroarenes: If amino groups are attached to heterocyclic aromatic compounds, $ISSP = 1.0$.

   Since the participation of a functional group in intramolecular hydrogen bonding reduces its ability to form intermolecular hydrogen bonds, the presence of n_{NO_2} in *ortho* positions or in positions close to n_{NH_2} groups has no appreciable effect, for example, in *o*-nitroaniline.

(2) The presence of certain functional groups: The value of $ISSP = 1.0$ if –COOH group or two –OH functional groups are attached to an aromatic ring.

(3) The existence of specific structural parameters: If the ⟨structure⟩ or –C=C– groups are attached to an aromatic ring, $ISSP$ equals 1.0.

(4) The presence of some specific groups in disubstituted benzene rings (in *para* position with respect to an –NO$_2$ group): If the –NO$_2$, R$_2$N–, or –N–C(=O)– groups are located in the *para* position with respect to the nitro group, ISSP equals 1.0.

3.2.2.2 *DSSP*

(1) The attachment of only alkyl or alkoxy groups to nitroaromatic rings: If $n_{NO_2}/n_{R,OR} \leq 1/3$ (where $n_{R,OR}$ is the number of –R or –OR groups attached to a nitroaromatic ring), then $DSSP = 0.0$. For $1/3 < n_{NO_2}/n_{R,OR} \leq 1$, the value of DSSP is equal to 1.0. The $DSSP$ value is also 0.5 for $n_{NO_2}/n_{R,OR} > 1$.

(2) The presence of alkyl nitramine groups attached to aromatic rings: For the presence of R–N–NO$_2$, $DSSP = 0.7 \times n_{RNNO_2}$, where n_{RNNO_2} is the number of alkyl nitramine groups.

(3) The presence of specific structural parameters: For the ⟨structure⟩ N$^+$–O$^-$ (or ⟨structure⟩) and ⟨structure⟩ rings, the values of $DSSP$ are 1.0 and 2.0, respectively.

(4) Mononitro-substituted carbocylic aromatic compounds: The value of DSSP is equal to 1.0 only for mononitro-substituted carbocylic aromatic compounds.

3.2.2.3 Different effects of *ISSP* and *DSSP* in polycylic nitroaromatic compounds

The presence of more than two nitroaromatic rings can result in different situations. If three aromatic rings are present and if $n_{NO_2}/n_{Ar} > 2.5$, then ISSP equals 2.0. However, if four aromatic rings are present and if $n_{NO_2}/n_{Ar} > 2.5$, then DSSP is 1.0.

Example 3.2: Consider the following molecular structure:

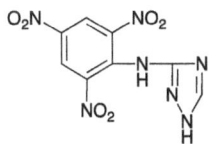

The calculated T_m by eq. (3.5) is given as follows:

$$T_m(K) = 355.0 + 3.33a - 5.63b + 14.57c + 90.83ISSP - 63.75DSSP$$

$$= 355.0 + 3.33(8) - 5.63(5) + 14.57(7) + 90.83(1) - 63.75(0)$$

$$= 546.3 \text{ K.}$$

The measured T_m value was reported to be 583 K [285].

3.2.3 Nitramines, nitrate esters, nitrate salts, and nitroaliphatics

The following equation is the simplest approach to predict the melting points of nitramines, nitrate esters, nitrate salts, and nitroaliphatics [286]:

$$T_m = 220.47 + 30.220c + 24.780d - 68.691C_{SFG} - 25.891n_{N-NO_2} \tag{3.6}$$

where n_{N-NO_2} is the number of nitramine groups in the energetic compound and C_{SFG} is the contribution of specific functional groups, which can be specified as follows:

(1) Nitrate salt: $C_{SFG} = 0$.
(2) Nitrate groups: C_{SFG} has the values 2.0, 2.5, and 3.5 for nitrated energetic compounds, which have one, two, and three or four $-O-NO_2$ groups, respectively.
(3) Hydroxyl group: $C_{SFG} = -1.0$ for nitro or nitrated energetic compounds that have at least one $-OH$ functional group.
(4) $-C(NO_2)_3$ group: C_{SFG} has the values 3.0 and 4.0 for the presence of one and more than one $C(NO_2)_3$ group, respectively.
(5) Mononitro compounds: $C_{SFG} = 1.5$ for mono nitro compounds which have the general formula $R'CH-RCH-NO_2$ (R or R'= $-H$, alkyl).

Example 3.3: For $HOCH_2C(CH_3)_2NO_2$, the predicted melting point using eq. (3.6) is calculated as follows:

$$T_m = 220.47 + 30.220c + 24.780d - 68.691C_{SFG} - 25.891n_{N-NO_2}$$

$$= 220.47 + 30.220(1) + 24.780(3) - 68.691(-1) - 25.891(0)$$

$$= 393.72 \text{ K.}$$

The measured T_m value was reported to be 361.7 K [199].

3.2.4 Nonaromatic energetic compounds

It was found that the elemental composition, as well as positive (T^+) and negative (T^-) correcting terms are important factors to include for nonaromatic energetic compounds, since this results in a more reliable correlation than eq. (3.5). The new correlation is given as follows [256]:

$$T_m = 281.7 + 28.97a - 12.08b + 29.75c - 9.966d + 102.92T^+ - 110.11T^-. \qquad (3.7)$$

The values of T^+ and T^- depend on some functional groups and structural parameters that are specified as follows.

(1) Nitroaliphatics and nitrate esters:
  (a) $C_nH_{2n+1}(NO_2 \text{ or } ONO_2)_{m=1 \text{ or } 2}$: The values of T^- depend on the number of carbon atoms in the alkyl group of mononitro or mononitrate alkanes:
    (i) If $n = 1$, then $T^- = 1.0$.
    (ii) If $n \geq 2$, then $T^- = 0.6$. For $m = 2$, the value of T^- depends on the molecular structure of the energetic compound. The value of T^- equals 0.40 in this case, except if two nitro groups are attached to one carbon, in which case $T^- = 0.0$.
  (b) The presence of the –OH group: If the hydroxyl group is present, T^+ is equal to 1.0.
(2) Energetic compounds with $-N-NO_2$, $-NH-NO_2$, and $-NHNO_3$ groups
  (a) The presence of $-N-NO_2$ groups: The ratio of the number of $-N-NO_2$ to $-CH_2$ (or $-CH_3$) groups has different effects:
    (i) If the ratio $n_{NNO_2}/n_{CH_2 \text{ or } 3} \geq 0.5$, then $T^+ = 0.5$.
    (ii) If the ratio $n_{NNO_2}/n_{CH_2 \text{ or } 3} \leq 0.2$, then $T^- = 0.6$. For other ratios of $n_{NNO_2}/n_{CH_2 \text{ or } 3}$, T^+, and T^- equal zero.
  (b) The presence of $-NH-NO_2$ and $-NHNO_3$ groups: The value of T^+ is equal to 1.0.

Example 3.4: Consider the following nonaromatic energetic compound:

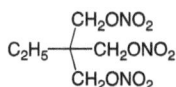

$$\text{C}_2\text{H}_5 \begin{array}{c} \text{CH}_2\text{ONO}_2 \\ \text{—CH}_2\text{ONO}_2 \\ \text{CH}_2\text{ONO}_2 \end{array}$$

The melting point is calculated as follows:

$$T_m = 281.7 + 28.97a - 12.08b + 29.75c - 9.966d + 102.92T^+ - 110.11T^-$$
$$= 281.7 + 28.97(6) - 12.08(11) + 29.75(3) - 9.966(9) + 102.92(0) - 110.11(0)$$
$$= 322.2 \text{ K}.$$

The measured melting point of this compound is 324.15 K [54]. If eq. (3.6) is used instead, a value of 293.7 K is obtained, which illustrates that the deviation of (3.6) from the experimental value is larger for such compounds.

3.2.5 Improved method for predicting the melting points of energetic compounds

For various aromatic and nonaromatic energetic compounds containing Ar–NO$_2$, C–NO$_2$, C–ONO$_2$, or N–NO$_2$ groups, an improved method has been introduced which expresses the melting points of these compounds as additive and nonadditive parts. The new correlation has the following form [287]:

$$T_\text{m} = 326.9 + 5.524T_\text{add} + 101.2T_\text{nonadd}, \tag{3.8}$$

where

$$T_\text{add} = a - 0.5049b + 2.643c - 0.3838d, \tag{3.9}$$

$$T_\text{nonadd} = T_\text{PC} - 0.6728T_\text{NC}. \tag{3.10}$$

Two correcting functions T_PC and T_NC were chosen based on the deviations of T_add from the measured values and are discussed in the following sections.

3.2.5.1 T_PC

(1) –NH$_2$ group: The presence of amino groups can not only enhance the thermal stability of energetic compounds [285], but also can increase their safety by decreasing their sensitivity to external stimuli such as impact [121, 122]. The effect of amino groups on the melting point of a compound can be classified as follows:

(a) The number of amino groups per aromatic ring in polynitroarenes or nonaromatic energetic compounds: If one –NH$_2$ group is present in this type of energetic compound (e.g., in 2,2′,4,4′,6,6′-hexanitrobiphenyl-3,3′-diamine), the value of T_PC is 0.5. One exception to this is o-nitroaniline, because intramolecular hydrogen bonding may cancel the effect of the amino group. The value of T_PC equals 2.0 if a larger number of amino groups is present in such compounds, for example, 1,1-diamino-2,2-dinitroethylene and 1,3,5-triamino-2,4,6-trinitrobenzene.

(b) Polynitroheteroarenes: The value T_PC equals 1.0 if amino groups are attached to heterocyclic aromatic compounds, for example, 4-amino-5-nitro-1,2,3-triazole.

(2) The presence of some specific polar groups and molecular fragments: The presence of certain special functional groups such as –COOH can increase the melting points of these compounds because of their ability to form reinforced intermolecular hydrogen bonds. The effects of these functional groups and molecular fragments in different energetic compounds can be classified as follows:

(a) Nitroaromatic: If –COOH, –NH–CO– and at least two –OH groups are attached to the aromatic ring, T_PC is 0.75. T_PC equals 0.75 if –NO$_2$, R$_2$N– or –N–C(=O)– are present in the para position relative to the nitro group. Moreover, the value of T_PC is 1.0 if the or –C=C–Ar groups are present.

(b) Nonaromatic energetic compounds: The value of T_{PC} is 1.0 for the existence of some specific polar groups or molecular fragments including $-NH-NO_2$, NH_4^+, more than one $-OH$, one cyclic ether or carbocyclic cage energetic compounds.

T_{NC}

For polynitroarenes and polynitroheteroarenes, the presence of alkyl and alkoxy groups attached to nitroaromatic rings can reduce the planarity of the molecules. The ratio of the number of nitro groups to the number of alkyl or alkoxy groups $(n_{NO_2}/n_{R,OR})$ may be important. For the nitramine polar group, the presence of an alkyl substituent can change its effect, because it can affect the nonplanarity of the nitroaromatic molecules. The attachment of certain heterocycles to a central aromatic ring in poynitro heteroarenes may result in a reduction of the symmetry and planarity of these compounds. Mononitro substitution in aromatic compounds may decrease the melting point compared to polynitro-substituted compounds. The effects of various structural parameters on T_{NC} are as follows.

(1) Nitroaromatics: Four different situations can be considered for polynitroarenes and polynitroheteroarenes.
  (a) Alkyl- or alkoxy-substituted nitroaromatics: For the ratios $n_{NO_2}/n_{R,OR} \leq 1$ and $n_{NO_2}/n_{R,OR} > 1$, the values of T_{NC} equal 1.0 and 0.5, respectively.
  (b) Alkyl nitramine groups attached to aromatic rings: The value of T_{NC} is equal to 0.7 times the number of alkyl nitramine groups.
  (c) Specific structural factors: If , , or are present, the value of T_{NC} is equal to 1.0.
  (d) Mononitro-substituted aromatic compounds: The value of T_{NC} is equal to 1.0.
(2) Nonaromatic energetic compounds: For those nitro and nitrate energetic compounds with general formula $-CH-(NO_2$ or $ONO_2)_n$, the values of T_{NC} are as follows:
  (a) If $n = 1$, then $T_{NC} = 2.0$.
  (b) If $n = 2$ or 3, then $T_{NC} = 1.0$.

3.2.5.2 Different behavior of some molecular structures

Due to the complex effects of symmetry, planarity, and local dipole moments, some specific molecular moieties can have different effects with respect to increasing or decreasing the melting point. The guidelines are as follows:

(1) Polycyclic nitroaromatic compounds: If three aromatic rings are present and if $n_{NO_2}/n_{Ar} > 2.5$, then $T_{PC} = 2.0$. However, if four aromatic rings are present and if $n_{NO_2}/n_{Ar} > 2.5$, then $T_{NC} = 1.0$.
(2) Cyclic nitramines containing methylene units: The ratio of n_{NNO_2} to the number of methylene units (n_{CH_2}) has different effects:

(a) if the ratio $n_{NNO_2}/n_{CH_2} \geq 1.0$, then $T_{PC} = 0.5$;

(b) if the ratio $n_{NNO_2}/n_{CH_2} \leq 0.2$, then $T_{NC} = 1.2$.

A summary of the above conditions is given in Tables 3.1 and 3.2.

Table 3.1: Summary of predicted values of T_{PC}.

Energetic compound	Specific groups or molecular moieties	T_{PC}	Comments
Polynitroarene or nonaromatic	$-NH_2$ group (the number of amino groups per ring in polynitroarene)	0.5	One amino group (except o-nitroaniline)
		2.0	More than one amino group
Polynitroheteroarene	$-NH_2$ group	1.0	–
Nitroaromatic	$-COOH$, $-NH-CO-$ and at least two $-OH$ groups	0.75	–
	$-NO_2$, R_2N- or $-N-C(=O)-$ in *para* position with respect to the nitro group		
		1.0	
Nonaromatic	$-NH-NO_2$, NH_4^+ and more than one $-OH$ as well as one cyclic ether or carbocyclic cage energetic compound	1.0	–
	Cyclic nitramine containing methylene units	0.5	$n_{NNO_2}/n_{CH_2} \geq 1.0$
Polycyclic nitroaromatic	Presence of three aromatic rings	2.0	$n_{NO_2}/n_{Ar} > 2.5$

Table 3.2: Summary of predicted values of T_{NC}.

Energetic compound	Specific groups or molecular moieties	T_{NC}	Comments
Nitroaromatic	Alkyl- or alkoxy-substituted nitroaromatics	1.0	$n_{NO_2}/n_{R, OR} \leq 1$
		0.5	$n_{NO_2}/n_{R, OR} > 1$
	Alkyl nitramine groups attached to aromatic rings	0.7 × the number of alkyl nitramine groups	–
		1.0	

Table 3.2 (continued)

Energetic compound	Specific groups or molecular moieties	T_{NC}	Comments
	Mononitro-substituted aromatic compound		
Nonaromatic	$-CH-(NO_2$ or $ONO_2)_n$	2.0	$n = 1$
		1.0	$n = 2, 3$
	Cyclic nitramine-containing methylene units	1.2	$n_{NNO_2}/n_{CH_2} \leq 0.2$
Polycyclic nitroaromatic	Presence of four aromatic rings	1.0	$n_{NO_2}/n_{Ar} > 2.5$

Example 3.5: Use eq. (3.8) to calculate the melting point of the following compound:

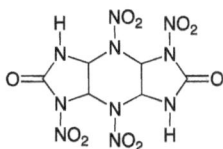

$$T_{add} = a - 0.5049b + 2.643c - 0.3838d$$

$$= 6 - 0.5049(6) + 2.643(10) - 0.3838(10)$$

$$= 25.56$$

$$T_m = 326.9 + 5.524T_{add} + 101.2T_{nonadd}$$

$$= 326.9 + 5.524(25.56) + 101.2(0)$$

$$= 468.1 \text{ K}.$$

Since the measured melting point of this compound is 498.15 K [120], the percent deviation of the calculated melting point from the measured melting point (−6.0 %) is much lower than the two group additivity methods of Joback–Reid [242] (1260.8 K, % Dev = 153.1) and Jain–Yalkowsky [194] (772.8 K, % Dev = 55.1).

3.2.6 Organic molecules containing hazardous peroxide groups

The study of various organic compounds containing peroxide groups has shown that it is possible to calculate the melting points of these compounds using core and correcting functions as follows [288]:

$$T_{m, peroxide} = 280.5 + 5.159 T_{core} + 38.90 T_{correcting}, \tag{3.11}$$

$$T_{core} = a - 0.556b + 2.064d, \tag{3.12}$$

$$T_{correcting} = T^+_{m, peroxide} - 1.345 T^-_{m, peroxide}, \tag{3.13}$$

where $T_{m, peroxide}$, T_{core}, and $T_{correcting}$ are the melting point of the peroxide compound, and the core and correcting functions, respectively; $T^+_{m, peroxide}$ and $T^-_{m, peroxide}$ are the positive and negative contributions of structural parameters in $T_{correcting}$, respectively. For the presence of several molecular moieties, the values of $T^+_{m, peroxide}$ and $T^-_{m, peroxide}$ can be specified as follows:

(1) $T^+_{m, peroxide}$: The values of $T^+_{m, peroxide}$ are 2.0 and 0.5 for the presence of more than one peroxy acid (without any functional groups) and –OH groups, respectively.

(2) $T^-_{m, peroxide}$: For some organic molecules including those containing –(CO)OO– or –O–C(O)–OO–(CO)–O– groups, the crystal packing efficiency of molecules may be reduced, which in turn, can decrease the interaction between local dipole moments of neighboring polar groups. If only one –(CO)OO– or –O–C(O)–OO–(CO)–O– group in the form R_1–(CO)OO–R_1 or R_1–O–C(O)–OO–(CO)–O–R_1 is present (in which R_1 should be the same on both sides of the organic molecule), the value of $T^-_{m, peroxide}$ is 1.0. For R–C(O)OOH, the value of $T^-_{m, peroxide}$ is 1.0 if the number of carbon atoms in the R group is less than five.

Example 3.6: Consider the following peroxide:

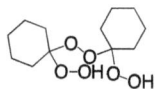

1,1′-peroxybis(1-hydroperoxycyclohexane)

$$T_{core} = a - 0.556b + 2.064d$$
$$= 12 - 0.556(22) + 2.064(6)$$
$$= 12.2$$
$$T_{m, peroxide} = 280.5 + 5.159 T_{core} + 38.90 T_{correcting}$$
$$= 280.5 + 5.159(12.2) + 38.90(0)$$
$$= 343.4 \text{ K}.$$

The measured melting point of this compound is 356 K [199]. The percent deviation of the calculated melting point using the above method is 3.5 %, which is much lower than that of the two group additivity methods of Joback–Reid [242] (619 K, % Dev = 74.2) and Jain–Yalkowsky [261] (454 K, % Dev = −27.4).

3.2.7 Organic azides

For organic azides, the following correlation has been introduced [289]:

$$T_m = 264.63 + 10.09a - 3.86b + 18.38c$$
$$-47.53n_{azide} + 45.76SPG - 65.58IPF,$$

(3.14)

where n_{azide} is the number of azide groups; SPG is the contribution of specific polar groups; IPF is the inefficient packing factor. Table 3.3 shows different SPG values containing a list of polar functional groups and molecular fragments. Table 3.4 shows different situations in which the effects of IPF can be considered.

Table 3.3: The values of specific polar groups (SPG).

Polar groups	SPG	Exception
–COOH	1.8	Ortho phenoxy molecular moiety (–O–Ar)
–CONH$_2$ or –SO$_2$NH$_2$	2	–
–OH	0.6 × the number of –OH groups	Disubstituted benzene containing only –N$_3$ and –OH or –CH$_2$OH
–SO$_3^-$ or 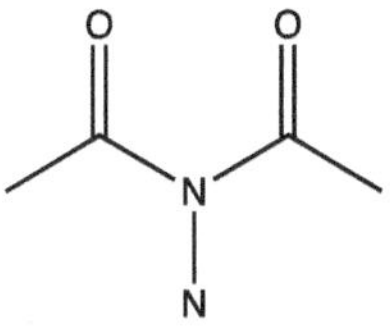 ![structure]	1	–
–NH–CO–NH–, –NH–CO–CH$_3$, cyclic –NH–CO, –CO–NH–N	1.2	–
Cyclic –NH–	0.8	–
Ar–NH$_2$	0.2 × the number of –NH$_2$	N–NH$_2$

Table 3.4: The values for the inefficient packing factor (IPF).

Compound	Illustration	IPF	Condition
Benzene and naphthalene derivatives	One or two separated benzene rings in form 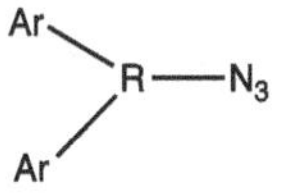 Ar—R—N$_3$ (Ar) or α-azidonaphthalene	1.2	No polar groups are present

Table 3.4 (continued)

Compound	Illustration	IPF	Condition
	β-azidonaphthalene derivatives, or three separated benzene rings in form 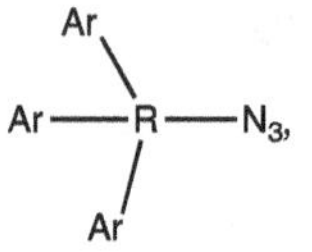 Ar \\ Ar —— R —— N$_3$, / Ar or two benzene rings in form Ar–Ar	0.6	
R–N$_3$	–	1.8	
Azides containing silicon	–	1.8	No polar groups are present and the silicon atom is directly attached to the azide group
		0.5	No polar groups are present

Example 3.7: The use of eq. (3.14) for 1-azidonaphthalene (C$_{10}$H$_7$N$_3$) gives

$$T_m = 264.63 + 10.09a - 3.86b + 18.38c - 47.53n_{azide} + 45.76SPG - 65.58IPF$$

$$= 264.63 + 10.09(10) - 3.86(7) + 18.38(3) - 47.53(1) + 45.76(0) - 65.58(1.2)$$

$$= 267.4 \text{ K}.$$

Since the measured melting point of this compound is 285.16 K [290], the percent deviation of the melting point calculated by this method from that of the measured melting point is 6 %.

3.2.8 General method for the prediction of melting points of energetic compounds including organic peroxides, organic azides, organic nitrates, polynitroarenes, polynitroheteroarenes, acyclic and cyclic nitramines, nitrate esters, and nitroaliphatic compounds

Investigations of the effects of different structural features in energetic compounds (including organic peroxides, organic azides, organic nitrates, polynitroarenes, polynitroheteroarenes, acyclic and cyclic nitramines, nitrate esters, and nitroaliphatic compounds) on the melting point of a compound has indicated that it is possible to express the melting points of these compounds as a function of additive and nonadditive parts in the following form [291]:

$$T_m = 323.0 + 5.511 T_{add,\,elem} + 101.2 T_{corr,\,struc}, \tag{3.15}$$

$$T_{add,\,elem} = a - 0.5251b + 2.402c, \tag{3.16}$$

$$T_{corr,\,struc} = T^+_{struc} - 0.6470 T^-_{struc}, \tag{3.17}$$

where $T_{add,\,elem}$ shows the contribution of the elemental composition as an additive part, whereas the parameter $T_{corr,\,struc}$ indicates the nonadditive part of the melting point. $T_{corr,\,struc}$ is a complex function that can be specified on the basis of the molecular structure because some specific polar groups such as –COOH, –NH$_2$, and –OH as well as other molecular moieties may enhance intermolecular interactions (and which are specified later). However, the presence of some specific molecular fragments can decrease the molecular attractions which are present. Large positive and negative deviations of the calculated $T_{add,\,elem}$ from the experimental data can be adjusted by the parameter $T_{corr,\,struc}$ in eq. (3.15) through the related variables T^+_{struc} and T^-_{struc} in eq. (3.17). The values of T^+_{struc} and T^-_{struc} are discussed in the following sections.

3.2.8.1 $T_{corr,\,struc}$

The contributions of different structural parameters of energetic compounds in $T_{corr,\,struc}$ can be specified through some specific polar groups and molecular fragments.

T^+_{struc}

(1) Peroxide group: The intermolecular hydrogen bonding effects of –OH, –COOH, –O–OH, and COOOH (peroxy acid group) without further functional groups results in the following conditions:
  (a) more than one –O–OH: The value of $T^+_{struc} = 0.5$;
  (b) two or more hydroxyl groups as well as an –O–O– group: $T^+_{struc} = 0.4$;
  (c) the presence of both COOH and COOOH groups: The value of $T^+_{struc} = 0.5$.
(2) Amino group: The melting points of different classes of energetic compounds may be increased by the introduction of amino groups. This can be accounted for as follows:
  (a) Nitroarenes and nonaromatic compounds: The value of $T^+_{struc} = 0.5$ for one –NH$_2$ group (e.g., 2,2′,2″,4,4′,4″,6,6′,6″-nonanitro-1,1′:3′,1″-terphenyl), except for o-nitroaniline. If more than one amino group is present in this category of compound (e.g., 2,2-dinitroethene-1,1-diamine) then the value of $T^+_{struc} = 2.0$.
  (b) Poly- and mono-nitro heteroarenes: The value of $T^+_{struc} = 1.0$ if amino groups are attached to heterocyclic aromatic compounds (e.g., 7-amino-4,6 dinitrobenzo[c][1,2,5]oxadiazole 1-oxide).
(3) Specific polar groups and molecular fragments: The influence of the molecular fragments and functional groups on different classes of energetic compounds can be classified as follows:

(a) Nitroaromatic energetic compounds: $T_{struc}^+ = 0.75$ if $-COOH$ or $-NH-CO-$ and at least two $-OH$ groups are attached to the aromatic ring. If $-NO_2$, R_2N- and $-N-C(O)-$ groups are attached to the ring in *para* positions with respect to the nitro group, then T_{struc}^+ is also equal to 0.75. The value of $T_{struc}^+ = 1.0$ if

the or $-C = C-Ar$ molecular moieties are present.

(b) Nonaromatic energetic compounds: if $-NH-NO_2$, more than one $-OH$ group, one cyclic ether or carbocyclic cage energetic compound is present, the value of $T_{struc}^+ = 1.0$. Furthermore, $T_{struc}^+ = 0.5$ if NH_4^+ is present in salts of nitro non-aromatic compounds.

3.2.8.1.1 T_{struc}^-

For some organic compounds which contain the $-(CO)OO-$ or $-O-C(O)-OO-(CO)-O-$ groups, the $T_{add, elem}$ values are higher than those from experimental data. These groups may reduce the packing efficiency of molecules in crystals because they can decrease the interactions between local dipole moments of neighboring polar groups. The presence of alkyl and alkoxy groups attached to nitroaromatic rings can also reduce the planarity of polynitroarenes and polynitroheteroarene molecules. This non-planarity not only decreases the packing efficiency of molecules in the crystal lattice, but it can also reduce the interactions between local dipole moments of neighboring nitro groups.

If polar nitramine groups are present, alkyl substituents can reduce the polar effect of these groups because alkyl groups can result in a reduction in the planarity of nitroaromatic molecules. The attachment of heterocycles to the central aromatic ring may result in a reduction of the symmetry in polynitroheteroarenes. Since there are larger attractive forces in polynitro-substituted compounds than in mononitro-substituted carbocyclic aromatic compounds, the latter have lower melting points. The effects of various structural parameters on T_{struc}^- may be categorized as follows.

(1) The presence of $-(CO)OO-$ or one $-O-O-$ group: If only one $-(CO)OO-$ or $-O-(CO)-OO-(CO)-O-$ is present in the form $R_1-(CO)OO-R_1$ or $R_1-O-(CO)-OO-(CO)-O-R_1$, the value of $T_{struc}^- = 1.0$ (R_1 should be the same on both sides of the organic molecule). The T_{struc}^- value is 1.0 and 0.5 for R–(CO)OOH and R′–(CO)OH (or R′–OO–R″), respectively, however, the number of carbon atoms in R and R′ should be less than five.

(2) The attachment of $-N_3$ to alkyl or aryl groups: The value of $T_{struc}^- = 1.4$.

(3) Nitroaromatic compounds: The following conditions can be considered for nitro arenes and nitro heteroarenes.

(a) Alkyl- or alkoxy-substituted nitroaromatics: For nitroaromatics containing alkyl or alkoxy groups with ratios of $n_{NO_2}/n_{R,OR} \leq 1$ and $n_{NO_2}/n_{R,OR} > 1$, the values of T^-_{struc} are 1.0 and 0.5, respectively.

(b) Alkyl nitramine groups attached to aromatic rings: $T^-_{struc} = 0.7 \times$ the number of alkyl nitramine groups.

(c) Specific structural factors in nitroaromatic compounds: If the N–O, or N–N molecular moieties are present, $T^-_{struc} = 1.0$.

(d) Mononitro-substituted aromatic compounds: $T^-_{struc} = 1.0$.

(e) Nitroheteroarenes: The value of $T^-_{struc} = 0.5$.

(4) Nitro and nitrate nonaromatic energetic compounds: For those nitro and nitrate energetic compounds with general formula $-CH-(NO_2$ or $ONO_2)_n$:

(a) if $n = 1$, then $T^-_{struc} = 2.0$;

(b) if $n = 2$ or 3, then $T^-_{struc} = 1.0$.

3.2.8.1.2 Polycyclic nitroaromatic compounds as well as cyclic nitramines with methylene units

In polycyclic nitroaromatic compounds as well as cyclic nitramines which contain methylene units, there are complex effects related to symmetry, planarity, and local dipole moments. Thus, the following conditions should be applied:

(1) Polycyclic nitroaromatic compounds: If three or four aromatic rings are present, the ratio of n_{NO_2} to n_{Ar} can estimate the contribution of T^+_{struc} and T^-_{struc}. If $n_{NO_2}/n_{Ar} > 2.5$, the value of T^+_{struc} is 2.0 if three aromatic rings are present, whereas if four aromatic rings are present, $T^-_{struc} = 1.0$ if $n_{NO_2}/n_{Ar} > 2.5$.

(2) Cyclic nitramines containing methylene units: Two situations are encountered based on the ratio of n_{NNO_2} to the number of methylene units (n_{CH_2}):

(a) if the ratio $n_{NNO_2}/n_{CH_2} \geq 1.0$, $T^+_{struc} = 0.5$;

(b) if the ratio $n_{NNO_2}/n_{CH_2} \leq 0.2$, $T^-_{struc} = 1.2$.

Tables 3.5 and 3.6 summarize the predicted values of T^+_{struc} and T^-_{struc} which are obtained by applying the above conditions.

Table 3.5: Summary of predicted values of T_{struc}^+.

Energetic compound	Specific groups or molecular moieties	T_{struc}^+	Condition	Example
Peroxide	O–OH group	0.5	More than one hydroperoxy group	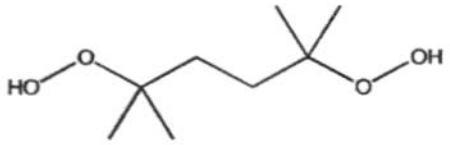 2,5-dihydroperoxy-2,5-dimethylhexane
	OH group	0.4	Two or more hydroxy groups	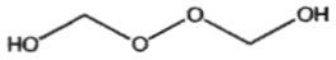 peroxydimethanol
	COOH and COOOH groups	0.5	The presence of two similar groups	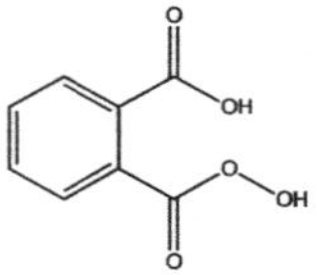 2-carboperoxybenzoic acid
Nitro arenes or nonaromatics containing amino groups	–NH₂ group	2.0	More than one amino group	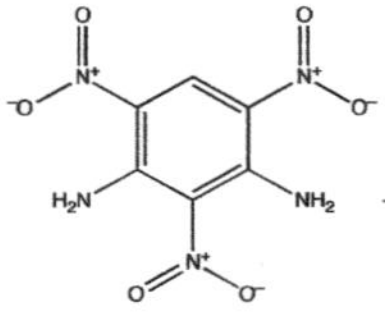 2,4,6-trinitrobenzene-1,3-diamine 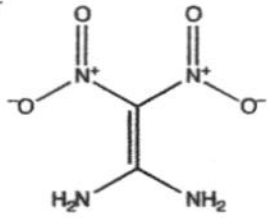 2,2-dinitroethene-1,1-diamine
	–NH₂ group	0.5	One amino group (except o-nitroaniline)	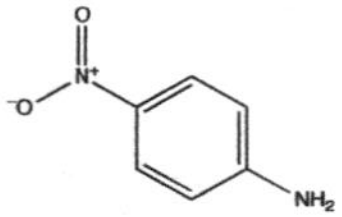 p-Nitroaniline

Poly and mononitro heteroarene	−NH₂ group	1.0	−

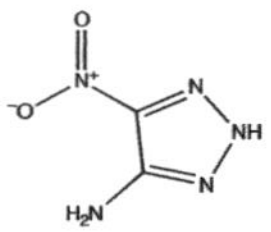

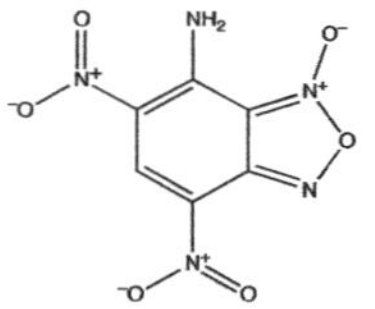

5-nitro-2H-1,2,3-triazol-4-amine 7-amino-4,6-dinitrobenzo[c][1,2,5]oxadiazole 1-oxide

Nitro aromatic	−COOH, −NH−CO, at least two −OH groups, −NO₂, R₂ N−, or −N−C(O)− in *para* position relative to the nitro group	0.75	−

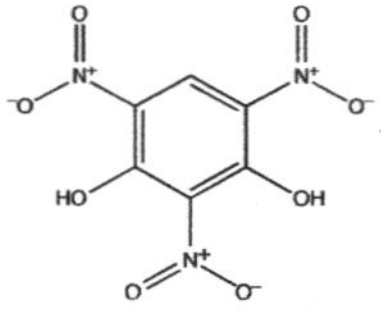

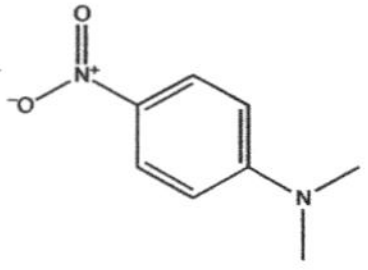

2,4,6-Trinitroresorcinol   Benzenamine, N,N-dimethyl-4-nitro-

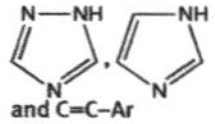

		1.0	−

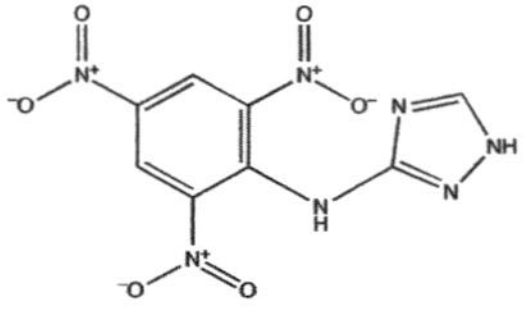

N-(2,4,6-trinitrophenyl)-1H-1,2,4-triazol-3-amine

Nitro nonaromatic	Cyclic nitramines containing methylene units	0.5	$n_{NNO_2}/n_{CH_2} \geq 1.0$

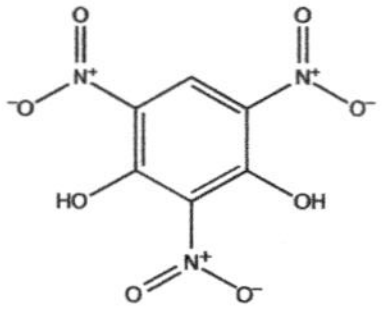

2,4,6-Trinitroresorcinol

(continued)

Table 3.5 (continued)

Energetic compound	Specific groups or molecular moieties	T_{struc}^+	Condition	Example
	$-NH-NO_2$ and more than one $-OH$, as well as one cyclic ether group	1.0	–	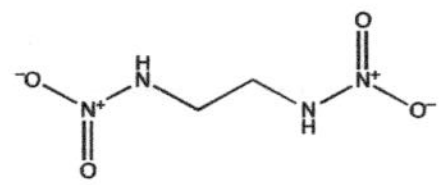 N,N'-(ethane-1,2-diyl)dinitramide
	NH_4^+	0.5	–	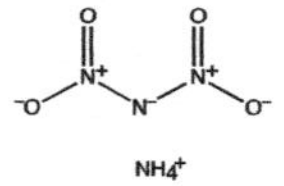 NH_4^+ N-nitronitramide, ammonium salt
Polycyclic nitroaromatics	The presence of three aromatic rings	2.0	$n_{NO_2}/n_{Aromatic\ ring} > 2.5$	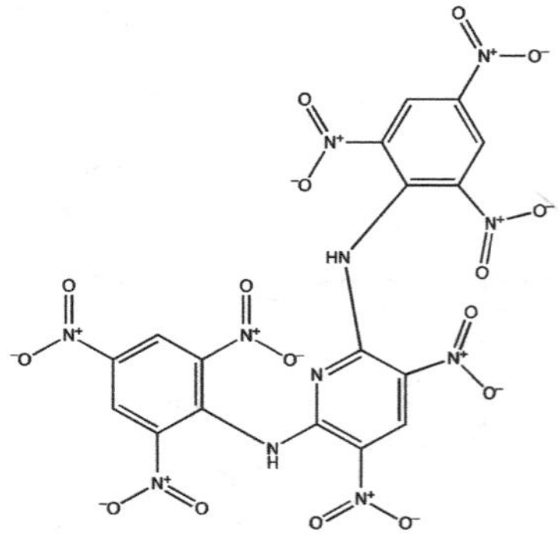 3,5-dinitro-N2,N6-bis(2,4,6-trinitrophenyl)pyridine-2,6-diamine

Table 3.6: Summary of predicted values of T^-_{struc}.

Energetic compound	Specific groups or molecular moieties	T^-_{struc}	Condition	Example
Peroxide	O–OH group	0.5	One hydroperoxy group	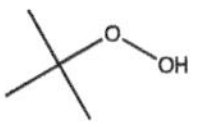 2-hydroperoxy-2-methylpropane
	O–O group	0.5	One group without any other group present	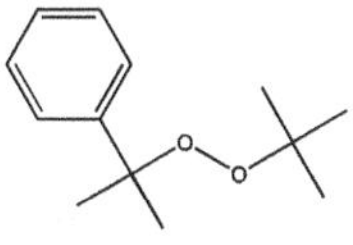 (2-(tert-butylperoxy)propan-2-yl)benzene
	(CO)OH, (CO)OO group	1.0	Only one group without other groups present	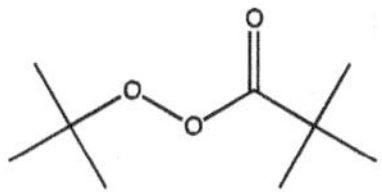 tert-butyl 2,2-dimethylpropaneperoxoate
Azide	N_3	1.4	The presence of N_3 group except ionic azides	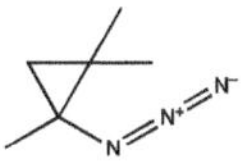 1-azido-1,2,2-trimethylcyclopropane
Nitro nonaromatic	–CH–(NO_2 or $ONO_2)_n$	2.0	$n = 1$	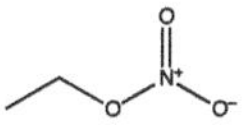 Ethyl Nitrate
		1.0	$n = 2, 3$	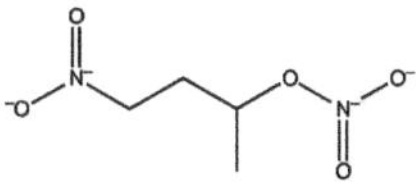 4-nitrobutan-2-yl nitrate
	Cyclic nitramines containing methylene units	1.2	$n_{NNO_2}/n_{CH_2} \leq 0.2$	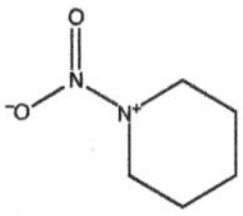 Piperidine, 1-nitro
Nitroaromatic	Alkyl- or alkoxy-substituted nitroaromatics	1.0	$n_{NO_2}/n_{R, OR} \leq 1$	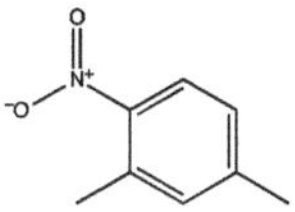 Benzene, 2,4-dimethyl-1-nitro
		0.5	$n_{NO_2}/n_{R, OR} > 1.0$	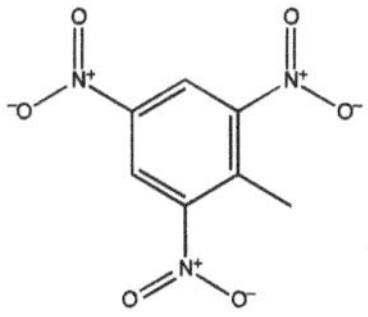 Benzene, 2-methyl-1,3,5-trinitro-

Table 3.6 (continued)

Energetic compound	Specific groups or molecular moieties	$\bar{T}_{struc}$	Condition	Example
	Alkyl nitramine groups attached to aromatic rings	$0.7 \times A$	A is the number of alkyl nitramine groups	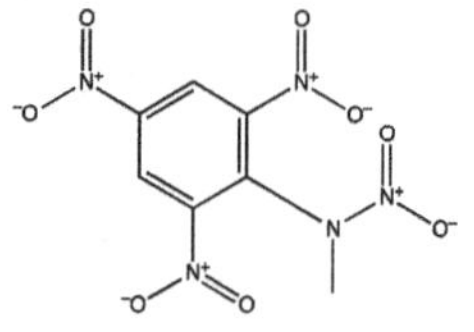 N-Methyl-N,2,4,6-tetranitroaniline
	Mononitro-substituted aromatic compounds	1.0	–	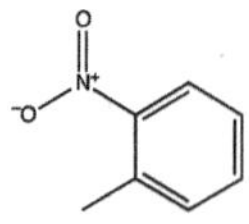 Benzene, 1-methyl-2-nitro-
	The presence of one NO_2 group in two ring-fused heteroarenes	0.5	–	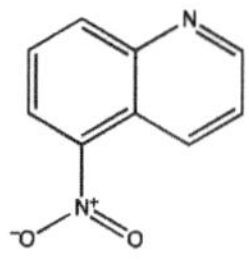 5-nitroquinoline
	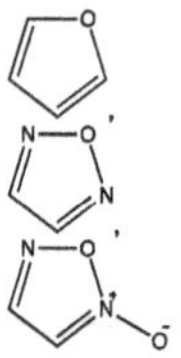	1.0		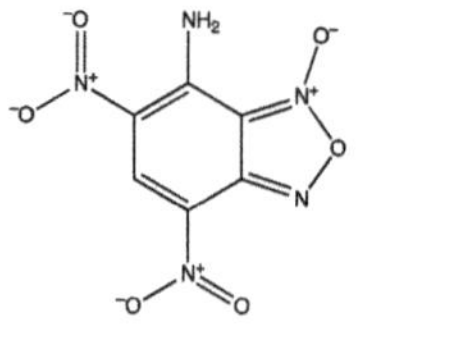 7-amino-4,6-dinitrobenzo[c][1,2,5]oxadiazole 1-oxide
Polycyclic nitroaromatic	$n_{NO_2}/n_{Aromatic\ ring}$ > 2.5	1.0	The presence of four aromatic rings	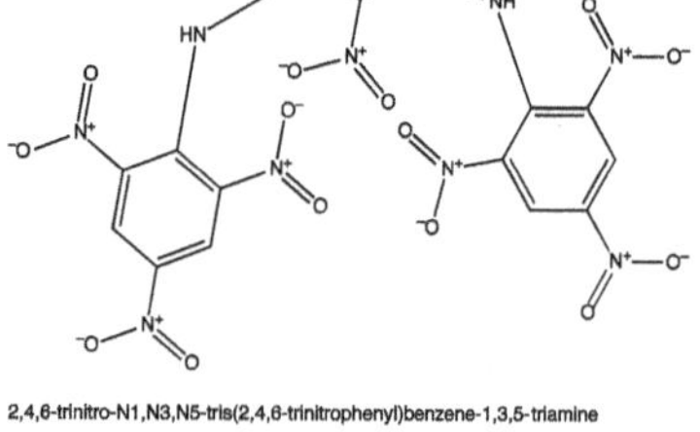 2,4,6-trinitro-N1,N3,N5-tris(2,4,6-trinitrophenyl)benzene-1,3,5-triamine

---

Example 3.8: The use of this method for the following compound

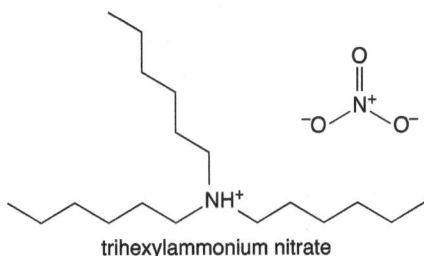

trihexylammonium nitrate

predicts the melting point as

$$T_{add, elem} = a - 0.5251b + 2.402c$$

$$= 18 - 0.5251(39) + 2.402(2)$$

$$= 2.325,$$

$$T_m = 323.0 + 5.511T_{add, elem} + 101.2T_{corr, struc}$$

$$= 323.0 + 5.511(2.325) + 101.2(0)$$

$$= 335.8 \text{ K}.$$

The predicted melting point of this compound is close to the experimentally determined value of 345 K [199].

3.2.9 Cyclic saturated and unsaturated hydrocarbons

Cyclic hydrocarbons or alicyclic hydrocarbons are a kind of closed chain hydrocarbons where they and their derivatives are cyclic hydrocarbons of alkane, alkene, and/or alkyne, which contain at least one ring. Polycyclic compounds are important cyclic hydrocarbons with more than one ring of carbon atoms whose rings share two or more same carbon atoms. The heats of combustion of cyclic hydrocarbons are much larger than we might expect by analogy with noncyclic compounds because of the existence of strain energy in cyclic hydrocarbons [6]. Thus, polycyclic saturated hydrocarbons can be used as liquid fuels, which are used in liquid bipropellants [1–3, 7]. For example, RJ-4 is a mixture of *endo-* and *exo-*tetrahydrodimethyldicyclopentadiene that is used as a synthetic jet fuel [7].

A general correlation has been developed for the prediction of the melting point of a wide range of cyclic saturated and unsaturated hydrocarbons with and without substituents including cyclic alkane, alkene, and/or alkyne, cage molecules, bridged cyclic and multicyclic hydrocarbon structures [292]. It depends on three variables as follows [292]:

$$T_m = 103.1 + 10.24a + 89.16T_{m,\,cyc\,hyd}^+ - 66.44T_{m,\,cyc\,hyd}^-, \quad (3.18)$$

where $T_{m,\,cyc\,hyd}^+$ and $T_{m,\,cyc\,hyd}^-$ are two correction terms, which correspond to increasing and decreasing values of the melting point corresponding to specific structural parameters given in Tables 3.7 and 3.8.

Table 3.7: Summary of the values of $T_{m,\,cyc\,hyd}^+$.

Type of cyclic hydrocarbon	Condition	$T_{m,\,cyc\,hyd}^+$	Example
Saturated (without substituents)	$5 < n_{C,\,ring} * < 19$ for monocyclic ring	1.0	
	Bicyclic ring	1.5 (for fused rings where each ring containing less than a five-membered ring)	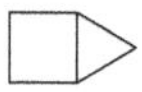
		2.0 (for bridged rings where each ring containing more than four-membered ring)	
	Tricyclic ring	1.5 (*Endo*)	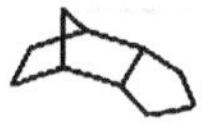
		3.0 (symmetric consecutive ring)	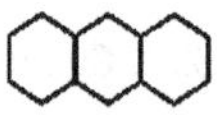
	Cage hydrocarbon with more than bicyclic ring	1.7	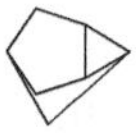
		3.5 (more than a four-membered ring)	
	Polycyclic containing more than three rings where each ring directly attached to another ring	1.0	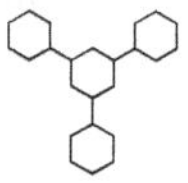
Saturated (with substituents)	$n_{C,\,ring} = 6$ with *trans*-dialkyl similar substituents or *cis, cis*-trialkyl with more than one similar substituents	1.0	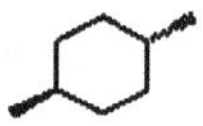
	The bridged bicyclic ring where each ring containing more than four-membered ring	1.7	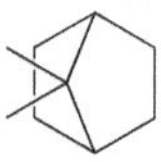

Table 3.7 (continued)

Type of cyclic hydrocarbon	Condition	$T^+_{m,\,cyc\,hyd}$	Example
		1.0 (one substituent)	
	Cage hydrocarbon containing more than two rings	1.7	
		2.2 (one substituent and more than a four-membered ring)	
	Each carbon of ring containing one or two methyl substituents	0.8	
Unsaturated (without substituents)	$n_{C,\,ring}$ = even number and number of double bonds = $\dfrac{n_{C,\,ring}}{2}$ or including only *trans* isomers with one double bond or allene	0.8	
	The bridged bicyclic ring where each ring containing more than four-membered ring	2.0 (one double bond)	
		1.0	
	Endo-tricyclic ring	1.5	
	Directly attachment of two rings with more than six carbon atoms in each ring	1.0	
	$n_{C,\,ring}$ > 11 in which all double bonds are in *trans* form	1.0	
	The existence of $-C\equiv C-$ in ring	1.2	

Table 3.7 (continued)

Type of cyclic hydrocarbon	Condition	$T^+_{m, cyc hyd}$	Example
Unsaturated (containing substituents)	Bridged bicyclic ring with an only methyl substituent	2.0	
	Attachment of substituents to ring through a double bond	1.0	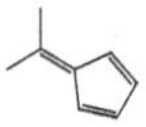

Table 3.8: Summary of the values of $T^-_{m, cyc hyd}$.

Type of cyclic hydrocarbon	Condition	$T^-_{m, cyc hyd}$	Examples
Saturated (without substituents)	$n^*_{C, ring} > 26$	1.0	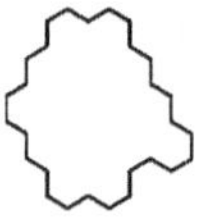
Saturated (containing substituents)	One substituent with a nonlinear chain with more than six carbons or the presence of a molecular fragment 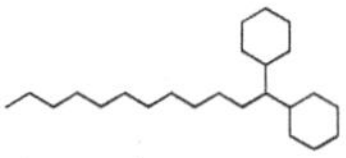 ![fragment] where R contains more than five carbon atoms	1.2	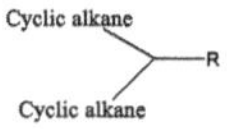 Cyclic alkane Cyclic alkane
	The existence of molecular moiety 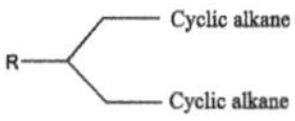 R—⟨ Cyclic alkane / Cyclic alkane	1.7	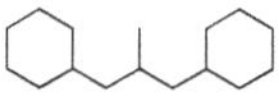

$^*n_{C, ring}$ represents the number of carbon atoms in the ring.

The present method can provide a suitable correlation for the quick reliable finding of novel cyclic hydrocarbon fuels with desirable melting points.

Example 3.9: 1,3,3-Trimethyl-bicyclo[2.2.1]heptane has the following molecular structure:

The use of eq. (3.18) gives

$$T_m = 103.1 + 10.24a + 89.16T_{m,\,cyc\,hyd}^+ - 66.44T_{m,\,cyc\,hyd}^-$$

$$= 103.1 + 10.24 \times 10 + 89.16 \times 1.7 - 66.44 \times 0$$

$$= 357.1\,K$$

The experimental value of the melting point of this compound is 331.0 K [199]. Meanwhile, the use of the method by Alantary and Yalkowsky [269] – as one of the best available predictive methods – gives 211.4 K.

3.3 Melting points of ionic liquids

Ionic liquids are low melting salts, which melt at or near room temperature. They have wide applications in the chemistry and chemical industries [293, 294]. Melting points of ionic liquids are essential for designs and applications but their melting-point data are scarce [295]. They show electrostatic attraction because larger sizes of cations and anions have a larger distance between two ions, which reduces their electrostatic attraction [296]. They depend on the arrangement of ions and the strength of the pairwise ion interactions in the crystal. The structure and interactions between the ions can provide ionic liquids with low melting points. Different approaches have been developed for the prediction of melting points of selected classes of ionic liquids – QSPR and group additivity methods are two common methods [297, 298]. Molecular dynamics (MD) and density functional theory (DFT) can predict melting points by the simulation of solid-to-liquid phase transitions but they are difficult as compared to other estimation methods [299, 300]. The high computational demands and the need for starting geometry are the main limitations of MD and DFT methods. Low et al. [297] investigated the effect of descriptor choice in machine learning models for ionic liquid melting point prediction from an experimental dataset of 2212 ionic liquid melting points consisting of diverse ion types. They introduced the best model, based on ECFP4 and molecular orbital energies, for the prediction of ionic liquid melting points with an average MAE of 29 K.

Chen and Bryantsev [299] used complex dispersion-corrected DFT method involving (semi)local (PBE-D3) and hybrid exchange-correlation (HSE06-D3) functionals to predict the lattice enthalpy, entropy, and free energy of 11 ILs containing imidazolium/pyrrolidinium cations and halide/polyatomic fluoro-containing anions.

Venkatraman et al. [299] used various machine learning methods to predict the melting points of structurally diverse 2,212 ILs based on a combination of 1,369 cations and 141 anions. They found that tree-based ensemble methods (cubist, random forest, and gradient boosted regression) can demonstrate slightly better performance over support vector machines and k-nearest neighbor approaches.

Mehrkesh and Karunanithi [300] used the quantum chemistry descriptors of the cation and anion radii, the density of ionic liquids, symmetrical value, the molecular weight of cation, and the dielectric energy of anion for the prediction of melting points of ILs based on the 37 experimental melting point data.

As datasets from scientific literature continue to grow, deep learning models [301] have proven more effective than traditional machine learning methods because they can handle vast amounts of data and extract meaningful patterns. This makes them a powerful tool for analyzing large collections of IL data through advanced data mining and machine learning. By developing robust modeling techniques, we can better identify and tackle key challenges in IL material design across various applications. This innovative approach could greatly accelerate the discovery and development of new ILs, offering a faster alternative to traditional methods like industrial-process modeling, thermodynamic analysis, and atomistic simulations – while still complementing these deeper, theory-driven studies. As a foundational step, Acar et al. [302] used a chemoinformatics-based deep learning model to predict the T_m of diverse ILs based on molecular descriptors. Their goal was to provide fresh perspectives that enhance existing theoretical models [303] and advance the field.

3.3.1 Group additivity approach

The group additivity methods can be used only for ionic liquids containing the specified cations and anions. Thus, they cannot be applied for many new ionic liquids where the contributions of cation and anion groups have not been considered. Lazzús [304] used experimental data of melting points of 40 ionic liquids to obtain the contributions of the cation-anion groups in a correlation set. Another data set including 23 ionic liquids was used to test the reliability of the method. The final equation for this model is [304]:

$$T_m = 98.599 + \sum_i n_{c,i} T_{c,i} + \sum_j n_{a,j} T_{a,i} \qquad (3.19)$$

where $n_{c,i}$ and $n_{a,j}$ are the occurrences of the groups i and j in the cation and the anion of the desired ionic liquid, respectively; $T_{c,i}$ and $T_{a,j}$ are the contribution of the cation and anion groups, respectively. Cation groups contain imidazolium, pyridinium, pyrrolidinium, phosphonium, and ammonium. Anion groups contain halides, pseudohalides, sulfonates, tosylates, imides, borates, phosphates, carboxylates, and metal complexes. The contribution values of the mentioned cation and anion groups are given in Table 3.9.

Table 3.9: The contribution of cation and anion groups for calculation of melting points of ionic liquids.

Cations

Name	Group	$T_{c,i}$	Group	$T_{c,i}$
Imidazolium		39.698	Z=−H	38.623
Pyridinium		82.227	Z=−CH$_3$	68.819
Pyrrolidinium		12.868	Z=−CH$_2$−	1.344
Ammonium		13.890	Z=−CH<	−79.375
Phosphonium		−448.738	Z=−N	21.626

Anions

Group	$T_{a,j}$	Group	$T_{a,j}$
=CH−	21.765	−Cl	10.923
>C<	9.910	−Br	10.213
−COO	13.484	−P [>P<]	33.726
−HCOO	13.531	−B [>B<]	−20.084
−O− [−O]	−9.850	−I	−3.753
−N− [>N−]	−5.493	>As<	−28.174
−NO$_3$	−4.482	−CB$_{11}$H$_6$	−5.603
−SO$_2$−	8.757	−CB$_{11}$H$_{12}$	66.553
−CF$_3$	−41.448	=CH− (ring)	8.067
−CF$_2$−	−1.811	=C< (ring)	3.132
−F	−4.930		

Example 3.10: Consider the following ionic liquid:

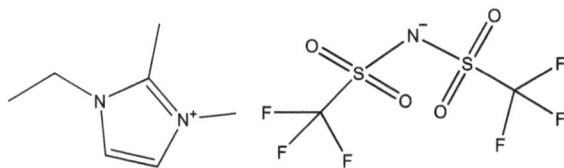

The melting point is calculated as follows:

$$T_m = 98.599 + \sum n_{c,i} T_{c,i} + \sum n_{a,j} T_{a,j}$$

$$= 98.599 + \text{Imidazolium} + 3(T_{c,-CH_3}) + T_{c,-CH_2-} + T_{a,-N-} + 2(T_{a,-SO_2}) + 2(T_{a,-CF_3})$$

$$= 98.599 + 39.698 + 3(68.819) + 1.344 - 5.493 + 2(8.757) + 2(-41.448)$$

$$= 275.22 \text{ K}$$

The measured melting point of this compound is 248.15 K [305].

Scientists have created several 'recipe books' (group additivity models) to predict ionic liquid properties, but they come with some frustrations. It's like comparing restaurant reviews when some critics tasted different dishes – the ratings don't line up properly. Many models also miss key ingredients, limiting what they can predict [307, 308].

Mital et al. [306] expanded on the method of Lazzús [304], focusing on the most versatile models for predicting melting points – a critical factor in designing advanced energy materials. After gathering the largest tasting menu yet (1,300+ data points across 933 unique ionic liquids), they upgraded one model by adding missing ingredients and fine-tuning the recipe as follows:

$$T_m = 210.97 + \sum_i n'_{c,i} T'_{c,i} + \sum_i n'_{a,j} T'_{a,i} \tag{3.20}$$

Table 3.10: Calculating melting points of ionic liquids through cation and anion group contributions.

Cations

Name	Group	$T'_{c,i}$	Group	$T'_{c,i}$
Imidazolium	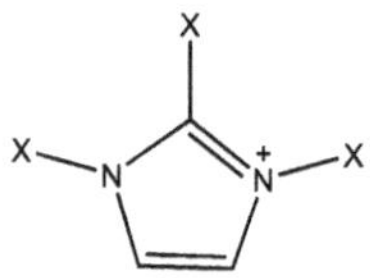	6.00	–H	24.52
Pyridinium	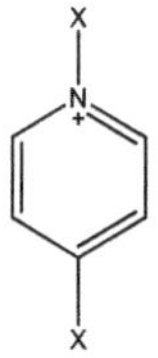	59.09	–CH$_3$	30.98

Table 3.10 (continued)

Cations

Name	Group	$T'_{c,i}$	Group	$T'_{c,i}$
Pyrrolidinium	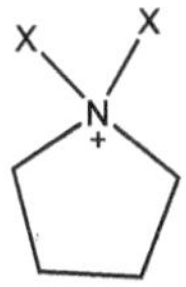	50.66	$-CH_2-$	−0.22
Pyrrolidinonium	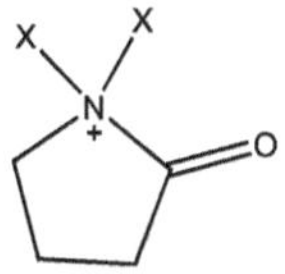	0.07	$-CH<$	−8.28
Piperidinium	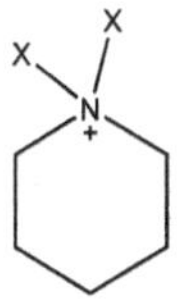	48.26	$-OH$	44.86
Morpholinium	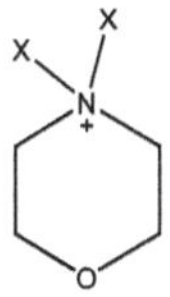	39.74	$-O-$	−18.99
Ammonium	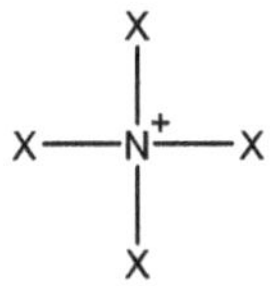	−4.38	$>C<$	12.71
Sulfonium	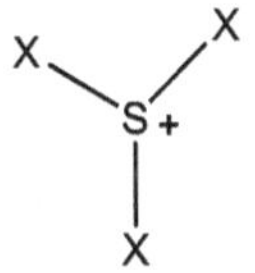	5.00	$-COO$	9.88
Phosphonium	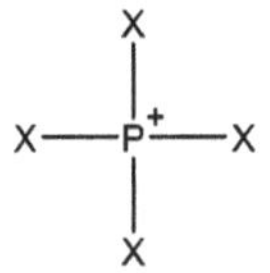	−10.01	$-CN$	40.01
Triazolium	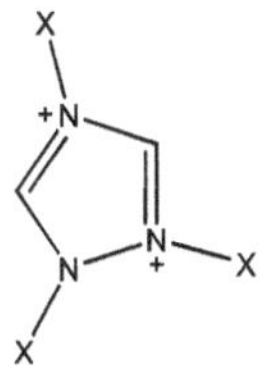	31.15	$-NH_2$	35.07

Table 3.10 (continued)

Cations

Name	Group	$T'_{c,i}$	Group	$T'_{c,i}$
Pyrazolium	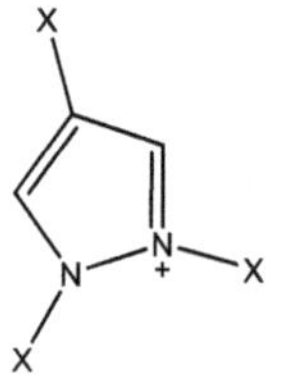	0.11	−F	22.38
Thiazolium	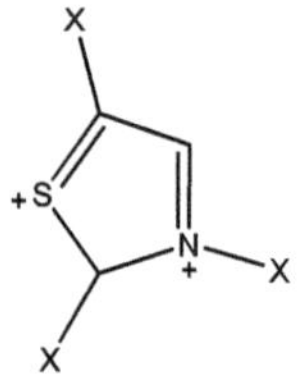	−0.19	−Br	22.24
			−Cl	0.11
			−SO$_2$−	5.95
			−Benzyl	52.26
			−Vinyl	0.86
			−CF$_2$−	10.05
			−CF$_3$	30.22
			−SF$_5$	−5.79
			>P=O	−36.95

Anions

Group	$T'_{a,i}$	Group	$T'_{a,i}$
−CH$_3$	1.55	−B	−11.05
−CH$_2$−	−2.71	−I	31.77
>CH−	1.06	−Al	−5.92
>C<	−0.77	−As	6.27
>CO	−7.63	−Nb	−2.74
−COO−	−20.94	−Ta	0.25
−HCOO	−24.97	−Sb	12.00
−OH	−0.72	−Sn	0.23
−O− [−O]	2.67	−W	−0.28
−CN	−16.50	−CB$_{11}$H$_{11}$− (carborane)	18.88
−N− [>N-]	−3.87	−CB$_{11}$H$_{12}$ (carborane)	18.19
−NO$_3$	9.21	−CB$_{11}$H$_6$ (carborane)	−16.68
−S−	0.42	CH− (ring)	0.18
−SO$_2$−	−4.82	=C< (ring)	17.02

Table 3.10 (continued)

Anions

Group	$T'_{a,i}$	Group	$T'_{a,i}$
$-CF_3$	−10.06	>CO (ring)	3.72
$-CF_2-$	4.27	−O− (ring)	3.34
−F	−4.74	PF6	28.32
−Cl	26.16	BF4	1.18
−Br	50.64	B(CN)4	−8.80
−P	4.77	Oleate	−12.14
B(CN)$_4$	−8.80	Phosphate	18.23

Example 3.11: The use of eq. (3.19) and Table 3.9 for the following IL gives:

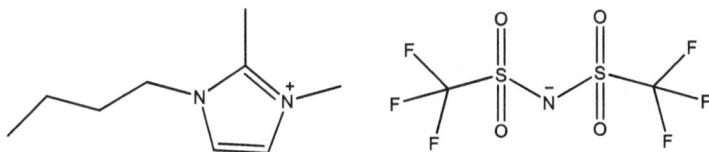

$$T_m = 210.97 + \sum_I n'_{c,i} T'_{c,i} + \sum_I n'_{a,j} T'_{a,j}$$

$$= 210.97 + \text{Imidazolium} + T'_{c,H} + 2(T'_{c,CH_3}) + 3(T'_{c,-CH_2-}) + T'_{a,-N-} + 2(T'_{a,-SO_2-}) + 2(T'_{a,-CF_3})$$

$$= 210.97 + 6.00 + 24.52 + 2(30.98) + 3(-0.22) + (-3.87) + 2(-4.82) + 2(-10.06)$$

$$= 269.16 \text{ K}$$

The predicted result is close to experimental value, that is, 267 K [307].

3.3.2 QSPR approaches based on complex descriptors

QSPR approaches usually need complex descriptors, expert users, and computer codes to evaluate the melting point of the desired compound. They correlate the requested chemical or physical properties with molecular descriptors, which are derived from the molecular structures of chemical compounds quantitatively [308]. They are based on different kinds of statistical tools with some restrictions, for example, application to similar compounds and the correct selection of molecular descriptors. They can predict the melting points of different classes of organic compounds. Wang et al. [309] used multiple linear regression and an artificial neural network for the estimation of the melting points of carbocyclic nitroaromatic compounds with six complex descriptors. Liu and Holder [310] have developed three QSPR models for organosilicon compounds containing a variety of silicon-containing organic compounds, silanes, and siloxanes, respectively. Liang et al. [311] have correlated the melting points of fatty acids with five complex descriptors, for example, the average valence connectivity index chi-5. Yan et al. [312] developed a QSPR model for the prediction of melting points of eight kinds of ionic liquid compounds including imidazolium, benzimidazolium, pyridinium, pyrrolidinium, ammonium, sulfonium, triazolium, and guanidinium based on the general topological index. Watkins et al. [313] used 49 complex descriptors to estimate the melting points of a wide range of persistent organic pollutants including chloro- and bromo-analogs of dibenzo-p-dioxins, dibenzofurans, biphenyls, naphthalenes, diphenyl ethers, and benzenes by GA-partial least squares (GA–PLS) modeling and the random forest method. Morrill and Byrd [314] developed some QSPR models to predict melting temperatures of energetic materials based on descriptors calculated using the AM1 semiempirical quantum mechanical method. Farahani et al. [315] developed a QSPR model using 12 complex descriptors such as descriptor X1A, which stands for average connectivity index of order 1, and genetic function approximation for the prediction of melting points of diverse ionic liquid compounds including sulfonium, ammonium, pyridinium, 1,3-dialkyl imidazolium, tri-alkyl imidazolium, phosphonium, pyrrolidinium, double imidazolium, 1-alkyl imidazolium, piperidinium, pyrroline, oxazolidinium, amino acids, guanidinium, morpholinium, isoquinolinium, and tetra-alkyl imidazolium.

Paduszyński et al. [316] reviewed existing QSPR models and developed new approaches to predict T_m and classify IL states at 300 K. Their study analyzed experimental data from 953 IL salts using multiple machine learning methods, including PLS regression, stepwise multiple linear regression, and several classification algorithms. They carefully examined how molecular descriptors and geometry optimization levels affect model performance while following rigorous validation protocols. The most effective models emerged as stepwise multiple linear regression model for regression and linear discriminant analysis for classification, both achieving about 80% accuracy – suggesting this may represent a current limit for IL melting point prediction. Unlike previous cation-focused models, their model accounted for both cation and anion effects, leading to better performance than established methods like Yan et al.'s [317]. Paduszyński et al. [316] made these models particularly practical by using accessible computational methods and providing all implementation details. Beyond theoretical improvements, they demonstrated real-world utility by predicting properties for over 35,000 potential IL combinations. This capability could significantly accelerate the design of ILs with specific phase behaviors. All supporting data and tools are openly available to help researchers apply and build upon our work in developing tailored ionic liquids for various applications.

Niu et al. [318] developed unified QSPR models using norm indices to estimate properties of organic compounds including Tm. Norm indices are like a molecule's ID badge – they use simple math to capture key structural features that help us predict how a compound will behave. By analyzing how atoms connect (like mapping out a social network of atoms), these descriptors translate molecular structure into useful property predictions – all through some clever number-crunching. Using datasets of 7,291 for T_m, they achieved good predictive performance ($R^2 = 0.837$).

Lotfi et al. [319] used a QSPR approach to predict the melting points of imidazolium-based ILs. Using the Monte Carlo algorithm in CORAL software, they developed a robust QSPR model for 353 imidazolium ILs. The model relies on a hybrid descriptor combining SMILES notation and hydrogen-suppressed molecular graphs. To ensure reliability, they validated the model internally and externally, dividing the dataset into four random splits: training (~33%), invisible training (~31%), calibration (~16%), and validation (~20%). The validation results showed strong predictability, with R^2 ranging between 0.74 and 0.90. They also analyzed which structural features influence melting points, helping identify what makes them rise or fall.

Building on previous research of Krossing et al. [320], Liu et al. [321] used key physicochemical parameters from a thermodynamic cycle model as descriptors for ILs. They analyzed how these descriptors relate to IL melting points while aiming for high accuracy, fewer descriptors, and lower computational costs – a key focus in current research. Specifically, they introduced a method using just 12 semi-empirical PM7-level descriptors to predict melting points of imidazolium-based ILs. This model helps quickly and accurately estimate melting points, simplifying IL selection for practical applications.

Dai et al. [322] analyzed melting points for 3,129 ionic liquids (–96 °C to 372 °C) and found their behavior depends on complex cation-anion interactions, not just individual components. While traditional methods like group additivity and QSPR offer practical predictions, conductor-like screening model for real solvents (COSMO-RS) provides the strongest theoretical foundation despite its current accuracy limitations. COSMO-RS is a computational chemistry method that predicts chemical properties (like melting points) by simulating how molecules interact in liquids. Instead of tracking individual atoms, it treats molecules as "electron clouds" and calculates their affinity for one another. The key challenge is incorporating transition thermodynamics properly – combining COSMO-RS with machine learning could bridge this gap, creating a model that's both theoretically sound and highly accurate.

3.3.3 Simple approach based on the structure of cations and anions

A simple correlation has been developed to predict melting points of important classes of ILs including imidazolium-, pyridinium-, pyrrolidinium-, ammonium-, phosphonium-, and piperidinium-based ILs and different types of anions. The experimental data of 195 different types of ILs were used to derive the new correlation as [323]:

$$T_m = 382.1 - 13.35a_{cat} + 4.796b_{cat} - 22.07c_{cat} - 11.24b_{ani} - 25.29c_{ani}$$
$$+ 70.69j_{ani} + 42.37f_{ani} - 205.4g_{ani} + 116.7T^+_{m, IL} - 121.8T^-_{m, IL} \tag{3.21}$$

where a_{cat}, b_{cat}, and c_{cat} are the number of carbon, hydrogen, and nitrogen atoms in cation, respectively; b_{ani}, c_{ani}, j_{ani}, f_{ani}, and g_{ani} are the number of hydrogen, nitrogen, bromine, chlorine, and aluminum atoms in anion, respectively; $T^+_{m,IL}$ and $T^-_{m,IL}$ are positive and negative adjusting functions, respectively. Table 3.11 shows the values of $T^+_{m,IL}$ and $T^-_{m,IL}$ for the presence of specific cations and anions in the desired IL.

Table 3.11: The values of $T^+_{m,IL}$ and $T^-_{m,IL}$ for the presence of specific cations and anions in the desired IL.

Cation	Anion	$T^+_{m,IL}$	$T^-_{m,IL}$	Condition
imidazolium structures	BF_4^-	0.5	0.0	$n=0$
		0.0	0.80	$2<n<10$
		0.5	0.0	$n>11$
	PF_6^-	0.0	0.5	$5<n<10$
		0.5	0.0	$n>11$
	tosylate ($-SO_3^-$)	11.0	0.0	$n<4$
	Cl^-	0.0	0.5	$2<n<10$
	trifluoroacetate	0.0	0.3	$1<n<10$
		0.5	0.0	$n>11$
	fluoroborate	0.0	0.5	$1\leq n<10$
	bis(trifluoromethylsulfonyl)imide	0.5	0.0	$n>11$
imidazolium	Cl^-	10.8	0.0	$0<n<4$
	trifluoroacetate	0.7	0.0	$n<3$
pyridinium	Br^-	1.1	0.0	$3<n<7$
	bis(trifluoromethylsulfonyl)imide	0.7	0.0	
pyridinium	PF_6^-	1.0	0.0	$11<n<18$
pyrrolidinium	bis(trifluoromethylsulfonyl)imide	0.5	0.0	$n<3$

Table 3.11 (continued)

Cation	Anion	$T^+_{m,IL}$	$T^-_{m,IL}$	Condition
	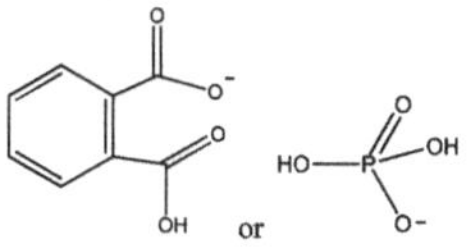 (phthalate anion) or (phosphite/phosphate anion)	0.6	0.0	$n < 4$
	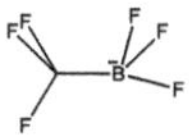 (fluoroborate/trifluoromethyl borate anion)	0.0	0.5	$n < 4$
	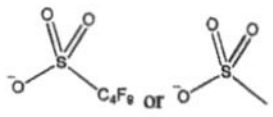 (acyl borate C_xF_x or acyl anion)	11.0 0.4	0.0 0.0	$n = 0$ $1 \leq n < 8$
	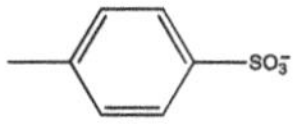 (benzenesulfonate, $-SO_3^-$)	11.0	0.0	$n < 6$
	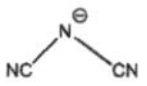 (dicyanamide, $NC{-}N{-}CN$)	11.0	0.0	$n = 1$
	 (methanesulfonate anion)	11.3	0.0	
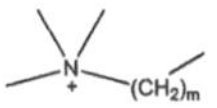 (ammonium cation with $(CH_2)_m$)	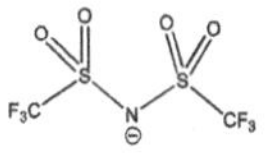 (bis(trifluoromethylsulfonyl)imide, $F_3C{-}SO_2{-}N{-}SO_2{-}CF_3$)	0.5 0.0	0.0 0.3	$m < 3$ $3 > m$
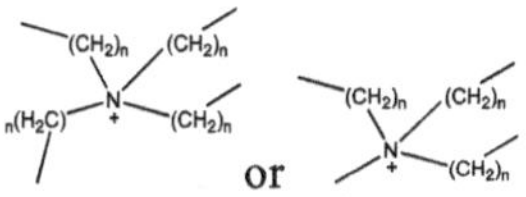 (diammonium cation with $(CH_2)_n$) or		0.6 0.2	0.0 0.0	$n < 5$ $n \geq 5$
	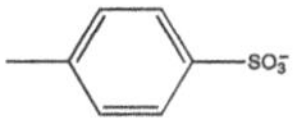 ($-SO_3^-$)	11.0	0.0	
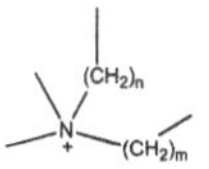	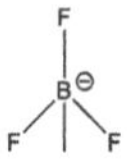 (BF_4^-)	0.7	0.0	m or $n < 4$
(ammonium cation with $(CH_2)_n$/$(CH_2)_m$)	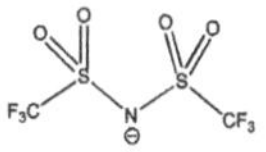 ($F_3C{-}SO_2{-}N{-}SO_2{-}CF_3$)	0.0	0.5	
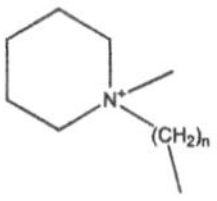 (piperidinium cation with $(CH_2)_n$)	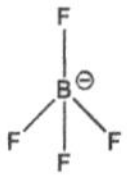 (BF_4^-)	0.5	0.0	$n < 9$
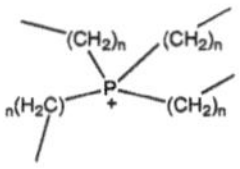 (phosphonium cation with $(CH_2)_n$)	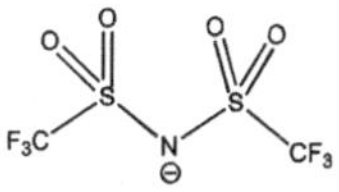 ($F_3C{-}SO_2{-}N{-}SO_2{-}CF_3$)	0.4	0.0	$n < 3$
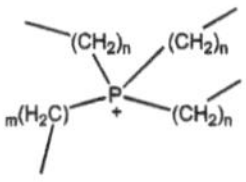 (phosphonium cation with $(CH_2)_n$)	Br^-	0.0	0.5	n or $m < 7$

Example 3.12: Consider the following ionic liquid:

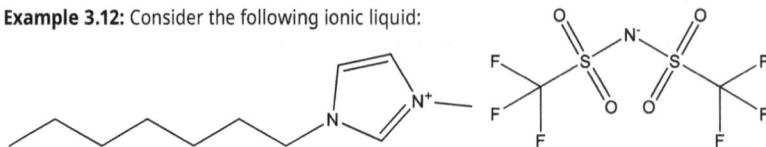

The use of eq. (3.21) gives:

$$T_m = 382.1 - 13.35a_{cat} + 4.796b_{cat} - 22.07c_{cat} - 11.24b_{ani} - 25.29c_{ani}$$
$$+ 70.69j_{ani} + 42.37f_{ani} - 205.4g_{ani} + 116.7T_{m,IL}^+ - 121.8T_{m,IL}^-$$
$$= 382.1 - 13.35 \times 11 + 4.796 \times 21 - 22.07 \times 2 - 11.24 \times 0 - 25.29 \times 1$$
$$+ 70.69 \times 0 + 42.37 \times 0 - 205.4 \times 0 + 116.7 \times 0 - 121.8 \times 0 = 266.5 \text{ K}$$

The measured melting point of this ionic liquid is 219.6 K [323].
The use of eq. (3.19) for this ionic liquid provides:

$$T_m = 98.599 + \sum n_{c,i} T_{c,i} + \sum n_{a,j} T_{a,j}$$
$$= 98.599 + \text{Imidazolium} + 2\left(T_{c,-CH_3}\right) + T_{c,-H} + 6\left(T_{c,-CH_2-}\right) + T_{a,-N-} + 2\left(T_{a,-SO_2}\right)$$
$$+ 2\left(T_{a,-CF_3}\right)$$
$$= 98.599 + 39.698 + 2(68.819) + 38.623 + 6(1.344) - 5.493 + 2(8.757) + 2(-41.448)$$
$$= 251.89 \text{ K}$$

3.4 Summary

This chapter shows how scientists predict melting points. Equation (3.1) is like basic math – just add up standard values for each part of the molecule. Equation (3.2) gets fancier, using four special measurements about the molecule's shape and electrical properties. Equation (3.3) simplifies things again, using just one key measurement that turns out to work surprisingly well. Together, they show how chemistry has moved from simple rules to smarter prediction methods. Several empirical methods have also been introduced for predicting the melting points of organic compounds containing energetic functional groups. For polynitroarenes and polynitroheteroarenes, the reliability of eq. (3.5) may be better than that of (3.4). For nonaromatic energetic compounds, the use of eq. (3.7) is recommended rather than eq. (3.6). Equations (3.8)–(3.10) can be applied to various aromatic and nonaromatic energetic compounds containing Ar–NO$_2$, C–NO$_2$, C–ONO$_2$, or N–NO$_2$ groups. Equations (3.11)–(3.14) can be used to predict the melting points of those energetic compounds which only contain peroxide and azide groups, respectively. Equations (3.15)–(3.17) provide a general approach for the prediction of melting points of various energetic compounds, including those of organic perox-ides, organic azides, organic nitrates, polynitroarenes, polynitroheteroarenes, acyclic and cyclic nitramines, nitrate esters and nitroaliphatic compounds, aromatic ring, ISSP equals 1.0. Equation (3.18) provides a reliable correlation for prediction of the melting points of cyclic hydrocarbons, which can be used as jet fuels. Several have also been introduced for prediction of melting points of ILs. Equation (3.19) is based on group addi-tivity approach, which can be applied for those ILs containing cations with imidazolium, pyridinium, pyrrolidinium, phosphonium, and ammonium groups as well as anions with halides, pseudohalides, sulfonates, tosylates, imides, borates, phosphates, carboxy-lates, and metal complexes. Equation (3.20) represents an extension of eq. (3.19) with broader applicability to diverse ILs. Equation (3.21) can be used for a wide range of ILs including imidazolium-, pyridinium-, pyrrolidinium-, ammonium-, phosphonium-, and piperidinium-based ILs and different types of anions.

4 Enthalpy and entropy of fusion

Thermal analysis of energetic compounds shows that they decompose at specific temperatures. Their exothermic decomposition can inhibit the explosive charge from being able to dissipate the applied heat [325]. The decomposition reaction begins usually above or during the melting process, so that energetic materials with higher melting points show high thermal stability. The enthalpy – or heat of fusion ($\Delta_{fus}H$) – is the enthalpy change which occurs in the transition from the most stable form of the solid to the liquid state of high energy compounds. It is related to the entropy of fusion ($\Delta_{fus}S$) and the melting point or fusion temperature, that is, $\Delta_{fus}H = T_m\Delta_{fus}S$. To measure the $\Delta_{fus}H$ of explosive materials, differential scanning calorimetry (DSC) can be used [326]. In this chapter, different approaches for the prediction of $\Delta_{fus}H$ and $\Delta_{fus}S$ will be reviewed.

4.1 Different approaches for the prediction of the enthalpy of fusion

Zeman et al. [327–330] have introduced some relationships between $\Delta_{fus}H$ and the impact sensitivity – as well as with the electric spark sensitivity – of nitramines and polynitro compounds. Therefore, the prediction of $\Delta_{fus}H$ provides a better insight into the intermolecular interactions and sensitivity of energetic molecules which have not yet been synthesized.

QSPR [331], quantum mechanics [332], group additivity methods [333, 334], artificial neural networks [335] and simple correlations based on molecular structures [336–342] are all suitable methods for predicting $\Delta_{fus}H$. QSPR methods need special computer codes and the databank set should cover a large number of compounds with different molecular structures in order to obtain suitable results for compounds with similar molecular structures as those included in the test library. Furthermore, they require complex molecular descriptors. However, such methods have often been used to predict the thermodynamic properties of particular families of compounds [343]. The group additivity methods have also been developed to predict the values of $\Delta_{fus}H$ for different types of organic compounds [187, 254, 344–346]. However, it has been shown that they may give $\Delta_{fus}H$ values which show large deviation from the expected values for some organic energetic compounds [225, 331]. Quantum mechanical methods were used to study the phase change properties of some energetic compounds [332, 347, 348], however, they require high-speed computers and specific computer codes. Some simple methods have also been developed for the prediction of $\Delta_{fus}H$ values [336–342] based on the molecular structures of energetic compounds. The group additivity method of Yalkowsky and coworkers [266–269], as well as simple methods based on molecular structures, will be demonstrated in the following sections.

https://doi.org/10.1515/9783112206768-004

Yan et al. [349] developed QSPR models to predict key thermodynamic properties of organic compounds, including standard vaporization enthalpy, enthalpy of fusion, entropy of fusion, and standard sublimation enthalpy. By incorporating molecular size and shape descriptors, they significantly improved prediction accuracy and model reliability, achieving excellent R^2. To handle cases where experimental $\Delta_{fus}H$ and $\Delta_{fus}S$ values fell below the model's average absolute error – which could produce unrealistic negative predictions in linear models – Yan et al. [349] successfully implemented nonlinear modeling approaches. When compared with existing methods, their models demonstrated superior predictive performance across a wide range of chemical compounds. These robust QSPR tools now provide valuable applications for chemical process design, reaction optimization, material property prediction, and pharmaceutical development, offering researchers and engineers reliable property estimation methods for both academic and industrial applications involving organic compounds.

4.1.1 Group additivity method for prediction of the enthalpy of fusion

The UPPER approach of Yalkowsky et al. [266–269] treats enthalpic interactions in additive constitutive properties in the general form:

$$\Delta_{fus}H = \sum n_i H_{fus,i}, \tag{4.1}$$

where $\Delta_{fus}H$ is the enthalpy of fusion of the whole molecule, n_i is the number of fragment i in the molecule, and $H_{fus,i}$ is the contribution of molecular fragment or group i to the enthalpy of fusion. Table 4.1 shows the values of $H_{fus,i}$ for different molecular fragments.

Example 4.1: Consider 2,4-dinitrobenzoic acid with the following molecular structure:

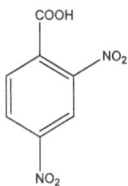

The enthalpy of fusion of this compound using eq. (4.1) and Table 4.1 is given as follows:

$$\Delta_{fus}H = \sum n_i H_{fus,i} = 3C_{ar} + 3CH_{ar} + 2YNO_2 + YCOOH$$

$$= 3 \times 1.777 + 3 \times 1.235 + 2 \times 4.584 + 11.785 = 29.99 \text{ kJ/mol}$$

The reported enthalpy of fusion for this compound is 24.6 kJ/mol [199].

As given in Table 4.1, the groups that participate in hydrogen bonds have higher values than those containing only dipole–dipole bonds because the latter only participates in dispersion forces.

Table 4.1: The contribution of i group or fragment in calculation of the enthalpy of fusion.

Group	Contribution	Group	Contribution	Group	Contribution
XCH_3	0.701	CH_{fus}	1.695	XNH_2	6.884
YCH_3	1.221	C_{fus}	1.332	XNH	1.181
ZCH_3	0.331	CH_{2RING}	1.054	YNH_2	5.681
XCH_2	1.408	CH_{RING}	1.046	YNH	3.799
$XCH_2{}^*$	-2.524	C_{RING}	0.757	YN	2.013
YCH_2	1.644	$=CH_{RING}$	0.883	YNO_2	4.584
$YYCH_2$	1.875	$=C_{RING}$	1.362	YNHCO	8.167
$YZCH_2$	-0.976	XF	-0.87	XCN	5.174
XCH	1.177	YF	0.409	YCN	6.558
YCH	-0.916	XCl	1.889	XCOO	9.488
XC	1.177	YCl	1.581	YCOO	6.208
YC	-1.076	XBr	4.674	XCOOH	14.287
$CH_2=$	0.454	2&6	-2.954	YCOOH	11.785
YCH=	1.691	YBr	2.911	YCHO	5.470
YYCH=	1.689	XI	4.034	XCO	8.037
YC=	2.250	YI	4.334	YCO	3.332
YYC=	0.655	XO	2.921	YOCO	7.568
CH≡	2.357	YO	3.162	YOCOO	5.335
ZC≡	3.853	YYO	-6.918	$YCONH_2$	12.814
YZC≡	-1.732	Ar–O	-0.922	YCONH	9.083
$C_{allenic}$	2.033	XOH	4.953	YNHCOO	6.929
C_{ar}	1.777	YOH	6.699	$YNHCONH_2$	14.865
CH_{ar}	1.235	YSH	2.635	YNHCON	16.721
C_{BIP}	2.602	YS	5.313	$YCONH_2$	13.418
C_{BR1}	1.329	YSO_2NH_2	10.642	Ortho	-0.282
C_{BR2}	-0.564	YSO_2N-X	6.739	IHB	-3.495

X, a group that is bonded to only sp^3 atoms.
Y, a group that is singly bonded to one sp^2 atom.
YY, a group bonded to 2 sp^2 atoms.
Z, a group that is bonded to a sp atom.
RING, a group within an aliphatic ring.
fus, an aliphatic bridge-head group.
ar, a group within an aromatic ring.
BR2, an aromatic carbon contained in 2 rings.
BR3, an aromatic carbons contained in 3 rings.
BIP, the central carbons of biphenyl rings.
Ortho, the number of ortho substitutions.
2 and 6, the number of halogen substitutions at the 2nd and 6th positions of a biphenyl ring systems.

4.1.2 Nitroaromatic carbocyclic energetic compounds

The study of the $\Delta_{fus}H$ values for various nitroaromatic carbocyclic compounds with general formula $C_aH_bN_c(O \text{ or } S)_{d''}$ has shown a new approach can be used to derive a useful equation for predicting $\Delta_{fus}H$ as follows [338]:

$$\Delta_{fus}H = 1.197 + 1.681a + 6.793c - 2.143d'' + 8.526C_{SPG}, \tag{4.2}$$

where $\Delta_{fus}H$ is in kJ/mol and C_{SPG} is the contribution of specific polar groups attached to an aromatic ring. As is seen in eq. (4.2), the C_{SPG} coefficient has a positive sign which can result in an increase in the value of $\Delta_{fus}H$. The value of C_{SPG} is determined as follows.

(1) Hydroxyl groups: C_{SPG} corresponds to the number of hydroxyl groups attached to an aromatic ring, for example, $C_{SPG} = 2.0$ for 1,3-dihydroxy-2,4,6-trinitrobenzene.

(2) Amino group: The value of C_{SPG} equals 2.0 for nitroaromatic carbocyclic compounds which have more than two amino ($-NH_2$) groups attached to the aromatic ring, for example, $C_{SPG} = 2.0$ for 1,3,5-triamino-2,4,6-trinitrobenzene (TATB).

(3) Other polar groups: The value of C_{SPG} is 1.0 if at least one $-C(=O)-O-C(=O)-$ or $-S(O)_2-$ (sulfone) is attached to an aromatic ring, whereas C_{SPG} is 2.0 if at least one $-COOH$ functional group is attached to an aromatic ring. For example, $C_{SPG} = 1.0$ for 4-nitrophthalic anhydride and $C_{SPG} = 2.0$ for 2-nitrobenzoic acid.

(4) Disubstituted nitroaromatic compounds: The value of C_{SPG} is 1.0 for disubstituted nitroaromatic compounds that contain only two nitro groups attached to aromatic rings, that is, 1,8-dinitronaphthalene.

Example 4.2: The use of eq. (4.2) for 2,2′,4,4′,6,6′-hexanitrobiphenyl with the following structure gives

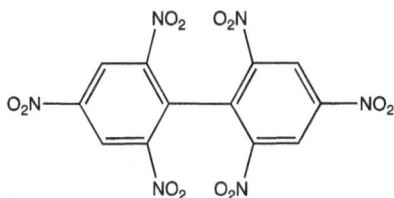

$$\Delta_{fus}H = 1.197 + 1.681a + 6.793c - 2.143d'' + 8.526C_{SPG}$$
$$= 1.197 + 1.681(12) + 6.793(6) - 2.143(12) + 8.526(0)$$
$$= 36.41 \text{ kJ/mol}.$$

The predicted value is close to the experimental value of 37.44 kJ/mol [328].

4.1.3 Nitroaromatic energetic compounds

Since eq. (4.2) may show relatively large deviations in the predicted results from the experimentally determined values for some halogenated and different isomers of nitroaromatics, a more reliable correlation of the following form has been introduced [340]:

$$\Delta_{fus}H = 3.817 + 1.196a + 5.8471c - 1.382d + 7.898C_{SSP}, \tag{4.3}$$

where the C_{SSP} factor can correct the predicted results on the basis of a, c, and d. For the presence of some polar groups and isomers, positive C_{SSP} values can increase the predicted $\Delta_{fus}H$ on the basis of a, c and d. In contrast, the attachment of tertiary and secondary amines as well as Ar–O– to nitroaromatic rings can result in the reverse situation being observed. The two opposite effects of the C_{SSP} can be specified according to the following conditions.

(1) Increasing effects of C_{SSP}: Increasing the number of hydroxyl (–OH) groups that are attached to an aromatic ring can increase the value for the enthalpy of melting because the strong hydrogen bonding which can result leads to a much more efficient crystal packing. Thus, the C_{SSP} value is equal to the number of –OH groups, except in the case of one hydroxyl group located between two nitro groups $O_2N\diagdown\!\!\!\diagup NO_2$.

    (a) Or if the hydroxyl group is *ortho* to the alkyl group, in which cases the value of C_{SSP} is 0.35.

    (b) Polar groups –C(=O)–C(=O)– and –S(O)$_2$– as well as –COOH: The values of C_{SSP} are 1.0 and 1.5 for the presence of –C(=O)–C(=O)– (or –S(O)$_2$–) and –COOH groups, respectively.

    (c) Amino (–NH$_2$) groups: if amino groups are present in nitroaromatic compounds, the predicted $\Delta_{fus}H$ can be higher. If one or two amino groups are present, $C_{SSP} = 0.5$, except if there is an amino group in the *ortho* position with respect to the nitro group in mononitro derivatives. If more than two amino groups are attached to the aromatic ring, $C_{SSP} = 2.5$.

    (d) Some polar groups in the *para* position relative to the nitro group in benzene rings: If –OH and –NH$_2$ groups are in the *para* position relative to a nitro group in disubstituted or halogenated benzene rings, the corresponding values for C_{SSP} from parts (a) and (c) should be replaced by 1.5 and 1.0, respectively.

    (e) Two nitro groups: in nitroaromatic compounds which contain only two nitro groups, the values of C_{SSP} are 0.50 and 1.5 for the *ortho* (or *meta*) and other positions, respectively.

    (f) If more than one alkyl group is attached to one benzene ring (or to two benzene rings that are not directly attached to each other): $C_{SSP} = 4/n_R$ in which n_R is the number of alkyl groups. Increasing n_R can decrease the planarity of the molecule and hence the molecular attractions.

(2) Decreasing effects of C_{SSP}: The value of C_{SSP} is −1.0 if tertiary, secondary amines or Ar–O– are attached to a nitroaromatic ring. This is because the presence of these groups can decrease the planarity of the molecules, as well as reduce the interactions between local dipole moments of neighboring nitro groups.

Example 4.3: Using eq. (4.3) for the following structure gives:

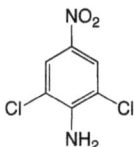

$$\Delta_{fus}H = 3.817 + 1.196a + 5.8471c - 1.382d + 7.898C_{SSP}$$

$$= 3.817 + 1.196(6) + 5.8471(2) - 1.382(2) + 7.898(1.0)$$

$$= 27.82 \text{ kJ/mol.}$$

The predicted value is closer to the value from the experimental data (32.64 kJ/mol [142]) than eq. (4.2) and group additivity method [194] are, which give values of 20.58 and 18.29 kJ/mol, respectively.

4.1.4 Nonaromatic energetic compounds containing nitramine, nitrate, and nitro functional groups

A simple correlation to predict the $\Delta_{fus}H$ of acyclic and cyclic nitramines, nitrate esters, and nitroaliphatic energetic compounds can be written as follows [337]:

$$\Delta_{fus}H = 16.81 + 1.896d + 4.186n_{EDNA}$$
$$+ 17.51\left(n_{NNO_2}^{>3,\,linear} - 2\right) - 11.52C_{-NO_2(-ONO_2)},$$

(4.4)

where n_{EDNA} is the number of N,N'-(ethane-1,2-diyl)dinitramide (EDNA) moieties $(O_2NNCH_xCH_xNNO_2)$ in cyclic nitaramines; $n_{NNO_2}^{>3,\,linear}$ is the number of $-NNO_2$ groups for those acyclic linear nitramines containing more than three nitramine groups, and $C_{-NO_2(-ONO_2)}$ is a parameter. The value of $C_{-NO_2(-ONO_2)}$ is 1.0 for those compounds which contain less than four $-NO_2$ and $-ONO_2$ groups, and is $C_{-NO_2(-ONO_2)}$ is − 0.75 for those compounds with four or more $-NO_2$ and $-ONO_2$ groups. The value of n_{EDNA} is zero for the simultaneous presence of $O_2NNCH_xNNO_2$ and $O_2NNCH_xCH_xNNO_2$ units in nitramine compounds (e.g., 1,3-dinitroimidazolidine (CPX)), which may be due to the lowering of the symmetry in these compounds.

Example 4.4: 1,4,5,8-Tetranitrodecahydropyrazino[2,3-b]pyrazine (TNAD) is a high performance explosive with the following molecular structure:

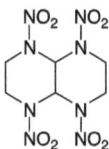

Using eq. (4.4) for TNAD gives

$$\Delta_{fus}H = 16.81 + 1.896d + 4.186n_{EDNA}$$
$$+ 17.51\left(n_{NNO_2}^{>3,\,linear} - 2\right) - 11.52C_{-NO_2(-ONO_2)}$$
$$= 16.81 + 1.896(8) + 4.186(3) + 17.51(0) - 11.52(0)$$
$$= 44.54 \text{ kJ/mol.}$$

The measured $\Delta_{fus}H$ for TNAD is 46.4 kJ/mol [326]. The predicted $\Delta_{fus}H$ using the group additivity method is 42.66 kJ/mol [261], which shows a larger deviation from the experimentally measured value than the value obtained using eq. (4.3).

4.1.5 Improved method for the reliable prediction of the enthalpy of fusion of energetic compounds

It was found that the following correlation can be used to obtain reliable predictions of the enthalpy of fusion of energetic compounds with the general formula $C_aH_bN_c(O \text{ or } S)_{d''}(halogen)_k$ [336]:

$$\Delta_{fus}H = 0.542a + 1.490b + 2.044c + 1.252d'' + 1.839k$$
$$+ 9.848\Delta H_{Inc,fus} - 11.675\Delta H_{Dec,fus},$$

(4.5)

where $\Delta H_{Inc,fus}$ and $\Delta H_{Dec,fus}$ are two correcting functions which are described in the following sections.

4.1.5.1 The presence of some specific polar groups
Polar groups such as $-COOH$, $-OH$, and $-NH_2$ may result in the presence of strong hydrogen bonding and consequently much more efficient packing as a result of the attractive forces. For acyclic nitamines, increasing the number of polar nitamine groups can result in enhancement of the electrostatic attractions. The contributions of $\Delta H_{Inc,fus}$ for these polar groups are:

1) The effects of –OH and –COOH groups: The value of $\Delta H_{\text{Inc,fus}}$ is 0.4 for the presence of these functional groups, except in the case of a single hydroxyl group located between two nitro groups $O_2N\diagup\overset{\overset{\text{OH}}{|}}{\diagdown}NO_2$, or in an *ortho* position relative to an alkyl group for which cases the value of $\Delta H_{\text{Inc,fus}} = 0.0$.

2) The influence of –NH$_2$ groups: If one or two amino groups are present, the value of $\Delta H_{\text{Inc,fus}}$ is 0.7, except if there is an amino group in an *ortho* position relative to the nitro group for mononitro derivatives in which case $\Delta H_{\text{Inc,fus}} = 0.0$. The value of $\Delta H_{\text{Inc,fus}}$ is 2.6 if more than two amino groups are attached to an aromatic ring.

3) The number of –NNO$_2$ groups in acyclic nitramines $n_{\text{NNO}_2}^{\text{acycl}}$: If more than three nitramine groups are present,

$$\Delta H_{\text{Inc,fus}} = n_{\text{NNO}_2}^{\text{acycl}} - 2.$$

4.1.5.2 Disubstituted nitroaromatics

Molecular interactions such as interactions between local dipole moments of neighboring atoms or groups in certain compounds can be responsible for the close proximity of molecules in the crystal. The value of $\Delta H_{\text{Inc,fus}}$ is 1.1 for dinitronaphthalene and also for disubstituted benzene derivatives in which polar groups such as –OH and –NO$_2$ are located *para* to the nitro group.

4.1.5.3 Structural parameters affecting $\Delta H_{\text{Dec,fus}}$

(1) Nitroaromatics with more than one benzene ring: The value of $\Delta H_{\text{Dec,fus}}$ is equal to 0.5, except if sulfur is present between two benzene rings.

(2) Cyclic nitramines with rings which are larger than six membered rings and which contain only carbon and nitrogen atoms $(m_{\text{cyc}}^{>6})$:

$$\Delta H_{\text{Dec,fus}} = \frac{m_{\text{cyc}}^{>6} - 6}{4} + 0.5.$$

4.1.5.4 Different effects of –NO$_2$ and –ONO$_2$ groups in nonaromatics

Intermolecular interactions may be increased if a larger number of –NO$_2$ and –ONO$_2$ groups are present. Furthermore, the presence of a lower number of –NO$_2$ and –ONO$_2$ groups can result in a reduction in the molecular packing. The values of $\Delta H_{\text{Dec,fus}}$ and $\Delta H_{\text{Inc,fus}}$ are 0.4 and 1.1 if three or less –NO$_2$ (or –ONO$_2$) groups and more than three –NO$_2$ (or –ONO$_2$) groups are present, respectively.

The predicted values of $\Delta H_{\text{Inc,fus}}$ and $\Delta H_{\text{Dec,fus}}$ using the above conditions are summarized in Tables 4.2, 4.3 and 4.4.

Table 4.2: Summary of the correcting function $\Delta H_{Inc,fus}$.

	Polar groups	$\Delta H_{Inc,fus}$	Condition	Exception
The presence of some specific polar groups	The effects of –OH and –COOH groups	0.4	–	One hydroxyl group between two nitro groups or *ortho* position to alkyl group
	The influence of –NH$_2$ groups	0.7	If one or two amino groups are present	If an amino group is in the *ortho* position relative to nitro group for mononitro derivatives
		2.6	If more than two amino groups are attached to an aromatic ring	–
	The number of –NNO$_2$ groups in acyclic nitramines $n_{NNO_2}^{acycl}$	$n_{NNO_2}^{acycl} - 2$	For more than three nitramine groups	–
Disubstituated nitroaromatics	–	1.1	For dinitronaphthalene or if polar groups such as –OH and –NO$_2$ are present in *para* positions to nitro group in disubstituated benzene derivatives	–

Table 4.3: Summary of the correcting function $\Delta H_{Dec,fus}$.

	$\Delta H_{Dec,fus}$	Condition	Exception
Nitroaromatics	0.5	More than one benzene ring	Sulfur is present between two benzene rings
Cyclic nitramines	$\dfrac{m_{cyc}^{>6} - 6}{4} + 0.5$	Rings larger than six membered rings which contain only carbon and hydrogen atoms ($m_{cyc}^{>6}$)	–

Table 4.4: Different effects of –NO$_2$ and –ONO$_2$ groups in $\Delta H_{Inc,fus}$ and $\Delta H_{Dec,fus}$.

	$\Delta H_{Inc,fus}$	$\Delta H_{Dec,fus}$	Condition
Non-aromatics	1.1	–	More than three –NO$_2$ (or –ONO$_2$) groups
	–	0.4	Three or less –NO$_2$ (or –ONO$_2$) groups

Example 4.5: The use of eq. (4.5) for the following molecular structure gives the following:

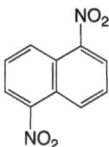

$$\Delta_{fus}H = 0.542a + 1.490b + 2.044c + 1.252d'' + 1.839k$$
$$+ 9.848\Delta H_{Inc,fus} - 11.675\Delta H_{Dec,fus}$$
$$= 0.542(10) + 1.490(6) + 2.044(2) + 1.252(4) + 1.839(0)$$
$$+ 9.848(1.1) - 11.675(0)$$
$$= 34.29 \text{ kJ/mol.}$$

The measured $\Delta_{fus}H$ for this compound is 33.03 kJ/mol [328]. The use of eq. (4.3) gives a value of 33.79 kJ/mol, whereas the predicted $\Delta_{fus}H$ by the group additivity method is 17.14 kJ/mol [261], which shows a much larger deviation from the measured value.

4.1.6 A reliable method to predict the enthalpy of fusion of energetic materials

The following improved simple approach has been introduced to enable prediction of the values of $\Delta_{fus}H$ in large classes of energetic compounds – including polynitroarenes, polynitroheteroarenes, acyclic, and cyclic nitramines, nitrate esters, nitroaliphatics, cyclic and acyclic peroxides, as well as nitrogen rich compounds [342]:

$$\Delta_{fus}H = 0.9781(\Delta_{fus}H)_{add} + 7.567(\Delta_{fus}H)^{Inc}_{nonadd} - 8.784(\Delta_{fus}H)^{Dec}_{nonadd}, \tag{4.6}$$

where

$$(\Delta_{fus}H)_{add} = 0.6047a + 0.6211b + 2.750c + 1.424d'' + 3.048k. \tag{4.7}$$

$(\Delta_{fus}H)^{Inc}_{nonadd}$ and $(\Delta_{fus}H)^{Dec}_{nonadd}$ are also nonadditive contributions corresponding to the increasing and decreasing effects of specific groups. The presence of polar groups results in an increase in the intermolecular attractions of a molecule with its neighboring molecules. The values of $(\Delta_{fus}H)^{Inc}_{nonadd}$ and $(\Delta_{fus}H)^{Dec}_{nonadd}$ are described in the following sections.

$$(\Delta_{fus}H)^{Inc}_{nonadd}$$

(1) –OH and –COOH groups in aromatic compounds: The value of $(\Delta_{fus}H)^{Inc}_{nonadd}$ is equal to 0.7, except if an –OH group is located between two nitro groups in an aromatic structure for which the value of $(\Delta_{fus}H)^{Inc}_{nonadd}$ is zero.

(2) $-NH_2$ group: The value of $(\Delta_{fus}H)_{nonadd}^{Inc}$ is 0.5 if one or two amino groups are present in aromatic or non-aromatic compounds. The value of $(\Delta_{fus}H)_{nonadd}^{Inc}$ equals 2.6 if more than two amino groups are attached to aromatic rings. For mononitro derivatives, if the nitro group is located in the *ortho* position with respect to the amino group, $(\Delta_{fus}H)_{nonadd}^{Inc}$ is zero.

(3) $>NH$ group: if the $>NH$ fragment is present, $(\Delta_{fus}H)_{nonadd}^{Inc}$ equals 0.5.

(4) $>N-NO_2$ group in acyclic nitramines: if more than three $>N-NO_2$ groups are present, $(\Delta_{fus}H)_{nonadd}^{Inc}$ is $\left(n_{NNO_2}^{acyclic} - 2\right)$ where $n_{NNO_2}^{acyclic}$ is the number of $>N-NO_2$ groups in acyclic nitramines.

(5) $-N(C=O)N-$ group: The value of $(\Delta_{fus}H)_{nonadd}^{Inc}$ is 1.5.

$$(\Delta_{fus}H)_{nonadd}^{Dec}$$

(1) Ar–X–Ar nitroaromatics: The value of $(\Delta_{fus}H)_{nonadd}^{Dec}$ is 0.5, except for Ar–S–Ar structures for which it is zero.

(2) Cyclic nitramines:

$$(\Delta_{fus}H)_{nonadd}^{Dec} = \frac{m_{cyc} - 6}{4} + 0.5$$

for cyclic nitramines in which the ring(s) are larger than six-membered rings and which contain only carbon and nitrogen atoms ($m_{cyc} > 6$).

(3) $-NO_2$ group in non-aromatic compounds: The value of $(\Delta_{fus}H)_{nonadd}^{Dec}$ is 1.0 for the presence of nitro groups in acyclic and cyclic alkanes.

(4) $-N-N=O$ group: The value of $(\Delta_{fus}H)_{nonadd}^{Dec}$ is 2.0 if the nitroso group is present.

Table 4.5 summarizes some functional groups and molecular fragments which can be used to determine the values of $(\Delta_{fus}H)_{nonadd}^{Inc}$ and $(\Delta_{fus}H)_{nonadd}^{Dec}$.

Table 4.5: Correcting functions which are used for the presence of different molecular fragments.

Molecular moieties	Effect on predicted $\Delta_{fus}H$		Comment
	$(\Delta_{fus}H)_{nonadd}^{Inc}$	$(\Delta_{fus}H)_{nonadd}^{Dec}$	
–OH and –COOH groups	0.7		In aromatic compounds
	0.0		In O_2N $\overset{OH}{\diagup}$ NO_2 fragments
–NH₂ group	0.5		For one or two $-NH_2$ groups
	2.6		For more than two $-NH_2$ groups
	0.0		In NO_2 / NH_2 fragments

Table 4.5 (continued)

Molecular moieties	Effect on predicted $\Delta_{fus}H$		Comment
	$(\Delta_{fus}H)_{nonadd}^{Inc}$	$(\Delta_{fus}H)_{nonadd}^{Dec}$	
>NH group	0.5		–
>NNO$_2$ group	$n_{NNO_2} - 2$		In acyclic nitramines
–N(C = O)N– fragment	1.5		–
Ar–X–Ar nitroaromatics		0.5	–
		0.0	where X is S
Cyclic nitramines		$\dfrac{m_{cyc} - 6}{4} + 0.5$	Rings which are larger than six membered rings
–NO$_2$ group		1.0	Nitroalkanes
–N–N=O group		2.0	–

Example 4.6: The use of eqs. (4.6) and (4.7) for different classes of energetic compounds, including acyclic nitramines, cyclic nitramines, nitrate esters, polynitroarenes, polynitroheteroarenes, nitroaliphatics, nitroaromatics, cyclic peroxides, acyclic peroxides, and nitrogen rich compounds, is given in Table 4.6.

4.1.7 A generalized predictive model for enthalpy of fusion in diverse hydrocarbons

A general correlation has been developed to predict the enthalpy of fusion for a broad variety of cyclic and acyclic hydrocarbons, covering both saturated and unsaturated aliphatic compounds as follows [350]:

$$\Delta_{fus}H = 5.18 + 0.09037ab - 0.119a^2 + 16.6\Delta_{fus}H_{corr}^{in} - 12.9\Delta_{fus}H_{corr}^{dec}, \tag{4.8}$$

where $\Delta_{fus}H$ represents the enthalpy of fusion (in kJ/mol), while $\Delta_{fus}H_{corr}^{in}$ and $\Delta_{fus}H_{corr}^{dec}$ are two correction terms. These terms account for additive and non-additive contributions from certain structural features. The optimized values for these corrections are provided in Table 4.7.

Table 4.6: Several examples for the use of eqs. (4.6) and (4.7) in some classes of energetic compounds.

Class of compound	Molecular structure	The calculated $\Delta_{fus}H$
Acyclic nitramine	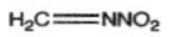	$\Delta_{fus}H = 0.9781[0.6047(1) + 0.6211(2)$ $+ 2.750(2) + 1.424(2) + 3.048(0)]$ $+ 7.567(0) - 8.784(0) = 9.97\text{kJ/mol}$
	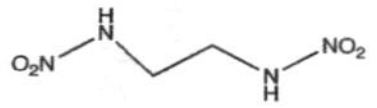	$\Delta_{fus}H = 0.9781[0.6047(2) + 0.6211(6)$ $+ 2.750(4) + 1.424(4) + 3.048(0)]$ $+ 7.567(0.5) - 8.784(0)$ $= 24.94\text{kJ/mol}$
Cyclic nitramine	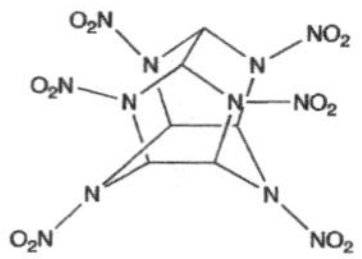	$\Delta_{fus}H = 0.9781[0.6047(6) + 0.6211(6)$ $+ 2.750(12) + 1.424(12)$ $+ 3.048(0)] + 7.567(0) - 8.784(0)$ $= 56.18\text{kJ/mol}$
	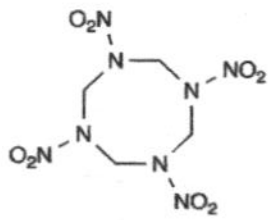	$\Delta_{fus}H = 0.9781[0.6047(4) + 0.6211(8)$ $+ 2.750(8) + 1.424(8) + 3.048(0)]$ $+ 7.567(0) - 8.784(1)$ $= 31.10\text{kJ/mol}$

Nitrate ester

$$\Delta_{fus}H = 0.9781[0.6047(4) + 0.6211(8)$$
$$+ 2.750(2) + 1.424(7) + 3.048(0)]$$
$$+ 7.567(0) - 8.784(0)$$
$$= 22.35 \text{kJ/mol}$$

$$\Delta_{fus}H = 0.9781[0.6047(5) + 0.6211(8)$$
$$+ 2.750(4) + 1.424(12) + 3.048(0)]$$
$$+ 7.567(0) - 8.784(0)$$
$$= 35.29 \text{kJ/mol}$$

Polynitroarene

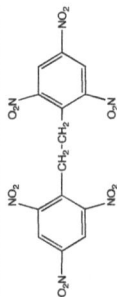

$$\Delta_{fus}H = 0.9781[0.6047(14) + 0.6211(8)$$
$$+ 2.750(6) + 1.424(12) + 3.048(0)]$$
$$+ 7.567(0) - 8.784(0.5)$$
$$= 45.99 - 4.39 = 41.60 \text{kJ/mol}$$

$$\Delta_{fus}H = 0.9781[0.6047(6) + 0.6211(5)$$
$$+ 2.750(5) + 1.424(6) + 3.048(0)]$$
$$+ 7.567(0.5) - 8.784(0)$$
$$= 32.17 \text{kJ/mol}$$

Polynitroheteroarene

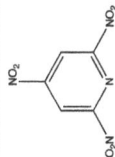

$$\Delta_{fus}H = 0.9781[0.6047(5) + 0.6211(2)$$
$$+ 2.750(4) + 1.424(6) + 3.048(0)]$$
$$+ 7.567(0) - 8.784(0) = 23.29 \text{kJ/mol}$$

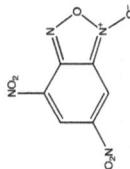

$$\Delta_{fus}H = 0.9781[0.6047(6) + 0.6211(2) + 2.750(4)$$
$$+ 1.424(6) + 3.048(0)] + 7.567(0) - 8.784(0)$$
$$= 23.88 \text{kJ/mol}$$

(continued)

Table 4.6 (continued)

Class of compound	Molecular structure	The calculated $\Delta_{fus}H$
Nitroaliphatic	H_3C-NO_2	$\Delta_{fus}H = 0.9781[0.6047(1) + 0.6211(3) + 2.750(1)$ $+ 1.424(2) + 3.048(0)] + 7.567(0) - 8.784(0)$ $= 7.89\,kJ/mol$
	(structure)	$\Delta_{fus}H = 0.9781[0.6047(2) + 0.6211(3) + 2.750(3)$ $+ 1.424(7) + 3.048(0)]$ $+ 7.567(0) - 8.784(1) = 12.04\,kJ/mol$
Nitroaromatic	(structure)	$\Delta_{fus}H = 0.9781[0.6047(6) + 0.6211(4) + 2.750(1)$ $+ 1.424(2) + 3.048(1)]$ $+ 7.567(0) - 8.784(0) = 14.43\,kJ/mol$
	(structure)	$\Delta_{fus}H = 0.9781[0.6047(7) + 0.6211(5) + 2.750(1)$ $+ 1.424(4) + 3.048(0)]$ $+ 7.567(0.7) - 8.784(0) = 20.73\,kJ/mol$
Cyclic peroxide	(structure)	$\Delta_{fus}H = 0.9781[0.6047(6) + 0.6211(12) + 2.750(0)$ $+ 1.424(4) + 3.048(0)]$ $+ 7.567(0) - 8.784(0) = 16.41\,kJ/mol$
	(structure)	$\Delta_{fus}H = 0.9781[0.6047(9) + 0.6211(18) + 2.750(0)$ $+ 1.424(6) + 3.048(0)]$ $+ 7.567(0) - 8.784(0) = 24.61\,kJ/mol$

Acyclic peroxide

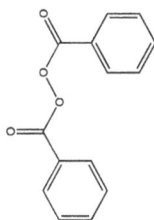

$$\Delta_{fus}H = 0.9781[0.6047(14) + 0.6211(10)$$
$$+ 2.750(0) + 1.424(4) + 3.048(0)]$$
$$+ 7.567(0) - 8.784(0) = 19.93\,kJ/mol$$

Nitrogen-rich compound

$$\Delta_{fus}H = 0.9781[0.6047(3) + 0.6211(6)$$
$$+ 2.750(4) + 1.424(0) + 3.048(0)]$$
$$+ 7.567(0) - 8.784(0) = 16.18\,kJ/mol$$

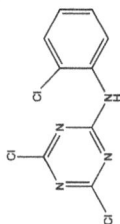

$$\Delta_{fus}H = 0.9781[0.6047(9) + 0.6211(5)$$
$$+ 2.750(4) + 1.424(0) + 3.048(3)]$$
$$+ 7.567(0.5) - 8.784(0.5) = 27.46\,kJ/mol$$

Table 4.7: The values of $\Delta_{fus}H^{in}_{corr}$ and $\Delta_{fus}H^{dec}_{corr}$.

Type of non-aromatic compound	Conditions	$\Delta_{fus}H^{in}_{corr}$	Example
Cyclic hydrocarbon	Saturated compound with one linear alkyl substituent and $n_C > 11$	$(n_C-11) \times 0.20$	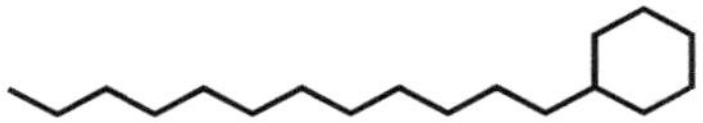
Acyclic alkane	Saturated linear and $n_C > 9$	1.40	
Acyclic alkyne	Saturated linear and $n_C > 4$	0.7	
 $H_2N{-}\overset{}{\underset{n}{}}{-}NH_2$	$n > 5$	1.5	
 $HOOC{-}\overset{}{\underset{n}{}}{-}COOH$	$n > 7$	0.9	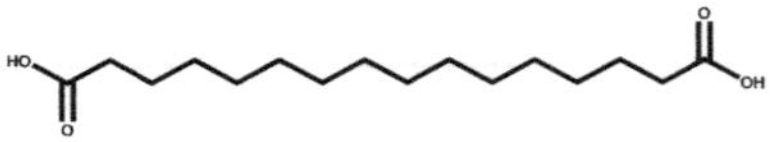
R-SH	R = linear alkyl substituent and $n_C > 6$	1.0	
R-OH	R = linear alkyl substituent and $n_C > 11$	1.5	
R-COOH	R = linear alkyl substituent and $n_C > 13$	1.0	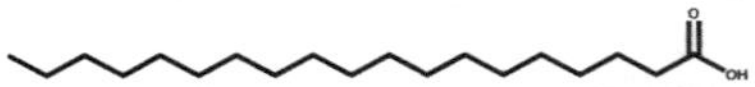
$R_1C(=O)R_2$	R_1 and R_2 are linear alkyl substituents and $n_C > 14$	2.2	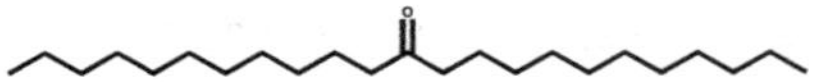

$R_1C(=O)OR_2$	R_1 and R_2 are linear alkyl substituents and $n_C > 18$		1.5
Cyclic hydrocarbon	Saturated monocyclic ring without substituent and $n_C > 11$		$(n_C - 11) \times 0.30$
Cyclic hydrocarbon	Saturated monocyclic ring with only methyl substituents where the numbers of cyclic carbon atoms >17		1.2
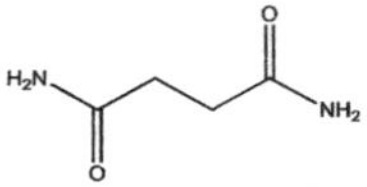	R_1 and R_2 are linear alkyl substituents or hydrogen atoms		1.5

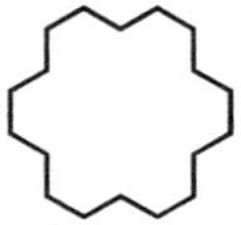

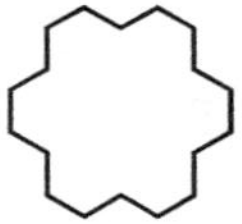

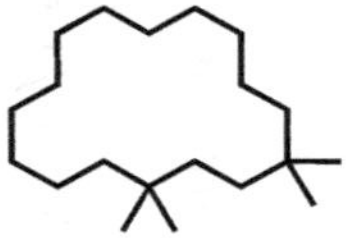

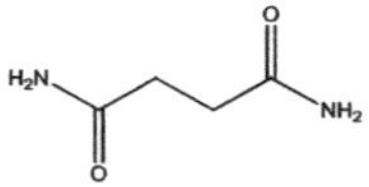

Example 4.7: The use of eq. (4.8) for 1,1,1-trifluoroethane ($C_2H_3F_3$) gives:

$$\Delta_{fus}H = 5.18 + 0.09037ab - 0.119a^2 + 16.6\Delta_{fus}H_{corr}^{in} - 12.9\Delta_{fus}H_{corr}^{dec}$$

$$= 5.18 + 0.09037(2)(3) - 0.119(3^2) + 16.6(0) - 12.9(0) = 4.65 \ \ kJ/mol.$$

The predicted value shows better agreement with the experimental data (6.19 kJ/mol [351]) compared to eq. (4.1), which yields a value of 2.5 kJ/mol.

4.2 Different methods to predict the entropy of fusion

Organic molecules with high symmetry may have lower entropy of fusion and a higher melting point as compared to their nonsymmetrical isomers. Their molecules have a higher probability of being in the right orientation to form the crystal. It was shown that some of the entropy determining parameters are not group additive [263, 352, 353]. There are more translational, rotational, and conformational constraints on molecules in the solid state than the liquid state. The entropy of fusion ($\Delta_{fus}S$) is an important property for predicting the melting point and solubility of organic compounds [187]. The value of $\Delta_{fus}S$ based on Walden's rule is constant with a value of 56.5 J/K mol [354] for aromatic compounds with little variation. The effect of molecular rotational symmetry (σ) can be incorporated into Walden's rule as follows [355]:

$$\Delta_{fus}S = 56.5 - R\ln\sigma. \tag{4.9}$$

The parameter σ is the rotational degeneracy of the molecule. It is the number of positions a molecule can be rotated into while maintaining the same atomic orientation of the original position. Figure 4.1 shows the symmetry number for some typical compounds. The value of σ is based upon rotation because it is the only symmetry operation that can physically be performed on a molecule. Thus, it does not enumerate mirror planes or operations that cannot be physically performed. Hydrogen atoms in OH, CH_3, and NH_2 groups as well as halogens of tri-homohalogenated carbons like trichloromethyl are treated as being radially symmetrical because they can be assumed to be freely and rapidly rotating. Carboxylic acids and nitro groups have bilateral symmetry. The molecules with geometry of cones, cylinders, and spheres have infinite symmetry for which σ is taken 20 to reflect their higher symmetry. Thus, these geometries are not considered as equal to unity as in chemistry. Rather they have $\sigma = 20$ because they are more symmetrical than benzene or neopentane as well as containing at least one infinite rotational axis. The value of σ is equal to 1.0 for all nonsymmetrical molecules.

The parameter σ' can be considered for those molecules with a slight structural difference, which causes an otherwise symmetrical molecule to be asymmetrical. The packing arrangement of this molecule is quite the same as that of its symmetrical homomorphy, that is, pseudo-symmetrical molecules have a higher probability of being

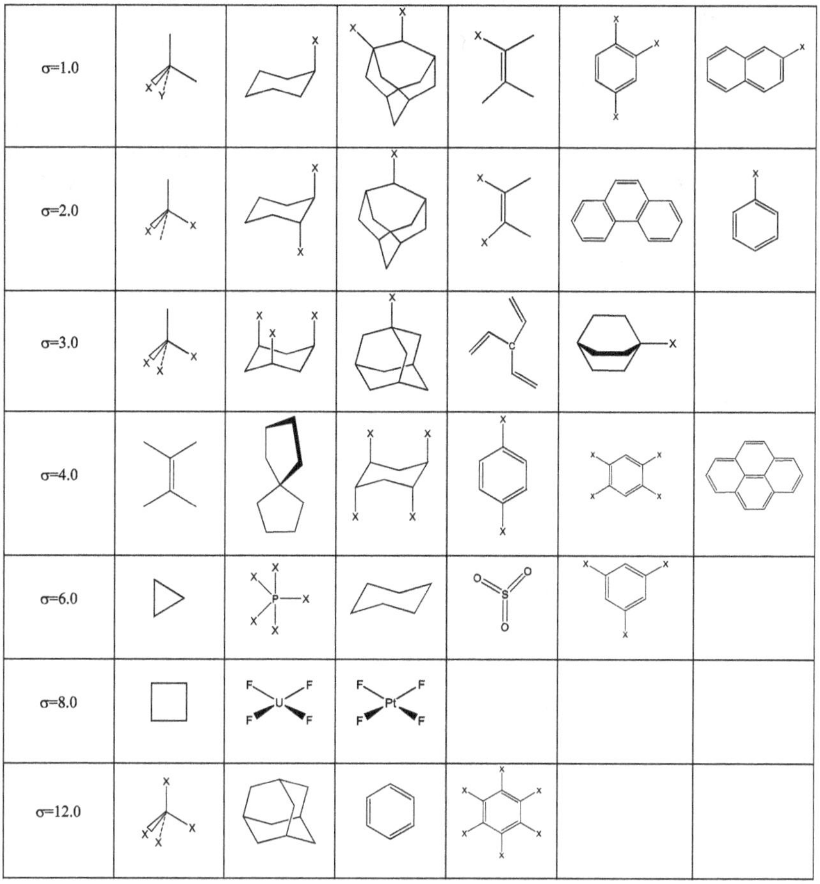

Figure 4.1: The value of σ for some representative compounds.

in the right arrangement required for a crystal lattice than nonsymmetrical molecules. Thus, halogens and methyl group are treated as being pseudosymmetrical atoms/group, that is, *p*-bromochlorobenzene is a pseudosymmetrical molecule ($\sigma' = 2$) as compared to its homomorph (*p*-dichlorobenzene or *p*-dibromobenzene, $\sigma = 2$).

Equation (4.9) can be modified to include the flexibility number (ϕ) as [356]:

$$\Delta_{fus}S = 50 - R\ln\sigma + R\ln\phi. \tag{4.10}$$

The parameter ϕ shows the internal conformational freedom of molecules. Flexible molecules, as compared to rigid molecules, tend to have a greater entropy change during melting. This parameter is calculated by an ad hoc expression uniting flexible segments as:

$$\phi = 0.3ROT + LINSP3 + 0.5(BRSP3 + SP2 + RING) - 1, \tag{4.11}$$

where *LINSP3* is the number of nonring, nonterminal, and nonbranched sp³ atoms; *ROT* is the extra entropy produced by freely rotating sp³ atoms and is calculated as *ROT* = *LINSP3*-4 (if *LINSP3* > 4 otherwise *ROT* = 0); *BRSP3* is the total number of nonring, nonterminal, and branched sp³ atoms; *SP2* is the number of nonring, and nonterminal sp² atoms; *RING* is the number of single, fused, or conjugated ring systems. If the calculated value of ϕ by eq. (4.3) is less than zero, the value of ϕ should be taken zero.

Example 4.8: The values of different parameters of eq. (4.11) are specified for 1-methoxy-2,4-dinitrobenzene with the following structure:

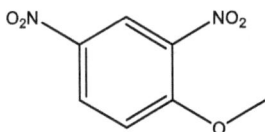

$$\phi = 0.3ROT + LINSP3 + 0.5(BRSP3 + SP2 + RING) - 1,$$

LINSP3 = 1 (the oxygen atom of methoxy group), *BRSP3* = 0, *SP2* = 2 (two nitrogen atoms of nitro groups), *RING* = 1 (one benzene ring), and *ROT* = 0 because *LINSP3* < 4.

The use of these parameters in eq. (4.11) gives ϕ = 1.5.

To predict $\Delta_{fus}S$ of nitroaromatic compounds, a suitable correlation based on the elemental composition has been used [352]. The reliability of this method is higher than that of the method of Jain and Yalkowsky [261], which is based on eq. (4.9) for 61 nitroaromatic compounds. This method has been restricted to nitroarmatic compounds because it may give values which show large deviation from the expected values for some of the other classes of energetic compounds. For nitroaromatics, Evan and Yallkowsky [263] have improved the reliability of the method of Jain and Yalkowsky [261] by introducing the molecular eccentricity (ε):

$$\Delta_{fus}S = 50 - R\ln\sigma + R\ln\phi + R\ln\varepsilon. \tag{4.12}$$

Yalkowsky and coworkers [269, 357] have improved various parameters to obtain the entropy of fusion for over 2000 organic compounds with a reasonable approximation as:

$$\Delta_{fus}S = 44.98 - 8.93\log\sigma - 2.17\log\sigma' + 11.36\log\varepsilon_{ar} + 8.26\log\varepsilon_{al} + 5.91\phi, \tag{4.13}$$

where σ is symmetry, σ' is pseudosymmetry, ε_{ar} and ε_{al} are aromatic and aliphatic eccentricity, respectively, and ϕ is flexibility.

The parameters ε_{ar} and ε_{al} are related to crystals of flat eccentric molecules, which have relatively less than average free volume due to their efficient packing. Flat eccentric molecules require more space to attain their free rotation as compared to spherical molecules of the same molecular weight. These parameters can be obtained by the number of atoms in aliphatic rings as well as the number of atoms in aromatic rings or directly connected to them or part of a conjugated system, that is, ε_{ar} is equal to 16 for 2,4,6-trinitrotoluene due to the existence of three nitro conjugated system (= 3 × 3), six carbon atoms of the benzene ring and one carbon atom of the methyl group. The values of ε_{ar} and ε_{al} equal to one for all nonring compounds.

Since the free energy of an organic compound at its melting point is equal to zero, the normal melting point is the temperature at which the solid and liquid are at equilibrium at one-atmosphere pressure. Thus, the use of eqs. (4.12) and (4.13) can give the normal melting point of an organic compound as:

$$T_m = \frac{\Delta_{fus}H}{\Delta_{fus}S} = \frac{\sum n_i H_{fus,i}}{44.98 - 8.93 \log \sigma - 2.17 \log \sigma' + 11.36 \log \varepsilon_{ar} + 8.26 \log \varepsilon_{al} + 5.91\phi}. \tag{4.14}$$

Example 4.9: 2,4,6-Trinitro-*N*-methyl-aniline has the following structure:

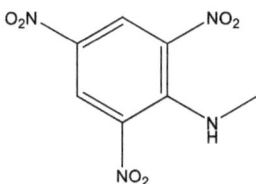

The value of $\Delta_{fus}H$ is calculated by eq. (4.12) and Table 4.1 as follows:

$$\Delta_{fus}H = \sum n_i H_{fus,i} = 4C_{ar} + 2CH_{ar} + 3YNO_2 + YNH + XCH_3$$
$$= 4 \times 1.777 + 2 \times 1.235 + 3 \times 4.584 + 3.799 + 0.701 = 27.83 \text{ kJ/mol},$$

The measured value of $\Delta_{fus}H$ for this compound is 25.9 kJ/mol [269].
    The value of $\Delta_{fus}S$ is calculated by eq. (4.13) as follows:

$$\Delta_{fus}S = 44.98 - 8.93 \log \sigma - 2.17 \log \sigma' + 11.36 \log \varepsilon_{ar} + 8.26 \log \varepsilon_{al} + 5.91\phi$$
$$= 44.98 - 8.93 \log 1 - 2.17 \log 1 + 11.36 \log 16 + 8.26 \log 1 + 5.91 \times 2.0$$
$$= 70.5 \text{ J/mol K}$$

The experimental value of $\Delta_{fus}S$ is 64.2 J/mol K [269].
    A general new method for predicting the entropy of fusion for various types of energetic compounds – including polynitroarene, acyclic, and cyclic nitramine, nitrate esters, and nitroaliphatic compounds – has been introduced as additive ($\Delta_{fus}S_{add}$) and nonadditive ($\Delta_{fus}S_{nonadd}$) functions [353]:

$$\Delta_{fus}S = \Delta_{fus}S_{add} - 23.86\Delta_{fus}S_{nonadd}, \tag{4.15}$$

where

$$\Delta_{fus}S_{add} = 39.99 + 5.88c + 1.22d. \tag{4.16}$$

The values of $\Delta_{fus}S_{nonadd}$ can be specified as follows.
(1) Ar–N(NO$_2$)– and –R–OH molecular moieties: The values of $\Delta_{fus}S_{nonadd}$ are 1.0 and 1.7, respectively, for the presence of these molecular fragmnents.
(2) 1,3,5-Trinitrobenzene derivatives: if 1,3,5-trinitrobenzene, 2,2′,4,4′,6,6′-hexanitrobiphenyl or two rings of 1,3,5-trinitrobenzene in the form:

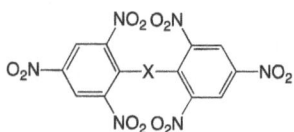

are present, the value of $\Delta_{fus}S_{nonadd}$ is 1.0 if X is a single atom or unsaturated molecular fragment (except –N=N– connecting two 1,3,5-trinitrobenzene rings).

(3) Cyclic nitramines containing more than three –NNO$_2$ groups:$\Delta_{fus}S_{nonadd} = 1.5$.

Example 4.10: The use of eqs. (4.15) and (4.16) for the following molecular structure

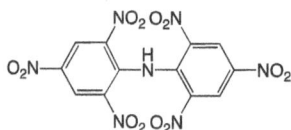

gives the following value:

$$\Delta_{fus}S_{add} = 39.99 + 5.88c + 1.22d$$
$$= 39.99 + 5.88(7) + 1.22(12)$$
$$= 95.79\,J/(K \cdot mol),$$
$$\Delta_{fus}S = \Delta_{fus}S_{add} - 23.86\Delta_{fus}S_{nonadd}$$
$$= 95.79 - 23.86(1)$$
$$= 71.93\,J/(K \cdot mol).$$

The measured $\Delta_{fus}S$ for this compound is 72.8 J/(K·mol) [54]. The predicted $\Delta_{fus}S$ using eq. (4.10) is 55.9 J/(K·mol), which deviates considerably from the measured value.

4.3 A group contribution approach incorporating cation-anion interactions and structural effects for ionic liquids

Group contribution methods remain popular as they provide quick estimates without complex computations [359], despite their inherent limitations – group definitions are somewhat arbitrary and accuracy depends heavily on the training dataset. Valderrama and Cardona [359] improved a new approach upon traditional group additivity methods by: (1) simplifying application, (2) differentiating cation/anion group contributions, (3) incorporating thermodynamic principles, (4) covering diverse ionic liquid (IL) types, and (5) allowing easy extension to new ILs.

Since the $\Delta_{fus}H$ of ILs depends on how molecular fragments of the cation and anion interact, Valderrama and Cardona [359] introduced eq. (4.17) to approximate it by summing contributions from each molecular group:

$$\Delta_{fus}H = 3.8315 + \sum_i n_{c,i}\Delta_{fus}H_{c,i} + \sum_j n_{a,j}\Delta_{fus}H_{a,j}, \tag{4.17}$$

where $n_{c,i}$ and $n_{a,j}$ count how often group "i" appears in the cation and anion, while $\Delta_{fus}H_{c,i}$ and $\Delta_{fus}H_{a,j}$ represent their specific enthalpy contributions. The model's optimized contribution values appear in Table 4.8.

Table 4.8: Group contributions for cations and anions in eq. (4.17).

Group	$\Delta_{fus}H_{c,i}$ (kJ/mol)	$\Delta_{fus}H_{a,j}$ (kJ/mol)
	Without rings	
–CH$_3$	1.7310	1.5040
–CH$_2$–	0.9808	0.4229
>CH–	0.7435	-0.0359
>C<	2.3858	0.7978
=CH$_2$	1.2630	-1.2514
=CH–	1.1151	-0.3161
=C<	-0.1434	-0.1434
=C=	0.3325	3.6680
=(-)CH	-1.7431	-1.7431
=(-)C-	3.6680	3.6680
–OH	3.1287	1.4754
–O–	1.3256	0.3047
>C=O	3.4444	3.4444
–CHO	1.7028	1.7028
–COOH	8.5964	8.5964
–COO–	2.8997	1.6170
HCOO–	10.2030	-0.0649
=O(other)	-0.7928	-0.7928
–NH$_2$	4.6890	3.8522
–NH$_3$	0.8072	4.0673
–NH–	2.8324	2.8324
>N–	0.5764	-0.0033
=N–	0.3478	1.5404
–CN	3.2063	-0.0373
–NO$_2$	1.8123	1.7907
–F	2.5688	0.7498
–Cl	0.0293	1.7157
–Br	2.1708	5.4581
–I	2.0461	3.9280
–P	0.4230	0.9264
–B	-1.8219	-1.5051

Table 4.8 (continued)

Group	$\Delta_{fus}H_{c,i}$ (kJ/mol)	$\Delta_{fus}H_{a,j}$ (kJ/mol)
	Without rings	
-S-	0.3026	0.5763
-SO$_2$	4.3331	1.5936
	With rings	
-CH$_2$-	0.2990	-0.3864
>CH-	0.4830	0.4830
=CH-	1.3369	3.7121
>C<	3.3692	3.3692
=C<	2.1579	-5.4406
-O-	0.7102	0.7102
-OH	4.9803	4.9803
>C=O	4.4901	4.4901
-NH-	1.8753	1.8123
>N-	0.2726	1.5968
=N-	0.1265	1.4521
Al	0	0.2990
As	0	5.3917
Fe	0	0.3238

For $\Delta_{fus}S$, Valderrama and Cardona introduced eq. (4.18):

$$\Delta_{fus}S = 9.774 \times 10^{-3} + 2.46 \times 10^{-4} Mw_{cat} + 1.458 \times 10^{-4} Mw_{ani}, \tag{4.18}$$

which separately accounts for cation (Mw_{cat}) and anion (Mw_{ani}) masses.

Combining these insights gives us eq. (4.19) for melting temperature:

$$T_m = \frac{\Delta_{fus}H}{\Delta_{fus}S} = \frac{3.8315 + \sum_i n_{c,i}\Delta_{fus}H_{c,i} + \sum_j n_{a,j}\Delta_{fus}H_{a,j}}{9.774 \times 10^{-3} + 2.46 \times 10^{-4} Mw_{cat} + 1.458 \times 10^{-4} Mw_{ani}}, \tag{4.19}$$

These models create a practical yet physically meaningful prediction tool that carefully distinguishes between cation and anion properties throughout.

Example 4.11: The use of eqs. (4.17), (4.18), and (4.19) for 1-(2-azidoethyl)-1,2,4-triazolium perchlorate (following structure) and Table 4.8 with Mw_{cat}= 139.1 g/mol and Mw_{ani}= 99.5 g/mol give:

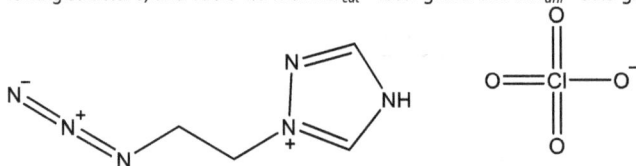

$$\Delta_{fus}H = 3.8315 + \sum_i n_{c,i}\Delta_{fus}H_{c,i} + \sum_j n_{a,j}\Delta_{fus}H_{a,j}$$

$\Delta_{fus}H = 3.8315 + 2(-CH_2-) + 3(=N-) + 2(=CH-) + 1(-NH-) + 1(>N-) + 1(=N-) + 4(-O-) + 1(-Cl)$

$\quad = 3.8315 + 2(0.9808) + 3(0.3478) + 2(1.3369) + 1(1.8753) + 1(0.2726) + 1(0.1265) + 4(0.3047)$

$\quad + 1(1.7157) = 14.719 \text{ kJ/mol}$

$$\Delta_{fus}S = 9.774 \times 10^{-3} + 2.46 \times 10^{-4}Mw_{cat} + 1.458 \times 10^{-4}Mw_{ani}$$

$$\Delta_{fus}S = 9.774 \times 10^{-3} + 2.46 \times 10^{-4}(139.1) + 1.458 \times 10^{-4}(99.5) = 0.0585 \text{ kJ/(K} \cdot \text{mol)}$$

$$T_m = \frac{\Delta_{fus}H}{\Delta_{fus}S} = \frac{3.8315 + \sum_i n_{c,i}\Delta_{fus}H_{c,i} + \sum_j n_{a,j}\Delta_{fus}H_{a,j}}{9.774 \times 10^{-3} + 9.46 \times 10^{-4}Mw_{cat} + 1.458 \times 10^{-4}Mw_{ani}}$$

$$T_m = \frac{\Delta_{fus}H}{\Delta_{fus}S} = \frac{14.719 \text{ kJ/mol}}{0.0585 \text{ kJ/(K} \cdot \text{mol)}} = 251.6 \text{ K}$$

Only experimental melting point of 1-(2-azidoethyl)-1,2,4-triazolium perchlorate is available, which is 217.1 K [360].

4.4 Summary

Different group additivity and empirical methods have been reviewed which can be used to predict the enthalpy and entropy of fusion. Equation (4.1) and Table 4.1 provide a suitable group additivity method for prediction of the enthalpy of fusion of organic compounds containing organic materials with energetic groups. Among the different methods for predicting the enthalpy of fusion, eqs. (4.6) and (4.7) provide a simple pathway for large classes of energetic compounds – including polynitroarene, polynitroheteroarene, acyclic and cyclic nitramines, nitrate esters, nitroaliphatic, cyclic and acyclic peroxides, as well as nitrogen-rich compounds. Equation (4.12) gives a suitable approaction for estimation of the entropy of fusion of organic compounds where it can be applied for organic energetic materials. Equation (4.8) establishes a comprehensive correlation for estimating fusion enthalpies across a wide range of hydrocarbon systems, encompassing cyclic/acyclic configurations and saturated/unsaturated aliphatic species. Equations (4.15) and (4.16) also provide a new general method for predicting the entropy of fusion for various types of energetic compounds including polynitroarenes, acyclic and cyclic nitramines, nitrate esters, and nitroaliphatic compounds. Equations (4.17) and (4.18) present new correlations for predicting the enthalpy and entropy of fusion for ILs.

5 Heat of sublimation

Sublimation – where solids transform directly into vapor without becoming liquid first – occurs below a material's triple point temperature [360], and understanding this process is crucial because it reveals fundamental insights about intermolecular forces while enabling key technologies, from semiconductor manufacturing [361] to pharmaceutical purification [362]. The energy required for this phase change, quantified as sublimation enthalpy, serves as both a practical design parameter and a window into non-bonding molecular interactions [360], making it essential for applications ranging from membrane separations [362] to crystal engineering [363].

Sublimation properties can help to understand the coarsening process which occurs [364, 365]. Knowledge of the heat of sublimation and the vapor pressure of organic explosives can also help to enable the design of new technology for the detection of explosive particles from concealed devices [366]. The sublimation properties of organic explosives have long-term effects on soil, water, and air. Since secondary explosives are widely used, understanding their sublimation properties is necessary to study their toxic effects on the environment during storage [367].

The heat – or enthalpy – of sublimation is the best parameter to characterize the strength of intermolecular interactions within a crystal. The gas-phase heat of formation ($\Delta_f H(g)$) and the heat of sublimation ($\Delta_{sub}H$) can be used to evaluate the solid-phase heats of formation ($\Delta_f H(s)$) as follows [368]:

$$\Delta_f H(s) = \Delta_f H(g) - \Delta_{sub}H. \tag{5.1}$$

The $\Delta_f H(s)$ value of an energetic compound is important in order to enable the prediction of its performance using various computer codes such as CHEETAH [369], ISPBKW [370], LOTUSES [371], EDPHT [87], and EMDB [69]. It can also be calculated by combining the predicted heat of sublimation and gas-phase heat of formation according to eq. (5.1).

The value of $\Delta_{sub}H$ for a specific compound can be considered to be the sum of its heat of fusion and its heat of vaporization, even if the liquid cannot exist at the pressure and temperature in question. Since the sublimation pressure at the melting point is only rarely known, it is difficult to use the Clausius–Clapeyron equation to obtain the heat of sublimation from the vapor pressure data [187].

While scientists can measure $\Delta_{sub}H$ directly using microcalorimeters or indirectly through vapor pressure measurements [372], these experiments remain notoriously challenging – they're expensive, time-consuming [373], and often yield inconsistent results between laboratories [374], particularly for less volatile compounds. There are a variety of approaches which can be used to predict the gas-phase heats of formation of energetic compounds [145, 161, 162, 187, 191, 348, 375], but few methods have been reported to predict the heat of sublimation of energetic compounds. Moreover, experimental data for the heats of sublimation of energetic compounds are rare because

https://doi.org/10.1515/9783112206768-005

these values have not yet been published for many energetic materials. In this chapter, some methods for the prediction of $\Delta_{sub}H$ for several classes of energetic compounds will be reviewed.

5.1 Group additivity method for prediction of the heat of sublimation

Mathieu [164] generated a model using 35 group contributions from a training set containing 814 compounds. Naef and Acree Jr. [346] introduced a suitable group additivity method for predicting the heat of sublimation of a wide range of organic compounds containing energetic materials. The general form of eq. (5.2) can be used for this purpose:

$$\Delta_{sub}H = 21.03 + \sum_i a_i A_i + \sum_j g_j G_j, \tag{5.2}$$

where a_i and g_j are the numbers of the ith atom group A_i and the special group G_j, respectively. For the atom groups, each group consists of a central atom (the backbone atom) and its immediate neighbor atoms. The central atom is bound to at least two other atoms. It is characterized by its atom name, its atom type being defined by either its orbital hybridization or bond type or its number of bonds, where required for distinction, and by its charge, if not zero. A term is used to collect the neighbor atoms, which are in the order $H > B > C > N > O > S > P > Si > F > Cl > Br > I$. The bond type of the neighbor bond with the backbone atom (if not single) encompasses its atom name and its number of occurrences (if >1). The symbol J is used instead of I because of the better readability of a neighbor term containing iodine. For the non-zero total net charge of the neighbor atoms, the charge "(+)" or "(−)" is appended to the neighbor term. The atom type "N sp^3" is used for N with three single bonds. The atom types "O" and "S" are used for O and S with two single bonds, respectively. If neighbor atoms are part of a conjugated moiety, the terms "(pi)", "(2pi)," or "(3pi)" supplement the neighbor term. The increased strength of a group's bonds in this situation is due to the π-orbital conjugation of the backbone atom's lone-pair electrons with conjugated neighbor moieties. Table 5.1 shows some examples for the backbone atom type (in boldface) and the term for its neighbors. Table 5.2 also gives atom groups and their contributions (in kJ/mol) for eq. (5.2).

The use of eq. (5.2) does not reflect any knowledge about the molecules' three-dimensional structure. It also depends on structural peculiarities such as buttressing effects, ring strains, gauche bond interactions, or internal hydrogen bonds.

Table 5.1: Group examples and their meanings.

Atom type	Neighbors	Meaning	Atom type	Neighbors	Meaning
C sp^3	H3C	C–CH3	N sp^3	H2C	C–NH2
C sp^3	H3N	N–CH3	N sp^3	H2C(pi)	C–N*H2
C sp^3	H2C2	C–CH2–C	N sp^3	C2N(2pi)	C–N*(N)–C
C sp^3	H2CO	C–CH2–O	N sp^2	H=C	C=NH
C sp^3	HC3	C–CH(C)–C	N sp^2	C=N	N=N–C
C sp^3	HC2Cl	C–CH(Cl)–C	N sp^2	=CO	C=N–O
C sp^3	HCO2	C–CH(O)–O	N(+) sp^3	H3C	C–NH$_3^+$
C sp^3	C3N	C–C(C)2N	N(+) sp^3	H2C2	C–NH$_2^+$–C
C sp^3	C2F2	C–CF2–C	N(+) sp^2	CO=O(–)	O=N$^+$(O$^-$)–C
C sp^2	H2=C	C=CH2	N aromatic	:C2	C:N:C
C sp^2	HC=C	C=CH–C	N(+) sp	=N2(–)	N=N$^+$=N$^{(-)}$
C sp^2	HC=N	N=CH–C	O	HC	C–OH
C sp^2	H=CN	C=CH–N	O	HC(pi)	C–O*H^c
C sp^2	HN=O	O=CH–N	O	Si2	Si–O–Si
C sp^2	C2=O	O=C(C)–C	P3	C3	C–PC)–C
C sp^2	C=CN	C=C(C)–N	P4	CO2=O	O=PO2)–C
C sp^2	=CNO	C=C(N)–O	P4	N2O=O	O=PO)(N)–N
C sp^2	N=NO	N=C(N)–O	S2	HC(pi)	C–S*H
C sp^2	NO=O	O=C(N)–O	S2	CS	C–S–S
C aromatic	H:C2^a	C:CH:C	S4	CO=O2	C–S(=O)2–O
C aromatic	H:C:N	C:CH:N	S4	O2=O	O–S(=O)–O
C aromatic	:CN:N	C:C(N):N	Si	C2Cl2	C–SiCl$_2$–C
C sp	H#C^b	C#CH	Si	OCl3	O–SiCl$_3$
C sp	C#N	N#C–C			
C sp	#CN	C#C–N			
C sp	=C2	C=C=C			
C sp	=C=O	C=C=O			

^{a}The symbol ":" represents an aromatic bond.
^{b}The symbol "#" gives a triple bond.
^{c}The symbol "*" shows lone-pair electrons forming π-orbital conjugated bonds with neighboring atoms.

Table 5.2: Atom groups and their contributions (in kJ/mol) for eq. (5.2).

Atom type	Neighbors	Contribution
B	C3	65.82
C sp^3	H3C	5.99
C sp^3	H3N	26.96
C sp^3	H3N(+)	98.98
C sp^3	H3O	28.51
C sp^3	H3S	30.06
C sp^3	H2C2	6.88
C sp^3	H2CN	21.98
C sp^3	H2CN(+)	27.46

Table 5.2 (continued)

Atom type	Neighbors	Contribution
C sp^3	H2CO	29.62
C sp^3	H2CS	23.29
C sp^3	H2CF	15.91
C sp^3	H2CCl	17.59
C sp^3	H2CBr	22.76
C sp^3	H2CJ	21.83
C sp^3	H2N2	43.95
C sp^3	H2NCl	36.29
C sp^3	H2O2	53.35
C sp^3	H2OS	54.78
C sp^3	H2S$_2$	47.45
C sp^3	HBC2	36.17
C sp^3	HC3	2.28
C sp^3	HC2N	14.28
C sp^3	HC2N(+)	21.01
C sp^3	HC2O	24.27
C sp^3	HC2S	17.59
C sp^3	HC2F	5.18
C sp^3	HC2Cl	11.49
C sp^3	HC2Br	0.95
C sp^3	HCN2	39.48
C sp^3	HCN2(+)	39.93
C sp^3	HCNO	34.73
C sp^3	HCNS	20.56
C sp^3	HCO2	39.96
C sp^3	HCF2	0.19
C sp^3	HCCl2	15.78
C sp^3	HN3(+)	37.31
C sp^3	HO3	72.23
C sp^3	C4	4.25
C sp^3	C3N	5.87
C sp^3	C3N(+)	18.44
C sp^3	C3O	15.18
C sp^3	C3S	6.40
C sp^3	C3F	1.89
C sp^3	C3Cl	8.06
C sp^3	C3Br	2.34
C sp^3	C2N2(+)	34.78
C sp^3	C2O2	39.73
C sp^3	C2S2	37.28
C sp^3	C2F2	7.07
C sp^3	CN3(+)	43.89
C sp^3	CN2F(+)	25.98
C sp^3	CO3	57.42
C sp^3	CF3	4.71
C sp^3	CCl3	16.10

Table 5.2 (continued)

Atom type	Neighbors	Contribution
C sp^3	N3F(+)	44.00
C sp^3	O4	73.43
C sp^2	H2=C	7.97
C sp^2	HC=C	5.10
C sp^2	HC=N	35.49
C sp^2	HC=N(+)	72.64
C sp^2	H=CN	32.79
C sp^2	HC=O	20.74
C sp^2	H=CO	16.89
C sp^2	H=CS	15.22
C sp^2	HN=N	55.52
C sp^2	HN=O	35.41
C sp^2	H=NO	40.91
C sp^2	H=NS	33.85
C sp^2	C2=C	3.91
C sp^2	C2=N	30.47
C sp^2	C2=N(+)	13.76
C sp^2	C=CN	26.81
C sp^2	C=CN(+)	41.65
C sp^2	C2=O	15.10
C sp^2	C=CO	22.08
C sp^2	C2=S	18.21
C sp^2	C=CS	15.64
C sp^2	C=CF	16.81
C sp^2	C=CCl	11.02
C sp^2	C=CBr	34.06
C sp^2	C=CJ	32.46
C sp^2	=CN2	64.94
C sp^2	=CN2(+)	60.65
C sp^2	CN=N	54.51
C sp^2	CN=N(+)	44.16
C sp^2	CN=O	39.66
sp^2	C=NO	42.74
C sp^2	CN=S	39.85
C sp^2	C=NS	34.89
C sp^2	=CNS(+)	41.29
C sp^2	=CNCl	38.14
C sp^2	CO=O	34.06
C sp^2	CO=O(−)	80.89
C sp^2	C=OCl	29.03
C sp^2	CS=S	56.97
C sp^2	N2=N	80.72
C sp^2	N2=N(+)	65.95
C sp^2	N2=O	59.57
C sp^2	N2=S	66.62
C sp^2	N=NS	51.62

Table 5.2 (continued)

Atom type	Neighbors	Contribution
C sp^2	NO=O	52.79
C sp^2	=NO2	61.12
C sp^2	N=OS	48.27
C sp^2	NO=S	58.04
C sp^2	=NOS	52.75
C sp^2	NS=S	60.83
C sp^2	=NS2	64.37
C sp^2	O2=O	41.40
C sp^2	=OS2	41.22
C sp^2	OS=S	73.06
C sp^2	S2=S	49.39
C aromatic	H:C2	5.36
C aromatic	H:C:N	18.20
C aromatic	H:C:N(+)	28.26
C aromatic	H:N2	23.27
C aromatic	B:C2	25.04
C aromatic	:C3	5.51
C aromatic	C:C2	3.12
C aromatic	C:C:N	11.10
C aromatic	C:C:N(+)	16.04
C aromatic	:C2N	22.21
C aromatic	:C2N(+)	28.67
C aromatic	:C2:N	17.03
C aromatic	:C2:N(+)	18.05
C aromatic	:C2O	20.46
C aromatic	:C2P	1.63
C aromatic	:C2S	16.31
C aromatic	:C2F	4.45
C aromatic	:C2Cl	12.48
C aromatic	:C2Br	14.66
C aromatic	:C2J	20.68
C aromatic	:C2Si	4.80
C aromatic	C:N2	28.80
C aromatic	:CN:N	29.72
C aromatic	:CN:N(+)	33.74
C aromatic	:C:NO	41.44
C aromatic	:C:NO(+)	33.50
C aromatic	:C:NCl	21.70
C aromatic	:C:NBr	31.31
C aromatic	N:N2	43.11
C aromatic	:N2O	39.92
C aromatic	:N2S	36.08
C aromatic	:N2Cl	35.90
C sp	=C2	6.39
C sp	C#C	3.24
C sp	C#N	16.49

Table 5.2 (continued)

Atom type	Neighbors	Contribution
C sp	C#N(+)	11.33
C sp	#CS	28.03
C sp	N#N	47.80
C sp	#NP	12.53
N sp^3	H2C	5.03
N sp^3	H2C(pi)	6.38
N sp^3	H2N	17.97
N sp^3	H2S	41.98
N sp^3	HC2	23.83
N sp^3	HC2(pi)	13.51
N sp^3	HC2(2pi)	20.10
N sp^3	HCN	0.15
N sp^3	HCN(pi)	6.71
N sp^3	HCN(2pi)	6.84
N sp^3	HCS(pi)	15.10
N sp^3	C3	51.07
N sp^3	C3(pi)	53.90
N sp^3	C3(2pi)	60.80
N sp^3	C3(3pi)	61.26
N sp^3	C2N(pi)	7.05
N sp^3	C2N(+)(pi)	5.52
N sp^3	C2N(2pi)	36.36
N sp^3	C2N(+)(2pi)	20.13
N sp^3	C2N(3pi)	54.74
N sp^3	C2S	49.13
N sp^3	C2F(2pi)	64.78
N sp^3	CN2(pi)	30.74
N sp^3	CN2(2pi)	49.40
N sp^3	CN2(+)(2pi)	3.72
N sp^3	CNF(2pi)	34.74
N sp^2	C=C	32.77
N sp^2	C=N	4.54
N sp^2	C=N(+)	15.43
N sp^2	=CN	4.63
N sp^2	=CN(+)	36.68
N sp^2	C=O	12.04
N sp^2	C=P	49.18
N sp^2	=CO	16.24
N sp^2	=CS	26.78
N sp^2	N=N	12.19
N sp^2	N=O	0.00
N sp^2	=NO	6.67
N aromatic	:C2	14.01
N aromatic	:C:N	4.98
N(+) sp^3	H3C	2.77
N(+) sp^3	H2C2	82.36

Table 5.2 (continued)

Atom type	Neighbors	Contribution
N(+) sp^2	C=CO(−)	68.61
N(+) sp^2	C=NO	26.37
N(+) sp^2	C=NO(−)	11.30
N(+) sp^2	CO=O(−)	4.38
N(+) sp^2	=CO2(−)	2.17
N(+) sp^2	NO=O(−)	0.15
N(+) sp^2	O2=O(−)	6.00
N(+) aromatic	H:C2	46.79
N(+) aromatic	:C2O(−)	7.10
N(+) sp	C#C(−)	14.36
N(+) sp	#CO(−)	0.00
N(+) sp	=N2(−)	19.14
O	HC	4.49
O	HC(pi)	8.19
O	HN(pi)	2.28
O	HO	29.95
O	C2	39.23
O	C2(pi)	31.33
O	C2(2pi)	24.06
O	CN(pi)	0.00
O	CN(+)(pi)	0.00
O	CN(2pi)	4.91
O	CO(pi)	27.16
O	CP(pi)	16.12
O	N2(2pi)	5.87
O	N2(+)(2pi)	6.27
P3	C3	16.70
P3	S3	66.68
P4	C3=N	0.00
P4	C3=O	30.50
P4	C3=S	46.30
P4	O3=O	0.00
S2	HC	2.58
S2	HC(pi)	18.47
S2	C2	22.69
S2	C2(pi)	15.86
S2	C2(2pi)	7.94
S2	CN(pi)	25.96
S2	CN(2pi)	6.82
S2	CS(pi)	6.16
S2	CP(pi)	0.00
S2	N2	2.00
S2	N2(2pi)	21.36
S2	NS	1.00
S4	C2=O	5.89
S4	C2=O2	4.26

Table 5.2 (continued)

Atom type	Neighbors	Contribution
S4	CN=O2	9.20
Si	C4	2.02
Si	C3Si	0.67
H	H Acceptor[a]	8.63
Alkane	No. of C atoms[b]	0.53
Unsaturated HC	No. of C atoms[c]	0.10

[a]Intramolecular H-bridge between acidic H (on O, N, or S) and basic acceptor (O, N, or F).
[b]Correction factor per carbon atom in pure alkanes.
[c]Correction factor per carbon atom in pure aromatics, olefins, and alkynes.

Example 5.1: The use of eq. (5.2) for (a) 1,3,5-trinitrobenzene, (b) hexanitroethane, and (c) 1,3-diamino-2,4,6-trinitrobenzene gives:

(a) According to Tables 5.1 and 5.2, 1,3,5-trinitrobenzene has three (Atomic type = C aromatic, Neighbors = H:C2, and Contribution = 5.36 kJ/mol), three (Atomic type = C aromatic, Neighbors = :C2N(+), and Contribution = 28.67 kJ/mol), and three (Atomic type = N(+) sp^2, Neighbors = CO=O(-), and Contribution = − 4.38 kJ/mol):

$$\Delta_{sub}H = 21.03 + \sum_i a_i A_i + \sum_j g_j G_j$$

$$= 21.03 + 3(5.36) + 3(28.67) + 3(-4.38) = 109.98 \text{ kJ/mol}.$$

(b) According to Tables 5.1 and 5.2, hexanitroethane has two (Atomic type = C sp^3, Neighbors = CN3(+), and Contribution = 43.89 kJ/mol), and six (Atomic type = N(+) sp^2, Neighbors = CO=O(-), and Contribution = − 4.38 kJ/mol):

$$\Delta_{sub}H = 21.03 + \sum_i a_i A_i + \sum_j g_j G_j$$

$$= 21.03 + 2(43.89) + 6(-4.38) = 82.53 \text{ kJ/mol}.$$

(c) According to Tables 5.1 and 5.2, 1,3-diamino-2,4,6-trinitrobenzene has one (Atomic type = C aromatic, Neighbors = H:C2, and Contribution = 5.36 kJ/mol), three (Atomic type = C aromatic, Neighbors = :C2N(+), and Contribution = 28.67 kJ/mol), two (Atomic type = C aromatic, Neighbors = C:C:N, and Contribution = 11.10 kJ/mol), and three (Atomic type = N(+) sp^2, Neighbors = CO=O(-), and Contribution = − 4.38 kJ/mol):

$$\Delta_{sub}H = 21.03 + \sum_i a_i A_i + \sum_j g_j G_j$$

$$= 21.03 + 1(5.36) + 3(28.67) + 2(11.10) + 3(-4.38) = 121.46 \text{ kJ/mol}.$$

The measured $\Delta_{sub}H$ for 1,3,5-trinitrobenzene, hexanitroethane, and 1,3-diamino-2,4,6-trinitrobenzene are 107.3, 70.7, and 143.5 J/K mol, respectively [199].

5.2 Quantum mechanical and complex approaches for predicting the heat of sublimation

Computational methods like DFT offer an alternative [376], but despite their promise, they struggle to achieve chemical accuracy (~5 kJ/mol) while remaining computationally affordable [377], creating a persistent gap in our ability to reliably predict this important property. Several quantum mechanical calculations have been introduced to predict the heats of sublimation of energetic compounds [173, 176, 178, 378–381], with which Politzer and coworkers have achieved significant success [178]. They used three quantum mechanical parameters in their calculations:

(1) the surface area of the 0.001 electron/bohr^3 isosurface of the electron density of the molecule,
(2) a measure of the variability of electronic potential on the surface,
(3) the degree of balance between the positive and negative charges on the isosurface.

Rice et al. [176] further improved this method to generate surface electrostatic potentials of individual molecules. Byrd and Rice [173] have modified previous methods by incorporating group additivity and by the use of the more complicated 6–311++G(2df,2p) basis set. Hu and coworkers [378] have also used the empirical relations of Politzer et al. [121] to predict the heats of sublimation of the condensed phases of energetic materials. There are also some relationships between the heats of sublimation of some polynitro compounds and lattice energies [382]. Suntsova and Dorofeeva [166] improved the electrostatic potential model of the Politzer approach by additional parameter, Π (average deviation of electrostatic potential), for estimating enthalpies of sublimation of nitrogen-rich energetic compounds based on experimental enthalpies of sublimation for 185 compounds. Meftahi et al. [165] have compared several QSPR methods based on complex descriptors for predicting the enthalpy of sublimation of organic compounds containing energetic materials.

Conventional QSPR approaches and machine learning-based QSPR (ML-QSPR) utilize chemical informatics-derived descriptors that eliminate the need for predefined substructure libraries while maintaining substantially faster computation times than quantum chemical QSPR calculations. Nevertheless, their predictive accuracy remains inherently dependent on training dataset quality and diversity. Although capable of extrapolating to new compound classes, such predictions may suffer from reduced reliability, underscoring the critical need for comprehensive datasets – particularly for energetic compounds – in developing robust $\Delta_{sub}H$ prediction models. To address these challenges, Liu et al. [382] augmented the DIPPR 801 dataset [383] with over 100 carefully selected energetic organic compounds. Their modeling strategy incorporated both quantum chemical descriptors (for physical interpretability) and topological descriptors (for computational efficiency), evaluated through four machine learning algorithms: SVR [384], RF [385], Extreme Gradient Boosting (XGBoost) [386], and Particle Swarm Optimization (PSO) [387]. Comparative analysis revealed XGBoost's superior performance, achieving remarkable prediction accuracy (MAE=2.8 kcal/mol) for ener-getic compounds. The PSO approach demonstrated comparable accuracy while offer-ing enhanced interpretability through its transparent functional form, making it par-ticularly suitable for practical implementation. These optimized ML-QSPR models combine quantum-chemical accuracy with computational efficiency, providing valu-able tools for rapid screening and development of novel energetic materials. Their balanced performance characteristics suggest promising applications in accelerating the discovery and optimization of high-performance energetic compounds. Wahler et al. [388] developed ML models that predict these enthalpies from simple structural formulas alone. Their goal was to provide more accurate predictions than traditional methods while maintaining simplicity – just sketch the molecule and get reliable re-sults. Liu et al. [389] created a validated DFT dataset for 845 diverse organic molecules containing C, H, O, N, F, and S elements, then building a machine learning model that not only predicts $\Delta_{sub}H$ from molecular structure but also identifies key influencing factors through feature analysis, with the model continuously improving via an active learning loop that strategically selects new calculations to maximize predictive perfor-mance while efficiently exploring chemical space, ultimately creating a versatile tool for materials design that could extend to hybrid materials and beyond.

5.3 The use of structural parameters

There are several simple methods for predicting $\Delta_{sub}H$ which are based on structural features [129, 168–170]. These methods can be applied for selected classes of energetic compounds and are demonstrated here.

5.3.1 Nitroaromatic compounds

For nitroaromatics, the following optimized correlation can be used to predict $\Delta_{sub}H$ according to [168]:

$$\Delta_{sub}H = 64.51 + 4.555a - 2.763b + 10.32c + 16.51C_{SG}, \qquad (5.3)$$

where $\Delta_{sub}H$ is in kJ/mol, and the variable C_{SG} shows the contribution of certain polar groups. The different values of C_{SG} for various polar groups attached to nitroaromatic rings are specified as follows:

(1) Alkoxy group (–OR) attached to a nitroaromatic ring: $C_{SG} = 1.0$, for example, 2-me-thoxy-1,3,5-trinitrobenzene.

(2) Carbonyl in the form of –C(=O)NRR′ or –C(=O)–R attached to an aromatic ring, in which R and R′ are alkyl groups: $C_{SG} = 0.75$, for example, 3-nitroacetophenone.

(3) Carboxylic acid functional group, two hydroxyl groups, or three amino groups: $C_{SG} = 2.0$. In the case of two hydroxyl groups, nitro groups should be separated from –OH by at least one –CH– group, for example, 4-nitrobenzene-1,2-diol.

Example 5.2: The use of eq. (5.1) for the following molecular structure gives

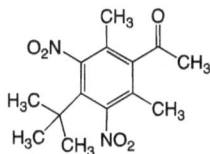

$$\Delta_{sub}H = 64.51 + 4.555a - 2.763b + 10.32c + 16.51C_{SG}$$

$$= 64.51 + 4.555(14) - 2.763(18) + 10.32(2) + 16.51(0.75)$$

$$= 111.6 \text{ kJ/mol}.$$

The measured $\Delta_{sub}H$ value for this compound is 107.9 kJ/mol [390].

5.3.2 Nitramines

It was found that the molecular weight and structural parameters are sufficient to establish a new correlation as follows [167]:

$$\Delta_{sub}H = 15.62 + 0.3911Mw + 10.36n_{O_2NNCH_2NNO_2}, \tag{5.4}$$

where Mw is the molecular weight of the nitramine and $n_{O_2NNCH_2NNO_2}$ is the number of –CH$_2$– groups between two nitramine functional groups in cyclic and noncyclic nitramines. Equation (5.4) cannot be used for cyclic nitramines with $n_{O_2NNCH_2NNO_2} \geq 5$.

Example 5.3: 4,10-Dinitro-2,6,8,12-tetraoxa-4,10-diazaisowurtzitane (TEX) has the following molecular structure:

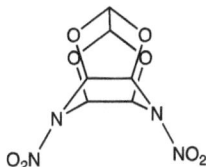

Equation (5.4) predicts $\Delta_{sub}H$ to be

$$\Delta_{sub}H = 15.62 + 0.3911Mw + 10.36n_{O_2NNCH_2NNO_2}$$
$$= 15.62 + 0.3911(262.13) + 10.36(0)$$
$$= 118.1 \, kJ/mol.$$

The measured $\Delta_{sub}H$ for TEX is 123.4 kJ/mol [329].

5.3.3 Nitroaromatics, nitramines, nitroaliphatics, and nitrate esters

For nitroaromatics, nitramines, nitroaliphatics, and nitrate esters, the molecular weight and the contribution of some specific functional groups, as well as structural features, can be combined by a general equation as follows [169]:

$$\Delta_{sub}H = 53.74 + 0.2666Mw' + 13.99C_{In} - 15.58C_{De}, \tag{5.5}$$

where Mw' is the molecular weight of the nitro compound (except halogenated nitro-aromatics and hydrogen-free nitro compounds, in which the contribution of halogen atoms in the calculation of the molecular weight should be neglected), C_{In} is the contribution of specific polar groups attached to aromatic rings, and C_{De} shows the presence of some molecular moieties. The values of C_{In} and C_{De} are specified according to the following conditions.

(1) Nitroaromatics:
  (a) Prediction of C_{In}:
    (i) –COOH and –OH functional groups: $C_{In} = 2.0$ for the compounds that contain the carboxylic acid functional group or two hydroxyl groups. Since the participation of a group in intramolecular hydrogen bonding can reduce its ability to form intermolecular hydrogen bonds, the presence of a nitro group in the *ortho* position relative to the –OH group can cancel its effect. If two hydroxyl groups are attached to the aromatic ring, nitro groups should be separated from –OH groups by at least one –CH– group, for example, 4-nitrobenzene-1,2-diol.
    (ii) Amino groups: the value of C_{In} is equal to the number of amino groups in such compounds, that is, $C_{In} = n_{NH_2}$.
    (iii) The presence of a carbonyl group in the form of an amide $\overset{O}{\underset{Ar}{\bigwedge}}_N$ or ketone $\overset{O}{\underset{Ar}{\bigwedge}}_R$: $C_{In} = 0.75$. Since carbonyl groups are in resonance with the aromatic ring, they can likely promote co-planarity and rigidity in some cases.
  (b) Prediction of C_{De}: The presence of alkyl groups – especially bulky groups such as *tert*-butyl – can decrease the intermolecular interactions for high ratios of n_R/n_{NO_2}. For $n_R/n_{NO_2} \geq 1$, the contribution of C_{De} should be considered. The values of C_{De} are 2.0 and 3.0 for the presence of one and more than one bulky

group, respectively. If only small alkyl groups such as methyl groups are present, $C_{De} = 1.0$.

(2) Nitramines: The contributions of C_{In} and C_{De} depend on the number of N–NO$_2$ groups in cyclic and acyclic nitramines. For five membered (or larger) cyclic nitramines that have only the fragments $\underset{H_2C-N-CH_2}{\overset{NO_2}{|}}$ and for acyclic nitramines, $C = 1.75 n_{NNO_2} - 4$. If $C < 0$ and $C > 0$, then C will become C_{De} and C_{In}, respectively. For nitramines with the molecular fragment $\underset{HN \quad NH}{\overset{\parallel}{\diagdown\diagup}}$, appreciable molecular interactions are present so that $C_{In} = 4.25$.

(3) Nitroaliphatic compounds: For nitroaliphatic compounds, $C_{De} = 3.0$.

Example 5.4: The use of eq. (5.5) for the following molecular structure gives

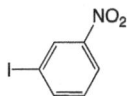

$$\Delta_{sub}H = 53.74 + 0.2666Mw + 13.99C_{In} - 15.58C_{De}$$
$$= 53.74 + 0.2666(122.10) + 13.99(0) - 15.58(0)$$
$$= 86.29 \text{ kJ/mol.}$$

The measured value of $\Delta_{sub}H$ is 83.0 kJ/mol [199].

5.3.4 General method for polynitroarenes, polynitroheteroarenes, acyclic and cyclic nitramines, nitrate esters, nitroaliphatics, cyclic and acyclic peroxides, as well as nitrogen-rich compounds

The following correlation can be used to predict $\Delta_{sub}H$ for a wide range of energetic compounds, including polynitroarenes, polynitroheteroarenes, acyclic and cyclic nitramines, nitrate esters, nitroaliphatics, cyclic and acyclic peroxides, as well as nitrogen-rich compounds [170]:

$$\Delta_{sub}H = 52.89 + 0.2689Mw' + 15.13F_{attract} - 13.29F_{repul}, \tag{5.6}$$

where $F_{attract}$ and F_{repul} are two parameters that take into account attractive and repulsive intermolecular forces. The values of $F_{attract}$ and F_{repul} are specified according to the following conditions, depending on the presence of various functional groups and molecular moieties:

(1) $-COOH$, $-OH$, ⬡N^+-O^-, $-NH_2$ (or $-NH-$) and $\underset{HN\quad NH}{\overset{O}{||}}$ polar groups: The values of $F_{attract}$ are as follows:

   (a) $F_{attract} = 2.0$ for compounds containing at least one $-C(=O)OH$, or two $-OH$ functional groups. For nitroaromatics containing two $-OH$ groups, nitro groups should be separated from $-OH$ at least by one$=CH-$ group, for example, 4-nitrobenzene-1,2-diol. The presence of a nitro group in the *ortho* position with respect to the $-OH$ group can cancel this condition so that $F_{attract} = 0.0$ in these compounds, for example, 2,4,6-trinitro-1,3-benzenediol (styphnic acid).

   (b) The value of $F_{attract}$ is equal to the number of $-NH_2$, $-NH-$ or ⬡N^+-O^- groups.

   (c) For nitramines or organic polynitrogen compounds that contain a $\underset{HN\quad NH}{\overset{O}{||}}$ molecular fragment, there is a large intermolecular attraction so that $F_{attract} = 4.7$.

(2) Nitramines:

   (a) Direct electrostatic interactions are dominant in polynitramine crystals which contain the $\underset{\quad C H_2}{N-N}\overset{NO_2\ NO_2}{}$ molecular fragment. Thus, $F_{attract}$ is 1.0 and 2.0 in acyclic and cyclic nitramines, respectively.

   (b) For other acyclic nitramines, $F_{repul} = 2.6$.

(3) Nitroaliphatic compounds: The value of F_{repul} is equal to 2.6 in these compounds.

(4) Alkylated nitroaromatics: $F_{repul} = 1.0$ in substituted nitroaromatics containing small alkyl groups such as methyl groups. However, $F_{repul} = 2.0$ for bulky groups such as the *t*-butyl group. These conditions can be applied if $n_R/n_{NO_2} \geq 1$.

(5) Peroxides: For acyclic and cyclic peroxides, $F_{repul} = 2.0$.

(6) Intramolecular hydrogen bonding (H-bonding): $F_{repul} = 0.5$, for example, 1,3-diamino-2,4,6-trinitrobenzene (DATB).

The conditions listed above are summarized in Table 5.3.

Example 5.5: 1,3,5-Trinitro-1,3,5-triazinane (RDX) has the following molecular structure:

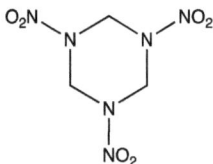

The use of eq. (5.6) for the following molecular structure gives

$$\Delta_{sub}H = 52.89 + 0.2689 M_{rev} + 15.13 Mw' - 13.29 F_{repul}$$
$$= 52.89 + 0.2689(222.12) + 15.13(2.0) - 13.29(0)$$
$$= 142.9 kJ/mol.$$

The deviation of the predicted value from the experimentally determined value (134.3 kJ/mol [391]), that is, 8.3 kJ/mol, is lower than that of the values obtained from the two complex quantum mechanical calculations of Rice et al. [176] ($\Delta_{sub}H = 102.5$ kJ/mol; Dev $= 31.8$) and [173] ($\Delta_{sub}H = 97.9$ kJ/mol; Dev $= 36.4$).

Table 5.3: Summary of correcting functions $F_{attract}$ and F_{repul}.

Molecular moieties	Effect on predicted $\Delta_{sub}H$		Comment
	$F_{attract}$	F_{repul}	
–OH and –COOH groups	2.0	–	(a) For one or more –COOH groups (b) For two –OH groups
–NH₂, –NH–, and N N⁺–O⁻ groups	No. of groups	–	–
HN NH structure	4.7	–	–
NO₂ NO₂ N–C–N structure H₂	1.0 2.0	– –	In acyclic nitramines In cyclic nitramines
–NO₂ and >NNO₂ groups	–	2.6	In nitroaliphatics and acyclic nitramines which are not included in the conditions two rows above
R– –NO₂ structure, $n_R/n_{NO_2} \geq 1$	– –	1.0 2.0	For small alkyl groups For bulky alkyl groups
C–O–O–C group	–	2.0	–
Intermolecular H-bonding	–	0.5	Intermolecular hydrogen bonding forms a 6-membered ring

5.4 Summary

This chapter has introduced different empirical methods for the prediction of the heats of sublimation of important classes of energetic compounds. Equation (5.2) can provide a group additivity method for prediction of the enthalpy of sublimation of organic compounds containing energetic materials. Equations (5.1) and (5.4) provide two simple and reliable approaches to estimate the heat of sublimation of nitroaromatics and nitramines, respectively. Equations (5.5) and (5.6) provide more complex empirical methods which can be applied to a wide range of organic compounds containing important energetic functional groups. Equation (5.6) is the best method because it can be used for a wide range of energetic compounds, including polynitroarenes, polynitroheteroarenes, acyclic and cyclic nitramines, nitrate esters, nitroaliphatics, cyclic and acyclic peroxides, as well as nitrogen-rich compounds.

6 Impact sensitivity

An organic energetic compound is a metastable molecule, which is capable of undergoing very rapid and highly exothermic reactions. Thus, the prediction of its sensitivity is a complex matter. Several properties contribute to a materials' response to the stimulus in a sensitivity test, which are a consequence of the kinetics and thermodynamics of the thermal decomposition of the explosive. They include:

(1) the ease with which a detectable reaction of any kind can be initiated in an energetic compound;
(2) the tendency for a small reaction to grow into destructive properties;
(3) the ease with which a higher-order detonation can also be established in an energetic compound.

There are several reviews in the literature which describe the calculation of the sensitivity of energetic compounds [121, 392–397]. Zeman and Jungová [396] have given an overview of the main developments in the study of the sensitivity of energetic materials to impact, shock, friction, electric spark, laser beams, and heat during the period 2006–2015.

The safe handling of an energetic compound is one of the most important issues to the scientists and engineers who handle energetic molecules. Some stimuli can cause the detonation of an energetic compound, including impact, shock, heat, electrostatic charge, and friction. Of these stimuli, impact is probably the most well-known among the various types of sensitivity because the drop-weight impact test is extremely easy to implement. Impact is one of the important factors in assessing an energetic compound since it provides information on the vulnerability of an energetic material to detonation due to accidental impact. Therefore, impact sensitivity is closely related to and highly relevant to many workplace accidents.

The drop hammer is one of the usual tests used for the evaluation of impact sensitivity. In this test, milligram quantities of an explosive material are placed between the flat tool steel anvil and the flat surface of the tool striker. It typically involves dropping a 2.5 kg mass from a predetermined height onto the striker plate. The impact drop height (H_{50}, cm) is the height from which there is a 50 % probability of causing an explosion, where 1 cm = 0.245 J (Nm) with a 2.5 kg dropping mass. Since the sensitivity is inversely proportional to H_{50}, impact sensitivity is shown in terms of the value of H_{50}. Although the impact sensitivity test itself is extremely easy to implement, obtaining reliable experimental data is known to be relatively difficult. Since there is some difficulty associated with the initiation mechanism of explosions caused by mechanical impact, it can be assumed that hot spots in the material contribute to initiation in the drop weight impact test. The results of impact sensitivity are often not reproducible because factors in the test that might affect the formation and growth of hot spots can strongly affect the measurements. Moreover, the experimental data are

https://doi.org/10.1515/9783112206768-006

extremely sensitive to the conditions under which the tests are performed. Despite all of the uncertainties associated with the impact sensitivity test, many different methods have been developed to correlate the impact sensitivity with other properties of energetic compounds, e.g., maximum heat of detonation [398], crystal lattice compressibility/free space [399], the available free space per molecule in the unit cell [400], ^{15}N NMR chemical shifts [401], nucleus-independent chemical shifts for aromatic explosives [402], and activation energy of thermal decomposition [403]. In recent years, some new correlations have been introduced to predict the impact sensitivity of different categories of energetic compounds, which are based on structural moieties [403–406], QSPR [407–412], artificial neural networks, and genetic algorithms [413, 414]. Some of these approaches are reviewed and illustrated in this chapter.

6.1 Complex computational methods

High-speed computers allow quantum mechanical calculations of impact sensitivities of different classes of energetic compounds. Since the molecular surface of electrostatic potentials of the nitroaromatic molecules has positively charged regions over the $C–NO_2$ bonds, some authors have used computed partial atomic charges [415, 416], heats of reaction [417], and heats of explosion [418] in order to estimate impact sensitivities of some classes of explosives.

Brinck et al. [419] introduced the term "polarity index" (Π), which can measure local polarity, and demonstrated its relationship to the dielectric constant. For nitroaromatics, there is a relationship between Π and their impact sensitivities. Xiao and coworkers [420] proposed the thermodynamic criteria of "the smallest bond order", "the principle of the easiest transition" and the kinetic criterion of "the reaction activation energy of pyrolysis initiation" to judge the impact sensitivity. These methods are only used to qualitatively compare the relative magnitudes of impact sensitivity.

Politzer et al. [421] have identified a few features of electrostatic potentials for $C_aH_bN_cO_d$ explosives that appear to be related to their sensitivity to impact. Owen et al. [303] investigated the electrostatic potential over the $C–NO_2$ bonding region, which reflects a degree of instability in the $C–NO_2$ bond. For 18 nitroaromatics (excluding hydroxynitroaromatic molecules), Murray et al. [416] also introduced a correlation between impact sensitivity measurements and an approximation of the electrostatic potential at the midpoint of the C–N bond. Rice and Hare [422] used approximations of the electrostatic potential at midpoints, statistical parameters of these surface potentials, and the property–structure relation method "generalized interaction property function" (GIPF) or computed heats of detonation to predict the impact sensitivity of $C_aH_bN_cO_d$ explosives. The impact sensitivities of $C_aH_bN_cO_d$ explosives have some dependence on the degree of internal charge imbalance within the molecule [423]. For nitramines, rupture of the $N-NO_2$ bond is a key step in the decomposition process initiated by heat, shock, and impact [424]. Edwards et al. [425] also used model IV of Rice and Hare [422] to calculate the heat of detonation of several nitramines using quantum mechanical theory, in which it was shown that there was a correlation between the exponential decrease of the HOMO and LUMO energies versus sensitivity at the DFT level of theory. Ren et al. [426] used seven models that related the features of molecular surface electrostatic potentials above the bond midpoints and rings, as well as statistical parameters of surface electrostatic potentials, to the experimental impact sensitivities of eight strained cyclic explosives with the $C-NO_2$ bonds at the DFT-B3LYP/6–311++G^{**} level. Oliveira and Borges Jr. [427] developed four mathematical models to correlate impact sensitivity to molecular charge properties using DFT.

Zhang et al. [428] derived some relationships between impact sensitivity and nitro group charges. They used the general gradient approximation (GGA) as well as the Becke hybrid functional and DNP basis set to calculate the Mulliken charges, which could be correlated with the impact sensitivity of nitro compounds. It was found that nitro compounds may be sensitive ($H_{50} \leq 40$ cm) when the nitro group has a negative charge of less than about 0.23. Since the charges on the nitro group can be used to estimate the bond strength, oxygen balance, and molecular electrostatic potential, compounds with higher Mulliken net charges at the nitro groups will be insensitive and show large H_{50} values. The method of Zhang et al. [428] can be applied to nitro compounds when the $C-NO_2$, $N-NO_2$, or $O-NO_2$ bond is the weakest in the molecule. Zeman and Jungová [396] have also reviewed some further publications which used quantum mechanical approaches. Bondarchuk [429] developed a theoretical approach for the prediction of the impact sensitivity of explosives based on the solid state-derived criteria, which include triggering pressure, the average number of electrons per atom, crystal morphology, energy content, and melting temperature. Cawkwell and Manner [430] demonstrated that chemical reactivity, rather than thermomechanical effects, is the dominant factor in explosive behavior in an impact test. They suggested that quantum-based molecular dynamics simulations may be a reliable computational tool for screening explosives for drop-height impact sensitivity. Mathieu [431] correlated linearly the impact sensitivity for several high explosives with their (detonation velocity)$^{-4}$ or equivalently with (detonation pressure)$^{-2}$ or (Gurney energy)$^{-1}$, which originated from the primary role of the amount of chemical energy evolved per atom for both performance and sensitivity.

Neural network architectures have been recently used as a prediction methodology for impact sensitivity. Cho and coworkers [432] utilized 17 molecular descriptors, which were composed of compositional and topological descriptors in an input layer and two hidden neurons in a hidden layer. Some structural parameters have also been used to predict impact sensitivity using an artificial neural network model by choosing only 10 molecular descriptors [433]. The final neural structure consists of the three layers input, output, and hidden. The network is composed of: 10 input nodes, fifteen hidden-layer neurons and a single output neuron corresponding to the impact sensitivity of explosive. The 10 structural descriptors include (1) a/MW; (2) b/MW; (3) c/MW; (4) d/MW; indicator variables for (5) aromaticity; (6, 7) heteroaromaticity (N and O); (8) N–NO$_2$; (9) α-hydrogen; (10) salt. The connection weights of the network were adjusted iteratively using back propagation algorithm. The predictive ability of the artificial neural network was checked with 275 experimental data. Impact sensitivities of 14 explosives in the test set were also compared with five quantum mechanical models of Rice and Hare [422]. It was shown that this model can provide better predictions compared to the quantum mechanical models of Rice and Hare [422].

Wang et al. [434] used the QSPR model by combining both the electronic and topological characteristics (ETSI approach) of the molecules under analysis. Since they used mixed data from very different structures, their predictions are, at best, just an indication. Xu et al. [412] performed a QSPR study for the entire set of 156 structurally different energetic compounds to estimate the impact sensitivity of new energetic materials. These QSPR approaches, however, do not allow an evaluation of the chemical physics of initiation.

Energetic materials must perform well in detonation while remaining stable and safe to handle, but achieving this balance has long been a challenge. Researchers are still working to understand the key factors that influence impact sensitivity and detonation performance, as well as how to design materials with the right properties. ML could help solve this problem by analyzing complex data and uncovering hidden relationships between a material's characteristics and its performance, potentially speeding up the development of better energetic compounds. Liu et al. [435] examined 222 different energetic materials, collecting data on their impact sensitivity. Using four ML models, they found that impact sensitivity is strongly linked to the heat of explosion, oxygen balance, decomposition products, and HOMO (highest occupied molecular orbital) energy levels, while detonation performance depends mostly on oxygen balance, decomposition products, and density.

Bao et al. [436] studied how quickly energy moves in seven explosive materials to see if it affects their sensitivity to impact. Energy transfer happens in two ways: inside the molecule (intramolecular) and between molecules (intermolecular). To measure this, Bao et al. [436] introduced a new parameter for the speed of energy flow within the molecule, while sound velocity helped track how fast energy escapes outward. The most sensitive explosives were those with fast internal energy transfer but slow energy dissipation between molecules. In other words, if energy gets trapped inside the material instead of spreading out, it's more likely to detonate from an impact.

Cawkwell et al. [437] developed a straightforward kinetic model to predict how sensitive organic explosives are to impact. Instead of relying on complex assumptions, it uses two key factors: (1) the heat released during explosion and (2) reaction barriers derived from gas-phase molecular dynamics simulations. These simulations – run for 24 diverse molecules – automatically account for all possible decomposition pathways, eliminating guesswork about how the explosives break apart. The results show that explosives with higher heat release tend to have lower reaction barriers, aligning with the Bell-Evans-Polanyi principle. The Bell-Evans-Polanyi principle shows a simple tradeoff – reactions that release more energy need less activation energy to start. For explosives, this means materials with weak trigger bonds and high energy output are most sensitive because their large heat release further lowers the reaction barrier. It's why powerful explosives often detonate easily – the huge energy payoff makes the initial trigger pull smaller. This explains why powerful explosives are often more sensitive. In short, impact sensitivity depends on two things: weak chemical bonds that break easily and high energy release that further speeds up reactions.

Duarte et al. [438] analyzed the electron distribution in 53 nitroaromatic explosives using advanced computational methods. By breaking down the molecular charge densities, they identified four key electronic features: (1) nitro group charge, (2) nitro group polarization, (3) carbon ring electron delocalization, and (4) the number of explosive nitro groups. These features were used to predict H_{50} through machine learning. After testing 42 different algorithms, four stood out as the most accurate: Extra Trees, Random Forests, Gradient Boosting, and AdaBoost. Their predictions matched experimental sensitivity measurements within a 19–28% error range. The analysis revealed that electron delocalization (39% importance) and nitro group polarization (35%) matter most, followed by nitro charge (16%) and count (10%).

Interestingly, these features affect sensitivity differently across the spectrum:
–   For highly sensitive explosives ($H_{50} < 50$ cm), all factors make the material more sensitive
–   For moderately sensitive ones (50–100 cm), nitro count and polarization increase stability, while other factors decrease it
–   For very stable explosives ($H_{50} > 200$ cm), all factors contribute to stability

These insights provide a clear molecular-level understanding of explosive sensitivity that can guide the design of safer materials.

Wu et al. [439] used four ML models – back propagation neural network (BPNN), multilayer perceptron, random forest, and support vector regression – to predict the impact sensitivity of energetic compounds. Key molecular descriptors, such as oxygen balance, nitro group charge, heat of release, and self-multiplying coefficient, were identified as the most influential factors. Among the models, BPNN demonstrated the highest accuracy in predicting impact sensitivity, outperforming the other three methods. This approach offers a promising way to improve sensitivity predictions for energetic materials using machine learning.

To improve accuracy, Peng et al. [440] used kernel methods combined with heuristic algorithms (Genetic Algorithm and Particle Swarm Optimization) to develop predictive models. The optimized model achieved an R^2 score of 0.871 for impact sensitivity – outperforming neural networks ($R^2 = 0.827$) and support vector regression ($R^2 = 0.822$). This approach provides higher accuracy without relying on experimental data, making it a safer and more efficient way to assess hazardous properties.

Deng et al. [441] analyzed 240 nitroaromatic compounds using machine learning to understand what drives impact sensitivity in energetic materials. ANN proved more accurate than sure independence screening and sparsifying operator methods at predicting sensitivity directly. Key molecular features emerged as major influencers – particularly oxygen-containing groups, how atomic properties are distributed, and a compound's hydrophilicity. Bulk modulus measures how resistant a material is to uniform compression (squeezing from all sides). It tells you how much pressure is needed to shrink a material's volume by a certain amount. An interesting relationship with bulk modulus was uncovered: when materials are less stiff (low bulk modulus), increasing stiffness dramatically reduces sensitivity, but this trend reverses for already-stiff materials, where greater stiffness actually makes them slightly more sensitive. These findings provide concrete guidance for designing safer explosives by highlighting which structural features matter most.

When designing new explosives, one of the most critical goals is reducing sensitivity – ensuring they won't detonate accidentally from something like an impact. Over the years, researchers have tried to predict impact sensitivity by linking it to molecular or crystal properties. A popular idea has been using the HOMO-LUMO gap (the energy difference between an explosive molecule's highest occupied and lowest unoccupied orbitals) as a potential indicator. Politzer and Murray [442] put that theory to the test with twelve nitroaromatic explosives, analyzing them using four different computational methods. They found that the HOMO-LUMO gap doesn't reliably predict impact sensitivity. Since detonation initiation is a complex, multi-step process, sensitivity likely depends on multiple factors rather than just one or two specific properties. Instead of chasing oversimplified correlations, a more promising approach may be identifying fundamental trends and key contributing factors – a strategy that's already shown some success.

For nearly 80 years, scientists have used the drop-weight impact test to measure how easily high explosives can accidentally detonate. While this test gives somewhat variable results, its simplicity and low material needs keep it widely used today. Marrs et al. [443] used an extensive collection of these test results, primarily from Los Alamos National Laboratory, combined with detailed chemical information about each explosive. This comprehensive dataset includes more than 500 different explosive compounds, over 1,000 repeated tests, and 100 distinct molecular characteristics – totaling about 1,500 data points. Using random forest machine learning techniques, Marrs et al. [443] developed a model that connects these chemical properties to handling sensitivity. The model successfully predicts sensitivity across many types of explosives, from very sensitive to relatively stable materials. Key factors like explosion heat, oxygen balance, and specific molecular groups strongly influence sensitivity, but interestingly, even simpler models without all these details still work reasonably well. These findings point to a fundamental truth about explosive sensitivity – it doesn't depend on just one or two simple factors, but rather emerges from many interconnected chemical properties working together in complex ways.

Salt formation offers an effective way to improve molecular stability against impact. Pallewela and Bettens [444] used quantum mechanical calculations to predict impact sensitivity trends for two nitrogen-rich energetic salts: 3-amino-1,2,4(4H)-oxadiazol-5-one and 4-nitramino-1,2,4-triazole. They evaluated several quantum-mechanical criteria – including the HOMO-LUMO energy gap, the ratio of bond dissociation energy to total molecular energy, electrostatic potential at bond midpoints, and bond topological parameters – and compared the results with experimental BAM fall hammer test data. After testing multiple DFT functionals and basis sets, Pallewela and Bettens [444] found that the CAMB3LYP/6–31G(d)/IEFPCM = water level of theory provided the best qualitative predictions for impact sensitivity. These findings show that quantum mechanical modeling is a powerful tool for designing more stable: 3-amino-1,2,4(4H)-oxadiazol-5-one – and 4-nitramino-1,2,4-triazole-based energetic salts before synthesis, helping to reduce costs and safety risks in the development process.

Impact sensitivity measures how easily energetic molecules react to mechanical forces like impacts, making it a crucial safety consideration. Nitro compounds, commonly used as explosives in military, industrial, and civilian applications, require careful handling due to this sensitivity. To improve safety predictions, Lotfi et al. [445] developed a QSPR model using the Monte Carlo algorithm in CORAL-2023 software to assess 404 nitro compounds. They represented molecular structures using SMILES notation and calculated correlation weight descriptors. They tested four modeling approaches: one using basic Monte Carlo optimization, two incorporating either the information index of correlation or the correlation index of information, and a fourth combining both the information index of correlation and the correlation index of information. Statistical analysis revealed the combined models performed best. These results demonstrate the value of incorporating both information measures for accurate sensitivity predictions.

Nitrobenzenic explosives offer an attractive combination of high energy density and low impact sensitivity, making them particularly valuable for practical applications. Siqueira Soldaini Oliveira and Roberta Borges Jr. [446] examined the charge density properties of 50 nitrobenzenic molecules using DFT combined with the distributed multipole analysis method to better understand their impact sensitivity characteristics. The distributed multipole analysis approach provides detailed insights through monopole, dipole, and quadrupole electric multipoles localized on molecular atoms, offering both precise charge density mapping and clear chemical interpretation. Using these distributed multipole analysis-derived multipoles, Siqueira Soldaini Oliveira and Roberta Borges Jr. [446] developed several models to correlate molecular charge properties with H_{50} values. They first evaluated three established models previously applied to just 17 nitroaromatic compounds: Model 1 relied solely on nitro group charge, Model 2 incorporated both nitro charge and benzene ring quadrupole values (measuring charge delocalization), while Model 3 added nitro group dipole moments (indicating site polarization) and average C–NO_2

bond distances (reflecting bond strength). For this expanded study, Siqueira Soldaini Oliveira and Roberta Borges Jr. [446] introduced two new models (Models 4–5) that additionally incorporated nitro group quadrupole values. All five models demonstrated strong predictive performance when applied to a larger dataset of 50 molecules, maintaining the quality seen in previous smaller-scale studies. Their analysis confirmed that two key distributed multipole analysis parameters remain crucial for accurate modeling, as identified in earlier work: the quadrupole values of ring atoms (quantifying electronic delocalization) and the total charge (monopole) values of the critical nitro explosophore groups. These findings validate and extend their understanding of the molecular charge characteristics governing impact sensitivity in nitrobenzenic explosives.

Understanding impact sensitivity through computational methods provides valuable insights for safely handling energetic materials during transport, synthesis, and storage. Guo et al. [447] investigated the connection between energy transfer rates and impact sensitivity using first-principles calculations. They analyzed 15 materials by calculating their phonon properties and determining the phonon bath range through frequency band gap analysis. To quantify energy transfer, Guo et al. [447] developed a new sensitivity index. Their results reveal a strong linear correlation ($R^2 \approx$ 0.89) between this index and experimental H_{50} values – the smaller new sensitivity index becomes, the more sensitive the material. While vibration mode band gaps and total up-pumped density help qualitatively identify sensitive materials, they don't provide precise sensitivity rankings. This limitation arises because impact sensitivity depends on multiple complex factors that can't be captured by any single parameter alone.

6.2 Advances in modeling impact sensitivity: phonon-vibration coupling and multi-phonon interactions in energetic materials

In solids, atoms aren't perfectly still – they constantly vibrate due to heat and quantum effects. These vibrations can be individual jiggles or coordinated waves moving through the lattice.

When many atoms vibrate together in a repeating pattern, we call these collective waves phonons. Think of them like sound waves in a crystal, but quantized – meaning their energy comes in tiny packets, like photons for light.

There are two main types:
- Acoustic phonons are like deep, rumbling sound waves (imagine shaking a spring).
- Optical phonons are higher-pitched, happening when atoms in a crystal's unit cell wiggle against each other (they can even interact with light).

Phonons aren't real particles, but they act like them, helping explain how heat travels, why some materials conduct electricity weirdly, and even how sound moves through solids. Without phonons, diamonds wouldn't stay cool, and speakers wouldn't work.

When a molecule vibrates (like a bond stretching or bending), that's called a vibron. Imagine a tiny molecule stuck to a surface, jiggling like a spring. That's its vibration – let's call it the molecule's "dance". Now, the atoms in the surface are also shaking (those are phonons), like a crowd doing the wave in a stadium. When the molecule's dance syncs up with the crowd's wave, they start influencing each other. The molecule's kicks might get weaker because the surface steals some energy, or the surface vibrations might get louder where the molecule sits. That push-and-pull is phonon-vibron coupling.

Bondarchuk [448] proposed a revised model for how mechanical energy converts to vibrational energy in explosive crystals under impact. The new approach fixes past inaccuracies in normalizing phonon-vibration couplings (ζ), which is crucial for comparing differently sized molecules. It also introduces two key damping factors, a^* and b^*. Factor a^* weights phonon overtones by coupling strength – weaker interactions mean stronger coupling. Factor b^* controls how quickly coupling weakens as phonons and vibrations diverge in energy. After testing 30 nitro-explosives, Bondarchuk [448] found optimal values: $a^* = 2.5$ and $b^* = 40$ cm^{-1}. The method was then applied to 21 crystalline energetic materials, including nitroexplosives and nitrogen-rich salts. The revised model eliminates errors for degenerate frequencies and cuts errors for split modes to under 8%. Impact sensitivity shows modest correlation ($R^2 = 0.53$–0.60) with many factors (trigger bond energy, crystal structure, electron transfer, etc.) influence sensitivity. The phonon up-pumping model is just one piece of the puzzle. For example, potassium pentazolate highlights the need to combine multiple mechanisms. This update improves accuracy but isn't a standalone solution.

Liu et al. [449] calculated ζ, a key factor in predicting sensitivity. Unlike previous approaches, their method accounts for interactions across a broader range of molecular vibrations (0–700 cm^{-1}), leading to more accurate results. By analyzing 45 different nitroexplosives, Liu et al. [449] refined the model's parameters and found that while ζ does influence sensitivity, it plays a secondary role compared to other factors, like how quickly decomposition spreads and the heat released during explosion. Liu et al. [449] also tested the relationship between ζ and impact sensitivity in 16 crystalline explosives and eight nitrogen-rich salts. The results confirmed a clear trend: as the phonon-vibron coupling increases, sensitivity tends to decrease.

Designing safer high-energy molecular crystals relies on accurate ways to predict sensitivity. One promising approach uses multi-phonon interactions. Bidault and Chaudhuri [450] put this method to the test with high-quality phonon calculations on 22 molecular crystals, using a physics-based criterion to define the phonon bath. The resulting shock sensitivity index was then compared to traditional impact sensitivity data. To keep things consistent, Bidault and Chaudhuri [450] focused on experiments with the same setup: a 2.5 kg hammer, grit, and 30–40 mg samples. The results were striking. The model predicted H_{50} values for single-molecule crystals with remarkable accuracy, even distinguishing between different polymorphs of HMX and CL-20. This success supports the idea that indirect vibrational up-pumping – where doorway modes interact with the phonon bath – plays a key role in sensitivity under these conditions. But there's a catch. While the method works well for single-molecule crystals, it struggles with cocrystals. The vibrational coupling between different molecules likely means we need a broader phonon bath to capture their behavior. Beyond theory, experimental factors matter too. Variations in sample density, granularity, and morphology can skew H_{50} measurements. To improve future models, Bidault and Chaudhuri [450] recommend standardizing and reporting these details alongside sensitivity data.

To understand how energetic materials respond to impact, Bao et al. [451] used phonon-upon transition theory to analyze the vibrational properties of seven different molecular crystals. Bao et al. [451] focused on "doorway modes", which they separated into two categories: low-energy phonon modes (collective lattice vibrations) and high-energy intramolecular vibrations. The key factor here is phonon-vibron coupling – the stronger this interaction, the faster energy transfers from the phonon modes into the molecules themselves. Using first-principles calculations, Bao et al. [451] quantified these energy transfer rates and compared them with experimental impact sensitivity data. The results revealed a clear trend: materials with faster energy transfer tend to be more impact-sensitive. This correlation suggests that phonon-vibron coupling plays a crucial role in determining how easily an energetic material can be triggered by impact.

6.3 Simple methods on the basis of molecular structure for neutral energetic compounds

In contrast to complex methods, simple empirical correlations have the advantages that neither complex quantum chemistry software, nor high speed computers need to be available for tedious computation. There are some simple relationships that relate impact sensitivities with measured and predicted molecular properties, e.g., the oxygen balance of the molecules [452, 453], molecular electronegativities [454, 455], vibrational states [456], parameters related to oxidation numbers [457], ^{15}N NMR chemical shifts and heat of fusion [457, 392, 458], as well as elemental composition and molecular structures [404, 406, 459–463]. Several simple correlations, which can be applied to different classes of energetic compounds, are reviewed in the following sections.

6.3.1 Oxygen balance correlations

Kamlet and Adolph [452, 453] found reasonable linear correlations between the oxygen balance and $\log H_{50}$ for some classes of high-energy molecules with similar decomposition mechanisms:

(1) Nitroaromatic:

$$\log H_{50} = 1.73 - 0.32 OB_{100} \tag{6.1}$$

(2) Nitroaromatic with α-CH linkage (e.g., TNT):

$$\log H_{50} = 1.33 - 0.26 OB_{100} \tag{6.2}$$

(3) Nitroaliphatic:

$$\log H_{50} = 1.74 - 0.28 OB_{100} \tag{6.3}$$

(4) Nitramine:

$$\log H_{50} = 1.37 - 0.17 OB_{100} \tag{6.4}$$

where $OB_{100} = 100(2d' - b' - 2a' - 2n'_{COO})$ in which d', b', a', and n'_{COO} are the number of oxygen, hydrogen, carbon, and carboxylate entities in the molecule, divided by the molecular weight of the explosive.

6.3.2 Elemental composition and molecular moieties

6.3.2.1 Polynitroaromatics (and benzofuroxans) and polynitroaromatics with α-CH and α-N–CH (e.g., tetryl) and nitramines

It was shown that the following simple equations are suitable for polynitroaromatics (and benzofuroxans) and polynitroaromatics with α-CH and α-N–CH (e.g., tetryl), as well as for nitramines [459]:

(1) Polynitroaromatics (and benzofuroxans):

$$\log H_{50} = 11.8a' + 61.72b' + 26.9c' + 11.5d'. \tag{6.5}$$

(2) Polynitroaromatics with α-CH and α-N–CH (e.g., tetryl) and nitramines:

$$\log H_{50} = 47.3a' + 23.5b' + 2.36c' - 1.11d'. \tag{6.6}$$

Example 6.1. Consider 2,3,4,5-Tetranitrotoluene (2,3,4,5-TetNT) with the following molecular structure:

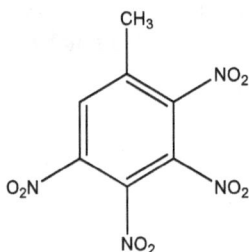

Since it is a nitroaromatic compound with a α-CH linkage, eq. (6.6) can be used, which gives

$$\log H_{50} = 47.3a' + 23.5b' + 2.36c' - 1.11d'$$
$$= (47.3(7) + 23.5(4) + 2.36(4) - 1.11(8))/272.13$$
$$= 1.565$$
$$H_{50} = 37 \text{ cm.}$$

The measured H_{50} for this compound is 15 cm [422].

6.2.2.2 Nitroaliphatics, nitroaliphatics containing other functional groups, and nitrate explosives

For nitroaliphatics, nitroaliphatics containing other functional groups, and nitrate explosives, the following correlation can be used to predict their impact sensitivity [461]:

$$\log H_{50} = 2.5 + 0.371[100(a' + b'/2 - d')]$$
$$- 0.485(100c') + 0.185n_{R-C(NO_2)_2-CH_2-},$$

(6.7)

where $n_{R-C(NO_2)_2-CH_2-}$ is the number of $R-C(NO_2)_2-CH_2-$ groups attached to the oxygen atom of carboxylate functional groups (R is an alkyl group).

Example 6.2. The use of eq. (6.7) for bis-(2,2-dinitropropyl) carbonate, with the following molecular structure gives

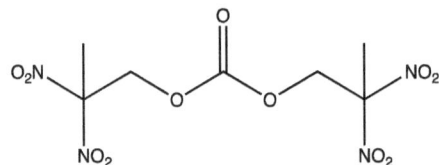

$$\log H_{50} = 2.5 + 0.371[100(a' + b'/2 - d')]$$
$$- 0.485(100c') + 0.185n_{R-C(NO_2)_2-CH_2-}$$
$$= 2.5 + 0.371[100(7 + 10/2 - 11)/326.17]$$
$$- 0.485(100 \times 4/326.17) + 0.185(2)$$
$$= 2.389$$
$$H_{50} = 228 \text{ cm.}$$

The measured H_{50} for this compound is 300 cm [464]. If eq. (6.3) is used, there is a large deviation between the predicted value and the experimental value of 121 cm (Dev = 179 cm).

6.3.2.3 Nitroheterocycles

For nitroheterocyclic energetic compounds, including nitropyridines, nitroimidazoles, nitropyrazoles, nitrofurazanes, nitrotriazoles, and nitropyrimidines, the following general equation can be used for various types of $C_aH_bN_cO_d$ nitro heterocycles [460]:

$$\log H_{50} = 46.29a' + 35.63b' - 7.700c' + 7.943d' + 44.42n'_{-CNC-} + 102.3n'_{-CNNC-}, \quad (6.8)$$

where n'_{-CNC-} and n'_{-CNNC-} are the number of –CNC– and –CNNC– moieties in the aromatic ring, divided by the molecular weight of the explosive.

Example 6.3 If eq. (6.8) is used for 4-methyl-3,5-dinitro-1,2,4-triazole with the following molecular structure,

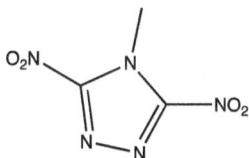

the value of H_{50} is calculated as follows:

$$\log H_{50} = 46.29a' + 35.63b' - 7.700c' + 7.943d' + 44.42n'_{-CNC-} + 102.3n'_{-CNNC-}$$
$$= (46.29(3) + 35.63(3) - 7.700(5) + 7.943(4)$$
$$+ 44.42(1) + 102.3(1))/173.09$$
$$= 2.229$$
$$H_{50} = 169 \text{ cm.}$$

The measured H_{50} for this compound is 155 cm [464]. It was found that the complex neural network [432] approach results in a larger deviation between the predicted value and experimental data, that is, 64 cm (Dev = 91 cm).

6.3.2.4 Polynitroheteroarenes

An improved correlation with respect to eq. (6.8) has been introduced to predict the impact sensitivity of different types of polynitroheteroarenes including nitropyridine, nitroimidazole, nitropyrazole, nitrofurazane, nitrooxadiazole, nitro-1,2,4-triazole, nitro-1,2,3-trazole, and nitropyrimidine explosives as [462]:

$$\log H_{50} = 52.13a' + 31.80b' + \frac{117.6 \sum SSP_i}{Mw},\tag{6.9}$$

where SSP_i are specific structural parameters that can decrease or increase impact sensitivity. The values of SSP_i are specified according to the molecular structures as follows:

(1) Amino derivatives as substituents in heteroarene: Amino derivatives (Ar–NH– or R–NH–) can decrease the impact sensitivity of some explosives [286]. For hetero-arenes containing tetrazole derivatives or three consecutive nitrogen atoms (e.g., 1,2,3-triazole derivatives) or the nitropyrimidine group attached to an aromatic ring via nitrogen (e.g., 1-picryl-2-picrylamino-1,2-dihydropyrimidine), the presence of amino groups has no notable effect, and the value of SSP_i is zero. For the presence of amino groups in nitropyridine, nitrofurazane (or nitrooxidazole), and nitro-1,2,4-triazole (or nitropyrimidine) explosives, the values of SSP_i are 0.5, 0.6, and 2.5 respectively. For other nitroheteroarenes which have only nitrogen as a heteroatom, SSP_i is 2.0 in the presence of amino derivatives.

(2) The attachment of an aromatic ring (e.g., picryl) to nitrogen and the presence of one nitro group in a specific position:

(a) The values of SSP_i are equal to 1.0 and 0.6 for the attachment of an aromatic ring to nitrogen in imidazole (or only in mononitro imidazol) and nitropyra-zole explosives, respectively.

(b) If only two aromatic substituents are attached to the heteroarene ring without further substituents, $SSP_i = 2$.

(c) If the polynitrophenyl group is attached to the nitrogen atom at the 4-position in 1,2,4-triazole explosives (e.g., 4-(2,4-dinitrobenzyl)-3,5–1,2,4-triazole), $SSP_i = 0.7$.

(d) The values of SSP_i are 0.6, 0.8 and 1.0 for the presence of one nitro group in positions of 2 in 3 (or 5) in nitropyrazole, nitroimidazole and in position 3 in nitro-1,2,4-triazole explosives, respectively. This condition is valid for nitroi-midazoles up to only disubstitueted nitroimidazole explosives.

As an illustrative example for this section, $\sum SSP_i$ is equal to 1.8 for 2-nitro-1-picryl-imidazole, whereas the $\sum SSP_i$ value is 1.0 for the isomer 4-nitro-1-picryl-imidazole.

Less sensitive materials could be designed if the "trigger linkage" could be identi-fied and avoided [465]. Ammonium salts are "unusually stable" because when an acid is converted to its ammonium salt. Thus, the ammonium salt will be less sensitive than the parent acid [464]. Since sensitivity of 1,2,4-triazoles are usually low, it appears that the ammonium counterion in ammonium 3,5-dinitro-1,2,4-triazolate provides no special insensitivity. For less sensitive derivatives of nitroimidazole, nitropyrazole, and nitro-1,2,4-triazole explosives, the insensitivity of the explosive is, in fact, a consequence of the chemistry preceding the rate determining step. Due to considerable charge delocalization through –N=N– and –C=C– double bonds, the insensitivity to impact may be accounted for in some isomers of polynitroheteroarenes, including nitroimidazole, nitropyrazole, and nitro-1,2,4-triazole explosives.

(3) Nitro-1,2,3-triazole explosives: For the attachment of an aromatic ring to a nitrogen atom in position 1, the value of SSP_i is equal to –1.0. As was mentioned in part (1), this condition is valid for compounds which are not amino derivatives. The sensitivity to impact and instability varies from isomer to isomer in nitro-1,2,3-triazole explosives. Since there are large differences in the impact sensitivities of 1-picryl-1,2,3-triazole compared to 2-picryl-1,2,3-triazole, and of 4-nitro-1-picryl-1,2,3-triazole compared to 4-nitro-1-picryl-1,2,3-triazole, it is possible that this is due to the facile loss of nitrogen in the 1-picryl isomers [466].

For the presence of another more active site to initiate decomposition, e.g., R–NO$_2$, the value of $\sum SSP_i$ is taken as zero.

Example 6.4. 5,5′-Dinitro-1H,1′H-3,3′-bi(1,2,4-triazole) has the following molecular structure:

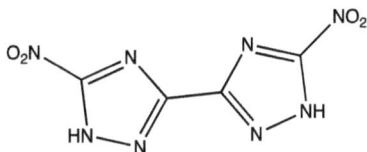

If eq. (6.9) is used for this compound, it gives

$$\log H_{50} = 52.13a' + 31.80b' + \frac{117.6 \sum SSP_i}{Mw}$$

$$= (52.13(4) + 31.80(2))/226.11 + \frac{117.6(2)}{226.11}$$

$$= 2.243$$

$$H_{50} = 175 \text{ cm}.$$

The measured H_{50} for this compound is 153 cm [464]. It was found that two complex neural networks, those of Cho et al. [432] and Keshavarz–Jafari [433], result in larger deviations between the predicted value and experimental data, that is, 73 cm (Dev = 80 cm) and 200 cm (Dev = 47 cm), respectively. The use of eq. (6.8) also results in a larger deviation, that is, 200 cm (Dev = 47 cm).

6.3.2.5 Nitroaromatics, benzofuroxans, nitroaromatics with α-CH, nitramines, nitroaliphatics, nitroaliphatics containing other functional groups, and nitrate energetic compounds

A simple correlation has been introduced to predict the impact sensitivity of nitroaromatics, benzofuroxans, nitroaromatics with α-CH, nitramines, and nitroaliphatics, as well as nitroaliphatics containing other functional groups and nitrate explosives as [463]

$$\log H_{50} = 48.81a' + 25.94b' + 13.73c' - 4.786d' + \frac{111.6DSSPH - 132.3ISSPH}{Mw}, \qquad (6.10)$$

where $DSSPH$ and $ISSPH$ are decreasing and increasing sensitivity structural parameters, respectively, which can be specified based on the molecular structures as follows.

(1) Prediction of $DSSPH$

(a) Nitroaromatics and bezofuroxanes: Since the presence of some special electron donating substituents which have an electron pair located on the atom which attaches to the aromatic ring (such as $-NH_2$ and $-OCH_3$), or the presence of double and triple bonds involving the carbon atom which is attached to an aromatic ring (e.g., $-C(=O)-$ and $-CN$) can decrease the sensitivity, the effects of these substituents can be predicted based on the molecular structure.

(i) $-NH_2$ group: $DSSPH$ equals 0.7, 1.2, and 1.7 for $n_{NH_2} = 1$, 2, and 3 per aromatic ring, respectively, if $n_{NO_2} \leq 3$ per aromatic ring.

(ii) Only $-OH$ groups: If $n_{NO_2} = n_{OH}$, then $DSSPH = 0.9$.

(iii) Benzofuroxans: $DSSPH = 0.6$.

(iv) $-OR$ and $-O^-$ groups attached to an aromatic ring: $DSSPH$ are 0.7 and 0.5, respectively. If an $-NH_2$ group is also attached to the same aromatic ring, the value of $DSSPH$ is 1.2.

(v) Double and triple bonds involving the carbon atom attached to the aromatic ring (e.g., $-C(=O)-$ and $-CN$): If $n_{NO_2} \geq 3$ in the aromatic ring, then $DSSPH = 1$. If the 2,2-dinitropropyl group is attached to the $-COO-$ functional group, $DSSPH = 2$.

(vi) Nitroaromatic explosives that contain one methyl group: For the presence of one phenyl or one $-OH$ group in the meta position with respect to the methyl group, $DSSPH = 0.8$.

(b) Nitramines: For nitramines which contain the $=C-N-NO_2$ group, $DSSPH = 0.45$.

(c) Nitroaliphatics: For nitroaliphatics containing the $-COO-$ functional group, the number of 2,2-dinitropropyl and nitroisobutyl groups attached to $-COO-$ can increase the value of $DSSPH$ since the value of $DSSPH$ equals the number of 2,2-dinitropropyl and nitroisobutyl groups.

(2) Prediction of *ISSPH*

    (a) Nitroaromatics

        (i) If an α-CH linkage is attached to a nitroaromatic ring, *ISSPH* = 0.5. It should be mentioned that condition (1) (a) (vi) cannot be used here.

        (ii) For those nitroaromatics containing azido or diazo functional group, *ISSPH* = 0.7.

    (b) Nitramines: *ISSPH* = 0.5. The presence of the molecular structure given in condition (1) (b) has the reverse effect.

Example 6.5. Consider ammonium nitrate with the following molecular structure:

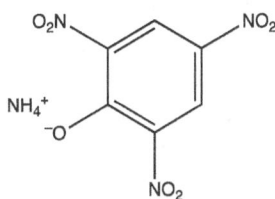

The use of eq. (6.9) gives

$$\log H_{50} = 48.81a' + 25.94b' + 13.73c' - 4.786d'$$

$$+ \frac{111.6 \, DSSPH - 132.3 \, ISSPH}{Mw}$$

$$= (48.81(6) + 25.94(6) + 13.73(4) - 4.786(7))/246.13$$

$$+ \frac{111.6(0.5) - 132.3(0)}{246.13}$$

$$= 2.136$$

$$H_{50} = 137 \text{ cm.}$$

The measured H_{50} for this explosive is 135 cm [464].

6.3.2.6 An improved, simple model for the prediction of the impact sensitivity of different classes of energetic compounds

An improved simple model, with respect to eq. (6.10) has been introduced to predict the impact sensitivity of nitropyridines, nitroimidazoles, nitropyrazoles, nitrofurazanes, nitrotriazoles, nitropyrimidines, polynitro arenes, benzofuroxans, polynitro arenes with α-CH, nitramines, nitroaliphatics, nitroaliphatics containing other functional groups, and nitrate energetic compounds as [404]:

$$(\log H_{50})_{core} = -0.584 + 61.62a' + 21.53b' + 27.96c' \tag{6.11}$$

$$\log H_{50} = (\log H_{50})_{core} + 84.47 \tfrac{F^+}{Mw} - 147.1 \tfrac{F^-}{Mw}. \tag{6.12}$$

The parameters F^+ and F^- in eq. (6.12) are two correcting functions, which are described in the following sections.

6.3.2.6.1 Molecular fragments affecting F^+

(1) $-NH_2$ groups and amino derivatives as substituents:
  (a) Polynitro heteroarenes: The values of F^+ are 0.4, 0.6 and 3.0 for the presence of amino derivatives (Ar–NH– or R–NH–) in nitropyridine, nitrofurazane (or nitrooxidazole), and nitro-1,2,4-triazole (or nitropyrimidine) explosives, respectively.
     (i)   For amino derivatives of other nitroheteroarenes which contain only nitrogen as heteroatoms, $F^+ = 2.3$.
     (ii) For heteroarenes which contain four nitrogen atoms (e.g., 1,2,3,4-tetrazole derivatives) or three nitrogens (e.g., 1,2,3-triazole derivatives) attached consecutively in one ring, or nitropyrimidine explosives in which an aromatic ring is attached to a nitrogen atom (e.g., 1-picryl-2-picrylamino-1,2-dihydropyrimidine), $F^+ = 0.0$
  (b) Polynitro arenes: For $n_{NO_2} \leq 3$ per aromatic ring, $F^+ = 0.7$, 1.6 and 2.2 for $n_{NH_2} = 1$, 2 and 3 per aromatic ring, respectively.
(2) Molecular fragments that increase insensitivity:
  (a) Polynitroheteroarenes: The attachment of an aromatic ring (e.g., picryl) to nitrogen, or if one nitro group is present in a specific position.
     (i)   The value of F^+ is equal to 1.1 and 0.6 for the attachment of an aromatic ring to nitrogen in imidazole (or only in mononitro imidazole) and nitropyrazole explosives, respectively.
     (ii) If only two aromatic substituents are present and attached to a heteroarene ring without further substituents, $F^+ = 2.8$.
     (iii) If the polynitrophenyl group is attached to the nitrogen atom at the 4-position in 1,2,4-triazole explosives (e.g., 4-(2,4-dinitrobenzyl)-3,5-1,2,4-triazole), $F^+ = 0.7$.
     (iv) The values of F^+ are 0.6, 0.9, and 1.0 if one nitro group is present in position 2 of nitroimidazoles, 3 (or 5) in nitropyrazoles, and 3 in nitro-1,2,4-triazole explosives, respectively. This situation is valid for nitroimidazoles up to only disubstituted nitroimidazole explosives. For the presence of one carbonyl group, or the attachment of two 5-nitro-1,2,4-triazole rings in nitro-1,2,4-triazole explosives, the value of F^+ is equal to 1.4.

(b) Polynitroarenes:
  (i) If only –OH groups are attached and if $n_{NO_2} = n_{OH}$, then $F^+ = 1.25$.
  (ii) If –OR and –O⁻ groups are attached to the aromatic ring, $F^+ = 0.7$ and 0.5, respectively. If an –NH$_2$ group is also attached to the aromatic ring, $F^+ = 1.2$.
  (iii) If the carbon attached to the aromatic ring participates in double or triple bonds (e.g., –C(=O)– and –CN), and $n_{NO_2} \geq 3$ for the aromatic ring, then $F^+ = 1.0$. The value of F^+ equals 2.0 for the attachment of the 2,2-dinitropropyl group directly to the –COO– functional group.
  (iv) For the attachment of one methyl group, if there is one phenyl or one OH group in the meta position with respect to the methyl group, $F^+ = 0.8$.
(c) Benzofuroxanes: $F^+ = 0.6$.
(d) Nitramines: For nitramines containing the group =C–N–NO$_2$, $F^+ = 0.7$.
(e) Nitroaliphatics: For nitroaliphatics containing the –COO– functional group, the value of F^+ is equal to the number of 2,2-dinitropropyl and nitroisobutyl groups attached to the –COO– group multiplied by 1.4. For 2,2-dinitropropanediol, $F^+ = 2.8$.

(3) Prediction of F^-:
(a) Nitro-1,2,3-triazole explosives: If an aromatic ring is attached to nitrogen in position 1, the value of F^- equals 1.0. This condition is valid for nonamino derivatives.
(b) Polynitro arenes:
  (i) The presence of an α-CH linkage attached to a carbocyclic nitroaromatic ring may increase the sensitivity, and therefore, $F^- = 0.5$. It should be mentioned that condition (2) (b) (iv) is an exception, that can decrease impact sensitivity.
  (ii) For those polynitro arenes that contain azido or diazo functional groups, $F^- = 0.7$.
(c) Nitramines: If the N–NO$_2$ functional group is present, $F^- = 0.5$. For nitramines that contain the =C–N–NO$_2$ group, condition (2) (d) does not apply.

The values of F^+ and F^- are summarized in Tables 6.1 and 6.2.

Table 6.1: Prediction of F^+.

Molecular moieties	Compound	F^+	Illustration	Exception
Presence of –NH$_2$ groups and amino derivatives as substituents	Polynitroheteroarenes	0.4	Nitropyridine	For central heteroarenes which contain four nitrogens (e.g., 1,2,3,4-terazole derivatives) and three nitrogens attached consecutively in one ring (e.g., 1,2,3-triazole derivatives), or nitropyrimidine explosives in which an aromatic ring is attached to nitrogen (e.g., 1-picryl-2-picrylamino-1,2-dihydropyrimidine), the presence of amino derivatives has no notable effect, and the value of F^+ can be taken as zero
		0.6	Nitrofurazane (or nitrooxidazole)	
		3.0	Nitro-1,2,4-triazole (or nitropyrimidine)	
		2.3	Other nitroheteroarenes which contain only nitrogen as hetero atoms	
	Polynitro arenes (if $n_{NO_2} \leq 3$ per each aromatic cycle)	0.7	$n_{NH_2} = 1$ per each aromatic cycle	
		1.6	$n_{NH_2} = 2$ per each aromatic cycle	
		2.2	$n_{NH_2} = 3$ per each aromatic cycle	
Molecular fragments which increase insensitivity	Nitroimidazole, nitropyrazole and nitro-1,2,4-triazole explosives	0.6	The attachment of an aromatic ring to nitrogen in nitropyrazole	
		1.1	The attachment of an aromatic ring to nitrogen in imidazole (or in mononitro imidazole)	
		2.8	If only two aromatic substituents are attached to a heteroarene ring without further substituents	
	1,2,4-Triazole explosives	0.7	If polynitrophenyl is attached to the nitrogen atom at the 4-position (e.g., 4-(2,4-dinitrobenzyl)-3,5-1,2,4-triazole)	–

Molecular fragments which increase insensitivity			
Nitroimidazole	0.6	Presence of one nitro group in position 2	This condition is valid for nitroimidazole and disubstituted nitroimidazole explosives
Nitropyrazole	0.9	One nitro group is present in position 3 (or 5)	
Nitro-1,2,4-triazole	1.0	One nitro group is present in position 3	One carbonyl group is present, or the direct attachment of only two 5-nitro-1,2,4-triazole rings, the value of F^+ is equal to 1.4
Polynitro arenes	1.25	Only one –OH group is attached: if $n_{NO_2} = n_{OH}$	–
	0.7	– –OR is present	If the –NH$_2$ group has been attached also to the aromatic ring simultaneously, the value of F^+ becomes 1.2
	0.5	– –O$^-$ is present	
	1.0	The carbon atom which is attached to the aromatic ring participates in double or triple bonds (e.g., –C (=O)– and –CN), and if $n_{NO_2} \geq 3$ in the aromatic ring	The value of F^+ is equal to 2.0 for the presence of 2,2-dinitropropyl attached to the –COO– functional group
	0.8	For the attachment of one methyl group, if there is one phenyl or one –OH group in the *meta* position with respect to the methyl group	
Benzofuroxanes	0.6		
Nitroaliphatics		For nitroaliphatics that contain the –COO– functional group, the value of F^+ is equal to the number of 2,2-dinitropropyl and nitroisobutyl groups attached to – COO– times 1.4	If the 2,2-dinitropropanediol group is present, $F^+ = 2.8$

Table 6.2: The value of F^- for specific molecular groups.

Compound	F^-	Illustration	Exception
Nitro-1,2,3-triazole	1.0	The attachment of an aromatic ring to nitrogen at position 1	This condition is valid in the absence of amino groups
Polynitro arenes	0.5	The presence of an α-CH linkage attached to a carbocyclic nitroaromatic ring	For the attachment of one methyl group, if there is one phenyl or one –OH group in the *meta* position with respect to the methyl group then $F^+ = 0.8$
	0.7	For those polynitro arenes that contain the azido or diazo functional group	
Nitramines	0.5		For nitramines that contain the group =C–N–NO_2, $F^+ = 0.7$

Example 6.6. For 2-picryl-1,2,3-triazole with the following molecular structure, the use of eqs. (6.11) and (6.12) gives

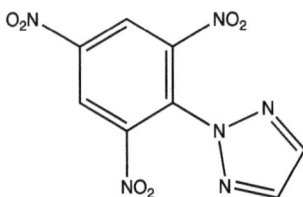

$$(\log H_{50})_{core} = -0.584 + 61.62a' + 21.53b' + 27.96c'$$
$$= -0.584 + (61.62(8) + 21.53(4) + 27.96(6))/280.15$$
$$= 2.082$$

$$\log H_{50} = (\log H_{50})_{core} + 84.47\frac{F^+}{Mw} - 147.1\frac{F^-}{Mw}$$
$$= 2.082 + 84.47\frac{0}{280.15} - 147.1\frac{0}{280.15}$$
$$= 2.082$$
$$H_{50} = 121 \text{ cm}.$$

The experimental value of H_{50} for this compound is 200 cm [464]. It was found that the complex neural network method of Keshavarz-Jafari [433] gives a value which showed a larger deviation between the predicted value and experimental data, i.e., 57 cm (Dev = 143 cm).

6.4 Impact sensitivity of quaternary ammonium-based energetic ionic liquids or salts

It was found that the impact sensitivity of quaternary ammonium-based energetic ionic liquids or salts can be correlated with the elemental composition of cations and anions, and two correcting functions as follows [467]:

$$IS_{IL}(J)$$

$$= 35.04$$

$$+ \frac{-1073a_{cat} + 728.9b_{cat} - 1761d_{cat} + 1032a_{ani} + 1061b_{ani} - 1261c_{ani} - 944.0d_{ani}}{Mw}$$

$$+ \frac{3663IS_{IL}^{+} - 4291IS_{IL}^{-}}{Mw}$$

$$(6.13)$$

where $IS_{IL}(J)$ is the impact sensitivity of a desired energetic of quaternary ammonium-based energetic ionic liquids or salts in J; a_{cat}, b_{cat}, and d_{cat} are the number of carbon, hydrogen, and oxygen atoms in the cation, respectively; a_{ani}, b_{ani}, c_{ani}, and d_{ani} are the number of carbon, hydrogen, nitrogen, and oxygen atoms in the anion, respectively; Mw is the molecular weight of the desired quaternary ammonium-based energetic ionic liquid or salt; IS_{IL}^{+} and IS_{IL}^{-} are two correction functions that depend on stabilizing or destabilizing structural parameters in cations or anions. The values of IS_{IL}^{+} and IS_{IL}^{-}, for the presence of specific cations or anions given in Table 6.3, are equal to 1.0.

Table 6.3: The contribution of IS_{IL}^{+} and IS_{IL}^{-} in ionic liquids or salts.

Cation	Anion
IS_{IL}^{+}	
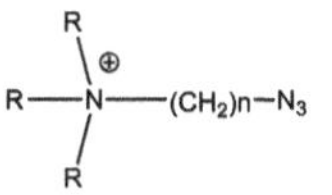 R—N⁺(R)(R)—(CH₂)n—N₃	$N(NO_2)^{2-}$
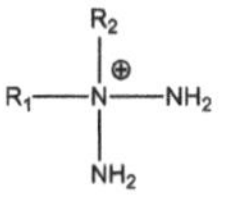 R₁—N⁺(R₂)(NH₂)—NH₂	$N(NO_2)(CN)^-$ or $C(NO_2)_2(CN)^-$ or NO_3^-
NH_4^+	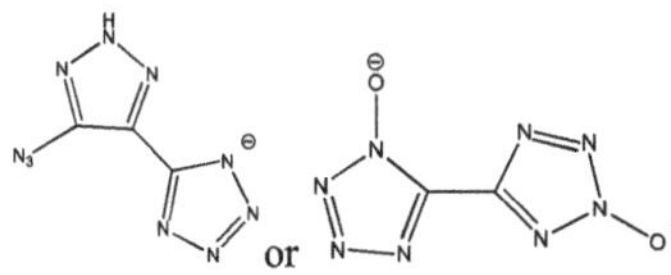

Table 6.3 (continued)

Cation	Anion
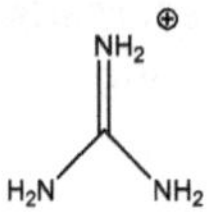	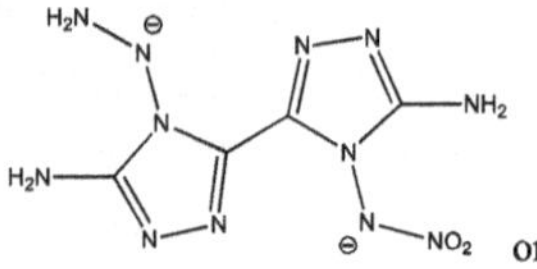 or 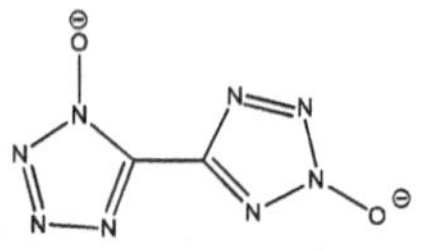
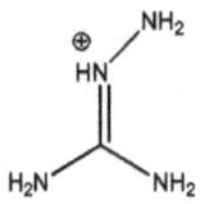	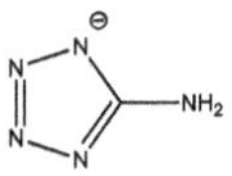
IS_{IL}^-	
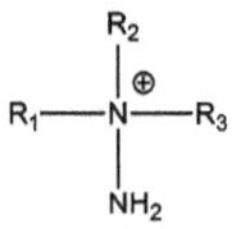 or 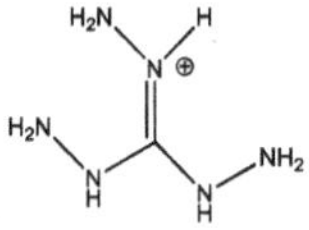	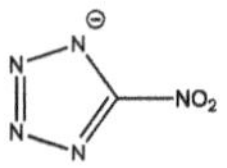 or 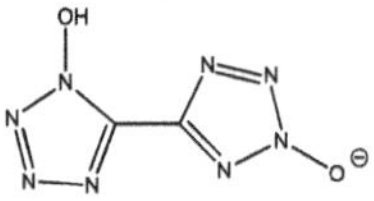
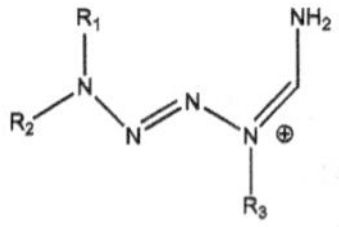	ClO_4^-
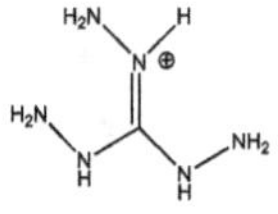	$N(NO_2)^{2-}$ or 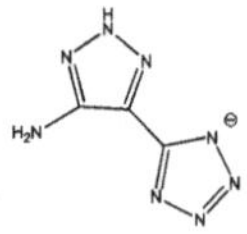
NH_4^+	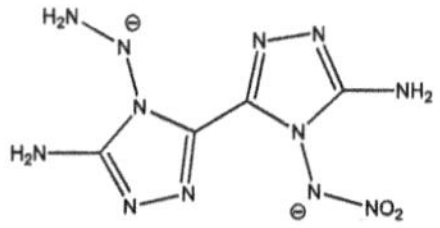
$N_2H_5^+$	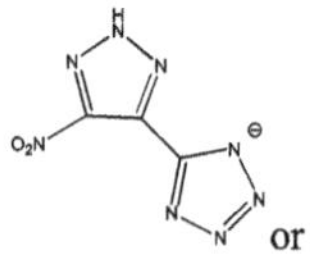 or 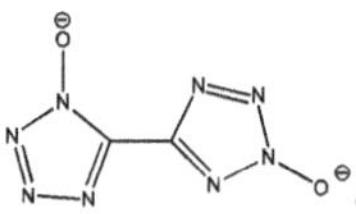 or 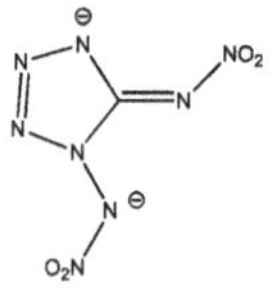

Example 6.7. The use of eq. (6.13) for the following compound

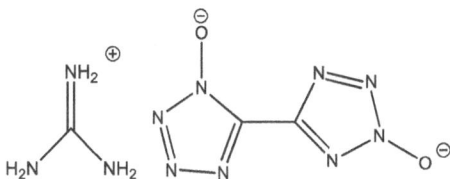

gives:

$$IS_{IL}(J)$$

$$= 35.04 + \frac{-1,073a_{cat} + 728.9b_{cat} - 1,761d_{cat} + 1,032a_{ani} + 1,061b_{ani} - 1,261c_{ani}}{Mw}$$

$$+ \frac{-944.0d_{ani} + 3,663IS_{IL}^{+} - 4,291IS_{IL}^{-}}{Mw}$$

$$= 35.04 + \frac{-1,073 \times 1 \times 2 + 728.9 \times 2 \times 3 - 1,761 \times 0 + 1,032 \times 2 + 1,061 \times 0 - 1,261 \times 8}{228}$$

$$+ \frac{-944.0 \times 2 + 3,663 \times 1 - 4,291 \times 0}{228} = 36.26 J.$$

The measured value of IS_{IL} for this compound is 40 J [468].

6.5 Quantitative impact sensitivity prediction: multiplicative incremental theory for energetic materials

Bondarchuk et al. [469] presented the first quantitative approach to predict impact sensitivity using a second-order incremental method based on explosive molecular structures. Their analysis revealed that H_{50} follows a multiplicative exponential relationship, where the exponents represent characteristic coefficients of structural features multiplied by their occurrence in each molecule. They developed this method using extensive experimental data from 450 diverse energetic materials, including nitro compounds, peroxides, nitrogen-rich salts, and heterocycles. The model was then validated on an independent set of 170 compounds, demonstrating significant correlation with experimental H_{50} values. The training set achieved $R^2 = 0.56$ (RMSE = 12.5 J), while the test set showed $R^2 = 0.63$ (RMSE = 18.8 J). The current implementation uses 53 structural parameters, but the framework allows for the expansion and refinement of these increments to improve prediction accuracy dynamically. Bondarchuk et al. [469] detailed the calculation algorithm with practical examples and validated the approach through machine learning techniques including genetic function approximation, multiple linear regression, and artificial neural networks. These analyses confirmed the robustness and informational value of the incremental theory of Bondarchuk et al. [469]. This innovative method advances the fundamental understanding of impact sensitivity while providing a practical tool so simple it could be implemented on a basic calculator. The approach bridges theoretical insights with real-world applicability for the safety assessment of energetic materials.

Bondarchuk et al. [469] formulated impact sensitivity through a multiplicative exponential relationship:

$$H_{50} = 2,000 \exp\left(\sum b_i x_i\right) \tag{6.14}$$

where x_i represents the count of specific structural features, and b_i are structure-specific parameters. This universal equation applies to all types of energetic materials when their corresponding b parameters are known.

Bondarchuk et al. [469] developed a comprehensive set of 53 structural parameters capable of describing an exceptionally diverse range of energetic compounds. Their dataset encompasses structurally varied materials including aliphatic and aromatic nitro/nitrato compounds, peroxides, hydroperoxides, nitramines, and nitrogen-rich heterocycles with both fused and isolated ring systems, including strained cycles. The complete list of these structural increments, along with their acronyms and parameters, appears in Table 6.4, with additional details provided later in the text.

Table 6.4: The structural components used in the impact sensitivity model by Bondarchuk et al. [469] are listed here, which shows each increment's acronym, description, and corresponding parameter value (b_i) for the multiplicative calculation.

No.	x_i	b_i	Discussion
1	NO_2	−0.88	A nitro group (NO_2) attached to a carbon-based ring structure, such as benzene or naphthalene
2	NH_2	0.65	An amino group (NH_2) connected to a carbon-based ring structure, like benzene or naphthalene
3	C(Ring)	−0.65	A carbon-based ring structure, like benzene or naphthalene
4	Alk-	0.35	An alkyl group attached to a conjugated ring system
5	(OH)OR	−0.15	A hydroxyl (−OH) or alkoxy (−OR) group bonded to a conjugated ring
6	N_3	−1.55	An azido group (−N_3) covalently attached to either a ring or a chain (cyclic or acyclic backbone)
7	C=N=N	−1.90	A diazo group (−N_2) covalently bonded to a cyclic structure
8	N–NO_2	−1.90	A nitramino group
9	O–O	−5.85	A peroxide bond
10	OOH	−2.00	A hydroperoxide bond
11	Azo, NH–NH	−2.00	An azo or hydrazo group

Table 6.4 (continued)

No.	x_i	b_i	Discussion
12	Het1	−1.10	A nitrogen-containing ring similar to pyridine, with no adjacent heteroatoms. For example: , and
13	Het1–NO₂	−1.15	A nitro group attached to Het1
14	Het2	−1.20	A *heterocyclic ring of the second kind* refers to: 1. Pyrrole-type systems: 5-membered rings with either: – One heteroatom (e.g., pyrrole, furan)*or* – Two adjacent heteroatoms (e.g., pyrazole, imidazole). 2. Pyridine-type systems: 6-membered rings with two adjacent heteroatoms (e.g., pyridazine, pyrimidine *only if N atoms are at the 1,2-positions*). For example: and
15	Het2–NO₂	−0.80	A nitro group attached to Het2
16	Salt	0.05	Is the compound a salt? This applies to all salts unless they are specifically listed elsewhere. Answer with '1' for 'yes' or '0' for 'no'
17	Het3	−2.80	A heterocyclic ring of the third kind refers to any conjugated ring that contains three or four heteroatoms in a consecutive sequence. For example: , and
18	Het3–NO₂	−1.95	A nitro group attached to Het3
19	C=O	−0.30	A carbonyl group connected to either cyclic (ring-shaped) or acyclic (non-cyclic) structures
20	N–OH	−0.95	A hydroxyl or $-O^-$ group attached to nitrogen atom
21	Het-NH₂	0.05	An amino group attached to any heterocyclic ring
22	NH₄⁺	0.35	Is the compound an ammonium salt? This is similar to the general 'Salt' category but specifically applies to ammonium salts
23	N₂H₅⁺	−0.05	Does this compound qualify as a hydrazinium salt? (Think of this as the 'Salt' category, but narrowed down specifically to hydrazinium salts.)

Table 6.4 (continued)

No.	x_i	b_i	Discussion
24	NH_3OH^+	−0.20	Is this compound a hydroxylammonium salt? (This is a specialized version of the general 'Salt' category, specifically tracking hydroxylammonium salts.)
25	G^+	1.50	Is this compound a guanidinium salt? (This is a specialized sub-category of 'Salt' that specifically tracks guanidinium salts – for example, An⁻.)
26	$G–NH_2^+$	0.65	Is this compound an aminoguanidinium salt? (This is a specialized category under 'Salt' that specifically tracks aminoguanidinium salts – like An⁻.)
27	$G–(NH_2)_2^+$	0.06	Is this compound a diaminoguanidinium salt? (This is a specialized subcategory of 'Salt' specifically for diaminoguanidinium compounds – examples include An⁻.)
28	$G–(NH_2)_3^+$	0.05	Triaminoguanidinium salt? (This counts as a 'Salt' entry, but specifically for triaminoguanidinium compounds – like An⁻.)
29	H_2O	0.65	Does the compound contain water of crystallization? (One water molecule per compound molecule)
30	$N–NH_2$	−0.90	An amino group attached to a nitrogen atom
31	$N(NO_2)_2^-$	−2.30	Is this compound a dinitramide salt? (This is a specialized category under 'Salts' that specifically identifies dinitramide compounds.)
32	≥5N	−1.25	Does the compound contain a chain of five or more nitrogen atoms in a row? (Answer with 1 for yes, 0 for no)

Table 6.4 (continued)

No.	x_i	b_i	Discussion
33	Stabil	1.50	This rule applies in three situations: When p-π conjugation appears after the trigger bond breaks (Case 1) When the trigger bond is already part of a π-conjugated system (Case 2) When this highly stable structural pattern is present (Case 3): 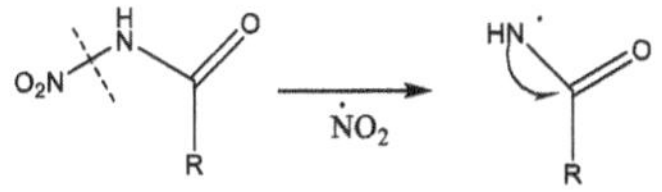 Case 1 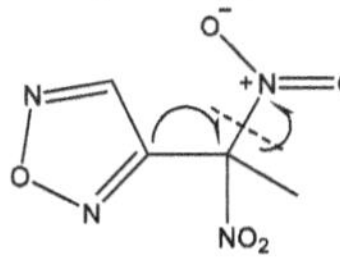 Case 2 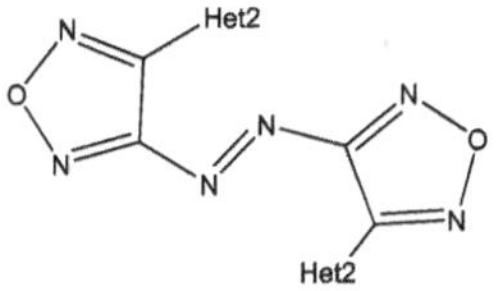 Case 3
34	EWG(ring)	−0.88	An electron-withdrawing group that doesn't participate in resonance effects. When the trigger bond is in an acyclic (non-ring) part of the molecule, this group can be any ring structure. Alternatively, it could be a non-conjugated chain substituent. For example: 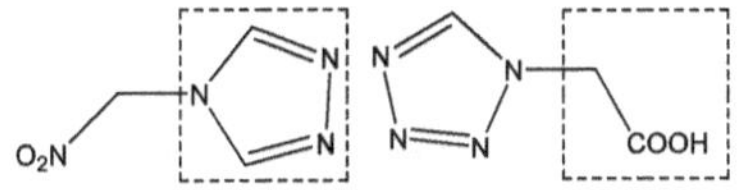
35	$CONH_2$	1.50	An amide group (–CONH–) or a structure where an amino group (–NH₂) sits right next to a carbonyl group (C=O)
36	NO_3^-, ClO_4^-	−0.75	Is this compound a nitrate or perchlorate salt? (This is a specialized category under 'Salts' that specifically identifies nitrate and perchlorate compounds.)
37	N_3^-	−1.75	Is this an azide salt? (This counts as a specialized type of 'Salt' – specifically for compounds containing the azide group.)
38	Het3N–Ph	0.85	This structural pattern significantly boosts stability. For example: 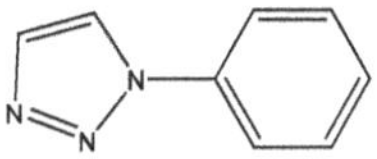
39	$G-CONH_2^+$	0.95	Is this a carbonylguanidinium salt? (A specialized type of salt where the guanidinium group is bonded to a carbonyl.)
40	NO_2_0	−1.21	The reactive NO_2 group and its symmetrical counterparts, all attached to the same carbon atom in a non-cyclic molecular framework

Table 6.4 (continued)

No.	x_i	b_i	Discussion
41	N-NO$_2$_0	−3.51	The reactive NO$_2$ group and its matching counterparts, all attached to the same nitrogen atom in an open-chain structure
42	O-NO$_2$_0	−3.21	The reactive NO$_2$ group connected to an oxygen atom in a chain-like structure (what chemists call an aliphatic nitrato group)
43	EWG_s_1	−0.37	An electron-withdrawing group located along a flexible carbon chain, positioned 1 to 10 bonds away from the reactive center
44	EWG_s_2	−0.30	
45	EWG_s_3	−0.27	
46	EWG_s_4	−0.25	
47	EWG_s_5	−0.17	
48	EWG_s_6	−0.15	
49	EWG_s_7	−0.11	
50	EWG_s_8	−0.08	
51	EWG_s_9	−0.05	
52	EWG_s_10	−0.03	
53	EDG_s_1	0.28	An electron-pushing group attached right next to the reactive bond in a carbon chain

Example 6.8. To demonstrate the model's application, consider pentanitroaniline as an example.

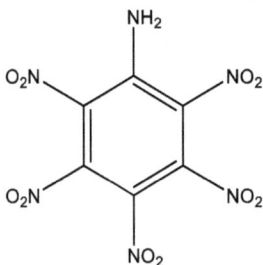

This compound contains three structural components: a benzene ring (1 occurrence), NO$_2$ groups (5 occurrences), and an NH$_2$ group (1 occurrence). Plugging the corresponding parameters from Table 6.3 into eq. (6.14):

$$H_{50} = 2{,}000 \exp\left(\sum b_i x_i\right)$$

$$H_{50} = 2{,}000 \exp(-0.65 \times 1 - 0.88 \times 5 + 0.65 \times 1) = 24.55 \, \text{cm}$$

This prediction closely matches the experimental value of 22 cm [469], demonstrating the model's practical utility.

When you smack or drop an energetic material, the physical shock doesn't just vanish – it gets converted into intense molecular vibrations through a process called phonon-vibron coupling [470, 471]. Think of it like a domino effect: First, low-frequency vibrations excite molecular motions, then this energy transfers to stronger valence vibrations – as these vibrations intensify, they eventually break the molecule's weakest bonds (called trigger bonds), creating free radicals that initiate explosive decomposition.

While the trigger bond concept is well-established [437, 472, 473], Bondarchuk et al. [469] introduced some key innovations. Bondarchuk et al. [469] represented molecules as two parts: a breaking fragment (detached group) and the remaining structure (residue). This makes impact sensitivity prediction deceptively simple in theory but complex in practice. Since trigger bond strength depends on electron density, electron-withdrawing groups (EWGs) increase sensitivity, while electron-donating groups (EDGs) decrease it. If we could precisely measure the residue's electronic effects, we could accurately predict sensitivity – crystal effects might cause minor deviations, but they're secondary [470, 471]. The complexity comes when we try to quantify these electronic effects. This seemingly obvious approach fails in practice. For example, calculations show heterocyclic azides require more energy to break N_3 than aniline needs to lose its amino group – yet aniline doesn't explode, while azides do [469]. For NH_4N_3, the problem worsens – there's no logical fragment to detach. Even dividing calculated triplet-state energies for two-bond cleavage doesn't work here [437, 472, 473].

This leaves us with one viable approach: semiquantitatively analyzing the residue's electronic effects through its structural components. The key factors are the immediate electron shifts inductive (I-effect) and mesomeric or resonance effect (M-effect) as:
- Inductive effect (I-effect): Electron shifts through σ-bonds (localized, distance-dependent).
- Mesomeric effect (M-effect): Electron delocalization through π-bonds or lone pairs (stronger, requires conjugation).

This method balances theoretical rigor with practical applicability, avoiding the pitfalls of direct bond energy calculations while capturing the essential chemistry governing sensitivity. The I-effect governs how electron density shifts through single bonds in molecules. But when we apply this to energetic materials, we see some fascinating patterns. Electron-donating groups (+I effect) actually make compounds less sensitive to impact (giving higher H_{50} values), while electron-withdrawing groups do the opposite. The M-effect follows similar principles but works through conjugated π-systems – a crucial distinction that explains why some explosives behave differently than we might initially expect.

For most explosives, the action centers on $R–NO_2$ bonds. Through careful studies, chemists have identified several reliable patterns:

- Symmetric molecules have equivalent trigger bonds:

- Central bonds weaken when pulled from both sides:

– C(sp^3)-NO$_2$ bonds break more easily than C(sp^2)-NO$_2$

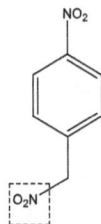

– More nitro groups on a carbon make its bonds weaker:

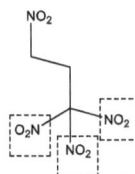

– N–NO$_2$ bonds are more fragile than C–NO$_2$

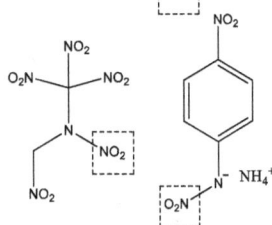

When molecules get really complex, modern computational methods come to the rescue. Even relatively simple quantum calculations (like AM1 or PM3) can reliably identify trigger bonds [469]. There's a clear hierarchy of bond weakness that experienced chemists keep in mind.

The difference in range between I-effects (short) and M-effects (long) becomes particularly important in aliphatic explosives.

Example 6.9 A striking case study comes from bis(2,2,2-trinitroethyl) 4,4,6,6,8,8-hexanitroundecanedioate:

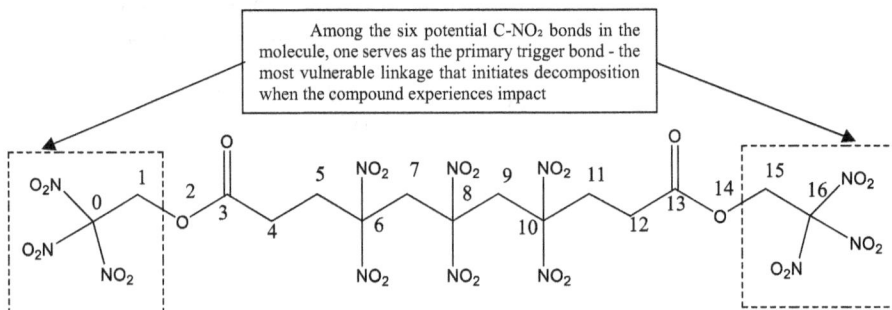

Among the six potential C-NO$_2$ bonds in the molecule, one serves as the primary trigger bond - the most vulnerable linkage that initiates decomposition when the compound experiences impact

With its 12 NO$_2$ groups, a naive calculation would predict extreme sensitivity, but reality tells a different story because:

1. Only terminal NO$_2$ groups serve as trigger sites
2. Other NO$_2$ groups contribute differently based on their positions
3. The I-effect's rapid attenuation with distance must be considered

The remaining NO_2 groups in the molecule act solely through their electron-withdrawing inductive effect (−I effect). However, since the inductive effect weakens rapidly with distance, each symmetrically distinct NO_2 group actually contributes differently to the overall sensitivity.

Breaking this down systematically (referring to Table 6.3):

- Three NO_2 groups at the primary position (NO₂_0)
- One oxygen atom two bonds away (EWG_s_2)
- A carbonyl group three bonds away (EWG_s_3)
- Two NO_2 groups six bonds away (EWG_s_6)
- Two NO_2 groups eight bonds away (EWG_s_8)
- Two NO_2 groups 10 bonds away (EWG_s_10)

We can ignore the two oxygen atoms and three additional nitro groups at positions 14 and 16 because they're simply too far away to meaningfully influence the trigger bond.

Important note about our numbering system: The numbers here represent the number of single bonds between each group and the trigger site – this is different from standard IUPAC nomenclature.

The key distinction lies in how we treat these groups:

- The primary NO_2 groups (at position 0) have special parameters (40–42 in Table 6.3) because they're directly involved in the X–NO_2 trigger bonds.
- All other electron-withdrawing groups, regardless of type, are treated similarly based on their distance (parameters 43–52).

Putting this all together, we calculate the impact sensitivity (H_{50}) as:

$$H_{50} = 2,000 \exp\left(\sum b_i x_i\right)$$

$$H_{50} = 2000 \exp((-1.21 \times 3)\left[\text{for } NO_0^2\right] + (-0.30 \times 1)[\text{for EWG\_s\_2}] + (-0.27 \times 1)[\text{for EWG\_s\_3}]$$

$$+ (-0.15 \times 2)[\text{for EWG\_s\_6}] + \exp(-0.08 \times 2)[\text{for EWG\_s\_8}](-0.03 \times 2)[\text{for EWG\_s\_10}]) = 18 cm$$

This step-by-step calculation shows how each group's electronic influence, weighted by its distance from the trigger site, combines to determine the compound's overall sensitivity. Experimental impact sensitivity of bis(2,2,2-trinitroethyl) 4,4,6,6,8,8-hexanitroundecanedioate is 32 cm [469].

6.5.1 Aromatic systems exhibit distinct behavior

Aromatic systems demonstrate fundamentally different characteristics compared to aliphatic compounds. In benzene derivatives, conjugation ensures all NO_2 groups contribute equally to the molecule's electronic properties. Although theoretical considerations might suggest the possibility of a benzene ring fully substituted with six NO_2

groups, practical limitations including ring strain, render this configuration improbable. Such a hypothetical compound would undoubtedly exhibit extreme sensitivity.

Example 6.10. The application of structural increments from Table 6.3 becomes clear when examining specific molecules like CL-12:

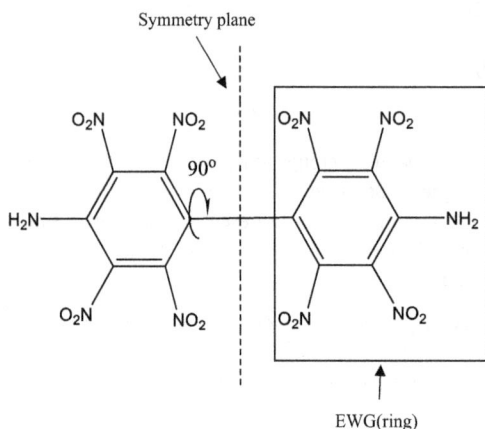

Initial structural analysis requires identification of molecular symmetry, which frequently simplifies subsequent calculations. CL-12 presents an interesting case study. A mirror plane in its structure enables treatment similar to tetranitroaniline, though with one critical distinction: the second aromatic ring functions as a fifth substituent. This configuration reveals an important structural feature – the near-perpendicular orientation of the two rings caused by steric hindrance prevents π-conjugation. Consequently:
1. The mesomeric (M) effect becomes negligible
2. The second ring can be classified as an electron-withdrawing group (EWG)
3. Specific assignment as an EWG(ring) increment proves appropriate

The impact sensitivity calculation for CL-12 proceeds as follows:

$$H_{50} = 2,000 \exp\left(\sum b_i x_i\right)$$

$$H_{50} = 2,000 \ \exp((-0.65 \times 1)[\text{amino group contribution}] + (-0.88 \times 4)[\text{four nitro groups}]$$

$$+ (0.65 \times 1)[\text{correction factor}] + (-0.88 \times 1)[\text{EWG(ring) contribution}]) = 24.55 \text{ cm}.$$

Comparison with experimental data shows excellent agreement:
- Calculated value: 24.55 cm
- Experimental measurement: 20 cm [469]
- Pentanitroaniline reference: 22 cm [469]

This strong correlation between calculated and experimental values confirms the validity of the analytical approach. Notably, CL-12 demonstrates behavior remarkably similar to pentanitroaniline when structural features are properly accounted for in the analysis.

6.5.2 Understanding symmetry in fused ring systems

Even when molecules share the same mirror symmetry, fused polycyclic systems like naphthalene require special treatment. Take 1,4,5,8-tetranitronaphthalene – we can't simply analyze it as two separate 1,4-dinitrobenzene units:

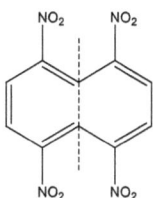

There's an important conversion factor to remember: a fused bicyclic system counts as about 1.5 monocyclic rings (3:2 ratio).

This leads to some interesting math:

- For the full molecule: 4 NO_2 groups ÷ 1.5 = 2.67 equivalent groups
- Analyzing one ring: 2 NO_2 groups ÷ 0.75 = same 2.67 value

Here's the key insight: symmetry in this context means equivalent numbers of substituents on each side, not necessarily identical positions. That's why the two isomers shown in 1,4,5,8-tetranitronaphthalene and 1,4,5,7-tetranitronaphthalene are treated the same way in this analysis:

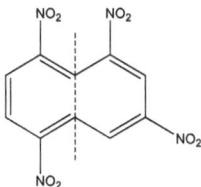

This principle applies broadly across molecular structures, not just to fused rings.

6.5.3 Identifying trigger bonds in symmetric structures

After accounting for symmetry, the next crucial step is locating the trigger bond. There are two main scenarios:

1. Conjugated ring attachment: The trigger bond connects directly to any conjugated ring system:

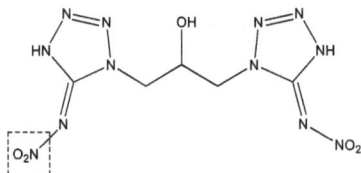

2. Non-conjugated attachment: The trigger bond is either:

    –   Part of an aliphatic chain:

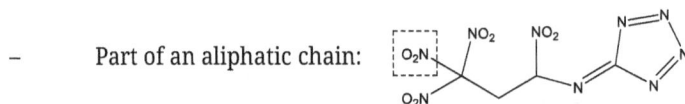

    –   In a saturated ring structure:

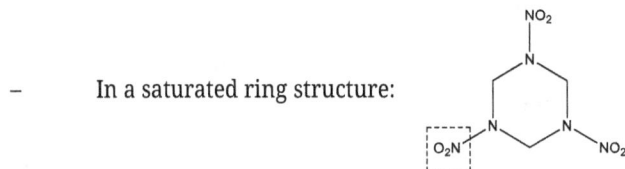

Once the trigger bond is identified, we analyze the remaining molecular framework (the "residue") using the same structural increment approach described previously. This systematic method ensures consistent evaluation of molecular sensitivity across different structural types.

6.5.4 Handling complex cases: when structures defy simple analysis

The trickiest scenarios arise when either:
1. The trigger bond can't be clearly isolated, or
2. The structure contains multiple independent segments, each contributing its own set of increments

This complexity frequently appears in heterocyclic salts with fused ring systems – though it certainly isn't limited to them. In such cases, we must analyze the entire structure holistically.

6.5.5 Breaking down the approach

Let's examine the salts in Figures 6.1 and 6.2 to understand how this works. The first step is normalizing all increments – essentially putting everything on a common scale. Here's the key: we can choose whether to account for molecular symmetry first, depending on which method proves more straightforward. Either path should lead us to the same final result.

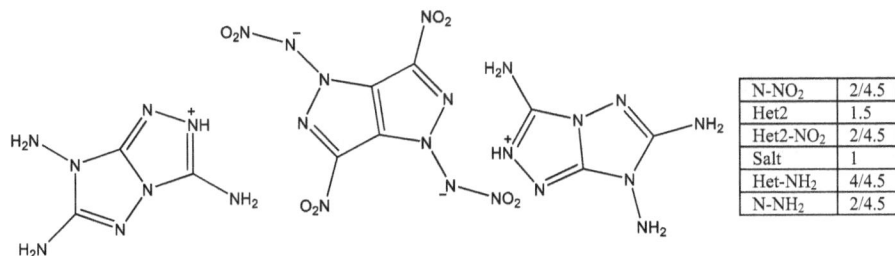

N-NO$_2$	2/4.5
Het2	1.5
Het2-NO$_2$	2/4.5
Salt	1
Het-NH$_2$	4/4.5
N-NH$_2$	2/4.5

Case Study 1:

Symmetry plane

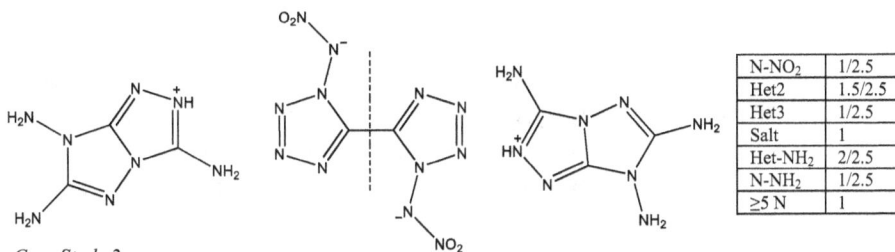

N-NO$_2$	1/2.5
Het2	1.5/2.5
Het3	1/2.5
Salt	1
Het-NH$_2$	2/2.5
N-NH$_2$	1/2.5
$\geq$5 N	1

Case Study 2:

Figure 6.1: Normalization of salt increments and molecular symmetry considerations.

Case study 1:
- Contains three identical fused bicycles (Het2 type)
- Symmetry consideration isn't the most efficient path here
- Normalization calculation:
  - Het2 increments: 4.5 total ÷ 3 units = 1.5 per unit
  - All other increments (except Salt) get normalized to 3 × 1.5 = 4.5
  - The Salt increment is binary (0 or 1) and doesn't scale
- Final normalized values appear in the accompanying table

Case study 2:
- Features different types of heterocycles
- Symmetry reduction works better here
- Breakdown:
  - One fused bicycle (Het2) = 1.5 units
  - One single cycle (Het3) = 1 unit
  - Total normalization factor: 2.5
- All increments (except salt and ≥ 5N groups) normalize relative to 2.5

This systematic approach ensures we don't miss critical contributions from any part of complex structures. The examples demonstrate how flexibility in method selection – whether using symmetry or not – leads to consistent results when properly applied.

6.5.6 When the rules don't apply: unexplained cases in impact sensitivity

While Bondarchuk et al.'s [469] approach successfully predicts impact sensitivity for most compounds, we've encountered some intriguing exceptions – materials that stubbornly refuse to follow the established rules. Figure 6.2 showcases these molecular rebels, whose behavior challenges Bondarchuk et al. [469] current understanding.

6.5.7 The puzzling case of compounds (I)–(IV)

These compounds feature a picryl group connected to a Het2 heterocycle, similar to compounds 484–486. Initial attempts to explain this pattern using a special Het3N-Ph increment fell short – reality turned out to be more complicated. The data show consistent anomalies across related systems, hinting that more than just additional empirical parameters may be needed.

The numbers tell a curious story:

- Compound (I): Experimental H_{50} = 312 cm (calculated predictions were half this value)
- Compound (II): 320 cm (a similar discrepancy)
- The same pattern holds for its pyrimidine variant (314 cm)

This consistency might suggest an unusually stable molecular configuration requiring its own descriptor. But then compound (III) disrupts the theory – adding just one NO_2 group causes sensitivity to plummet to 46 cm, a dramatic shift current methods can't fully explain.

6.5.8 Nitrogen's subtle influence: compounds (II) vs. (V)

These nearly identical compounds differ only in the position of one nitrogen atom (creating Het2 vs. Het3 configurations). Since neither middle ring serves as the trigger site – with multiple NO_2 groups offering alternative trigger bonds – minimal sensitivity differences might be expected. Yet, the observed variations highlight how subtle structural changes can have outsized effects.

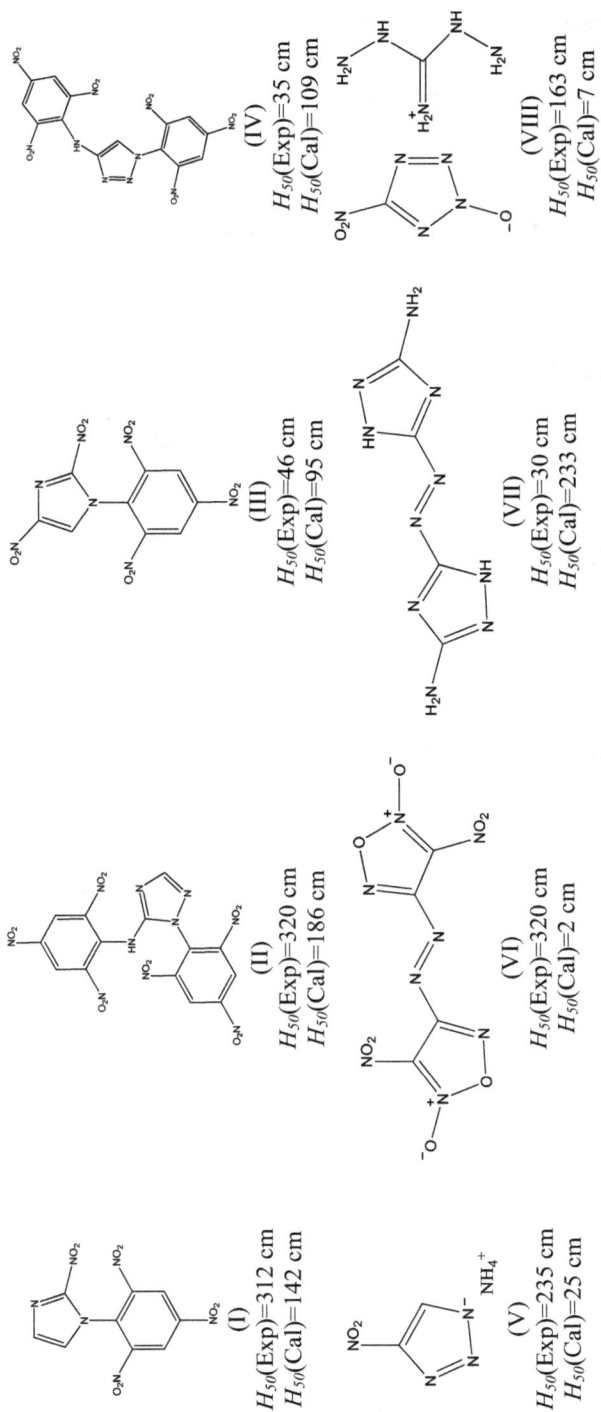

(I)
$H_{50}(Exp)=312$ cm
$H_{50}(Cal)=142$ cm

(II)
$H_{50}(Exp)=320$ cm
$H_{50}(Cal)=186$ cm

(III)
$H_{50}(Exp)=46$ cm
$H_{50}(Cal)=95$ cm

(IV)
$H_{50}(Exp)=35$ cm
$H_{50}(Cal)=109$ cm

(V)
$H_{50}(Exp)=235$ cm
$H_{50}(Cal)=25$ cm

(VI)
$H_{50}(Exp)=320$ cm
$H_{50}(Cal)=2$ cm

(VII)
$H_{50}(Exp)=30$ cm
$H_{50}(Cal)=233$ cm

(VIII)
$H_{50}(Exp)=163$ cm
$H_{50}(Cal)=7$ cm

Figure 6.2: Unexplained exceptions to Bondarchuk's sensitivity rules.

6.5.9 The azo-linked surprise: compounds (VI)–(VII)

At first glance, an experienced chemist might predict:
- Compound (VI): Sensitive
- Compound (VII): Insensitive

But the experimental data completely flip these expectations, proving once again that molecular behavior can defy intuition.

6.5.10 The curious case of salt (VIII)

This molecule combines multiple "sensitive" features (Het3, Het3–NO_2, and N–OH groups) with a diaminoguanidinium cation (a known stabilizer, $b = 0.06$). It exhibits unexpectedly low sensitivity that current calculations struggle to capture. The presence of N^+–O^- coordination bonds in both (VI) and (VIII) suggests stabilization mechanisms that haven't yet been fully accounted for.

6.5.11 Looking ahead

While the current method successfully models over 620 energetic materials, these exceptions reveal important gaps. Simply adding more specialized increments would only patch the problem. Digging deeper to uncover fundamental structural relationships could lead to more universal predictive tools. These puzzling cases aren't failures – they're exciting opportunities to refine the understanding of molecular sensitivity. Each anomaly points toward new chemical insights waiting to be discovered.

6.6 Summary

This chapter demonstrates several simple empirical methods for the prediction of the impact sensitivity of important classes of energetic compounds. Equations (6.5), (6.6), and (6.7) are suitable correlations for predicting the impact sensitivity of
(1) Polynitroaromatics (and benzofuroxans),
(2) Polynitroaromatics with α-CH and α-N–CH linkages (e.g., tetryl) and nitramines,
(3) Nitroaliphatics, nitroaliphatics containing other functional groups, and nitrate explosives.

For nitroheterocyclic energetic compounds containing nitropyridines, nitroimidazoles, nitropyrazoles, nitrofurazanes, nitrotriazoles, or nitropyrimidines, eq. (6.8) is a simple method for estimating the impact sensitivity of these classes of nitro heterocycles. Equation (6.9) is an improved correlation of equation (6.8), which can be used not only for the previously mentioned classes of polynitroheteroarenes, but also for two further important classes of energetic compounds, namely nitro-1,2,4-triazole and nitro-1,2,3-trazole derivatives. Equation (6.10) is an extended correlation for the prediction of the impact sensitivity of nitroaromatics, benzofuroxans, nitroaromatics with α-CH linkages, nitramines, and nitroaliphatics, as well as nitroaliphatics containing other functional groups and nitrate explosives. Equations (6.11) and (6.12) introduce a general correlation that can be used for a wide range of the above-mentioned classes of energetic compounds, including: nitropyridines, nitroimidazoles, nitropyrazoles, nitrofurazanes, nitrotriazoles, nitropyrimidines, polynitro arenes, benzofuroxans, polynitro arenes with α-CH linkages, nitramines, nitroaliphatics, nitroaliphatics containing other functional groups, and nitrate energetic compounds. Equation (6.13) provides a simple correlation for the prediction of the impact sensitivity of quaternary ammonium-based energetic ionic liquids or salts. Bondarchuk et al. [469] developed eq. (6.14) with 53 structural features (e.g., nitro groups, rings) and x_i representing their counts, validated on 620 compounds with $R^2 \sim 0.6$. The model's simplicity – calculable on a basic calculator – bridges theory and practical safety screening, though outliers reveal unexplored chemistry.

7 Electric spark sensitivity

The electrostatic or electric spark sensitivity (E_{ES}) of an energetic compound is an important aspect for estimating its safety in an electrostatic discharge (ESD) environment – and could be helpful in reducing accidents. It can be defined as the degree of sensitivity to an ESD. It represents the ease with which an explosion can be initiated by an electrostatic spark.

7.1 Measurement of electric spark sensitivity

Electric spark sensitivity is determined by subjecting an energetic compound to a high-voltage discharge from a capacitor. The required energy is calculated from the known capacitance C (in F) of the circuit and voltage U (in V) at the condenser as follows:

$$E_{ES} = 0.5CU^2. \tag{7.1}$$

Since the value of E_{ES} depends on the configuration of the electrodes and structure of the circuit, it can be expected that various results should be obtained by different test specifications of the electrode energy used by different authors [474]. Some of the important parameters for the determination of the E_{ES} include:

(1) the surface area of the tip of the electrode can affect the energy density of the spark;

(2) the structure of the electrical circuit will affect the shape and duration of the electrical pulse, which in turn influences the rate and duration of the delivery of energy to the sample; and

(3) confinement of the sample (or lack thereof) can have a dramatic effect on the measured electrostatic sensitivity because of the tendency of some samples to form dust clouds when unconfined.

To determine the values of the electric spark sensitivity of some secondary explosives, Zeman and coworkers [474, 475] used an instrument marked as RDAD, which was constructed in the R&D Department of Zbrojovka Indet, Inc., Vsetín, Czech Republic [474] for this purpose. Furthermore, they have measured the electric spark sensitivity for a large set of polynitro-secondary explosives. The capacitance of the capacitor in the RDAD instrument is chosen so as to allow measurements in the voltage range of 8 to 14 kV. If the initiation of the explosive is successful, the next measurement is carried out at a voltage which is 0.2 kV lower. Whereas, if initiation was unsuccessful, the voltage is increased by the same value (up and down method). Fortunately, there is a linear relationship between the measured data obtained using the RDAD instrument with some of those obtained by other experimental methods used to determine the

https://doi.org/10.1515/9783112206768-007

electric spark sensitivity in other recognized laboratories [474]. Since the RDAD instrument is not suitable for determining the sensitivity of primers and pyrotechnics [63], another instrument marked as ESZ KTTV has also been developed, with financial support from Czech Ministry of Industry and Commerce [475]. The ESZ KTTV instrument is suitable for both primary and secondary explosives. Details of the apparatus and procedures for the RDAD and ESZ KTTV instruments were described elsewhere [474–476].

7.2 Different methods for predicting electric spark sensitivity

For some classes of the secondary explosives, there are some correlations between the electric spark sensitivity and some characteristics, such as the detonation velocity and the Piloyan activation energy of decomposition which is obtained from differential thermal analysis [477–481]. Zeman et al. [477, 481, 482] have indicated that there is a linear relationship between the electric spark sensitivity and the square of the detonation velocity for some specific categories of explosives. Zeman and Liu [483] have shown that the electric spark sensitivities of some nitramines are directly proportional to the crystal lattice-free volumes but there were several nitramines with the opposite course of this relationship. For 28 polynitroarenes and their derivatives, Zeman [484] demonstrated some relationships between their impact and electric spark sensitivities with the volume heats of their explosion and their enthalpies of formation. Tan et al. [485] correlated the electrostatic spark energy with four parameters – the standard deviation of the negative electrostatic potential, the minimum surface electrostatic potential, the minimum ionization energy, and the detonation pressure using genetic function approximation. Wang et al. [486, 487] used quantum chemistry methods to optimize the molecular geometries and electronic structures for some explosives. They have shown that there are quantitative relationships between the experimental electrostatic spark sensitivity values and the predicted detonation velocity and pressure for some special groups of explosives. Some new correlations have also been developed which use the maximum obtainable detonation pressure (or velocity) as well as some specific molecular fragments to predict the electric spark sensitivity [488–491]. These methods have the advantage that there is no need to use the crystal density and heat of formation of an explosive to predict its detonation pressure and velocity. A simple method was also introduced to correlate the electric spark and impact sensitivities of nitroaromatic compounds [492].

Türker [493] has used quantum chemistry to derive some correlations between the computational data obtained from ionic nitramine salts and their electric spark sensitivity values. Zhi et al. [494] used the lowest unoccupied molecular orbital energy and Mulliken charges of the nitro group, as well as the number of the aromatic rings and certain substituents on polynitroaromatic compounds to predict their electric spark sensitivity. Yan and Zeman [395] as well as Zeman and Jungová [396] have reviewed some predictive methods for the prediction of the electric spark sensitivity of some classes of explosives.

Machine learning approaches for predicting electrostatic spark sensitivity remain rare, with only a few studies employing theoretical methods like DFT [494]. Wu et al. [495] expanded the descriptor set to include structural (oxygen balance, molecular volume), electronic (dipole moment), and energetic parameters (theoretical density, heat of formation, detonation velocity, and pressure. Among the seven parameters, oxygen balance, theoretical density, heat of formation, detonation velocity, and pressure showed inverse correlations with sensitivity, whereas molecular volume and dipole moment displayed positive trends – with detonation velocity, and pressure demonstrating particularly strong negative relationships due to nitro-group electron-withdrawing effects. Wu et al. [495] compared four algorithms: SVR, RF, multilayer perceptron , and BPNN, with BPNN achieving superior performance after 16,000 training iterations. The BPNN model yielded R^2 values of 0.984 (train) and 0.932 (test), with errors (MSE = 0.35/1.14) significantly lower than other methods.

7.3 Simple methods for predicting electrostatic spark sensitivity based on the RDAD instrument

There are several simple methods for predicting the electric spark sensitivity of polynitroaromatic and nitramine explosives based on the RDAD instrument, which are described in the following sections.

7.3.1 Polynitroaromatic compounds

For polynitroaromatic compounds, the following equation can be used to predict the electric spark sensitivity from structural parameters as [496]:

$$E_{ES} = 4.60 - 0.733a + 0.724d + 9.16(b/d) - 5.14C_{R,OR}, \tag{7.2}$$

where E_{ES} is in J; $C_{R,OR}$ represents the presence of alkyl (–R) or alkoxy (–OR) groups according to the following conditions:

(1) The value of $C_{R,OR}$ is 1.0 for alkyl groups attached to a nitroaromatic ring, that is, 2-methyl-1,3,5-trinitrobenzene.
(2) The value of $C_{R,OR}$ equals –2.0 for the attachment of an alkoxy group to a nitroaromatic ring, that is, 2-methoxy-1,3,5-trinitrobenzene.

Example 7.1: The use of eq. (7.2) for 2-methoxy-1,3,5-trinitrobenzene (TNA) with the following molecular structure

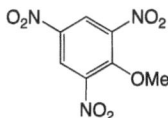

gives

$$E_{ES} = 4.60 - 0.733a + 0.724d + 9.16(b/d) - 5.14C_{R, OR}$$
$$= 4.60 - 0.733(7) + 0.724(7) + 9.16(5/7) - 5.14(-2)$$
$$= 21.36 \, J.$$

The measured value of E_{ES} is 28.59 J [481].

7.3.2 Cyclic and acyclic nitramines

It was found that the ratio of carbon to oxygen atoms, the presence of methylenenitramine units ($-CH_2NNO_2-$) in cyclic nitramines, as well as $-COO-$ (or amide) groups can be used to predict the electrostatic sensitivity of cyclic and acyclic nitramines as follows [497]:

$$E_{ES} = 3.460 + 6.504(a/d) - 4.059C_{CH_2NNO_2 \geq 3, C(=0)(O \, or \, NH)}, \tag{7.3}$$

where $C_{CH_2NNO_2 \geq 3, C(=0)(O \, or \, NH)}$ is either the number of methylenenitramine groups greater than two in cyclic nitramines or the presence of $-COO-$ (or $-CONH-$) functional groups.

Example 7.2: If eq. (7.3) is used for 2,4,5-trinitro-2,4,6-triazaheptane (ORDX) with the following molecular structure,

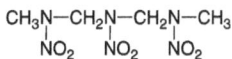

$$\underset{NO_2 \quad NO_2 \quad NO_2}{CH_3N-CH_2N-CH_2N-CH_3}$$

it gives

$$E_{ES} = 3.460 + 6.504(a/d) - 4.059C_{CH_2NNO_2 > 2, C(=0)(O \, or \, NH)}$$
$$= 3.460 + 6.504(4/6) - 4.059(0)$$
$$= 7.80 \, J.$$

The measured value of E_{ES} equals 8.08 J [498].

7.3.3 General correlation for polynitroaromatics as well as cyclic and acyclic nitramines

For various nitroaromatic and nitramine compounds, it was shown that the following correlation can be used to prediction the electric spark sensitivity [394]:

$$E_{ES} = 5.12 + 2.323\left(\tfrac{a}{d}\right) + 1.513\left(\tfrac{b}{d}\right) + 7.519E_{ES}^{+} - 3.637E_{ES}^{-}, \tag{7.4}$$

where E_{ES}^{+} and E_{ES}^{-} are correcting functions that can increase and decrease the predicted results, based on the ratios of the number of carbon and hydrogen to oxygen atoms, respectively. The values of E_{ES}^{+} and E_{ES}^{-} can be given as

(1) Prediction of E_{ES}^{+}:
  (a) for nitroaromatic energetic compounds, the values of E_{ES}^{+} are 2.0, 1.0 and 0.75 in the presence of –OR, three –NH$_2$ and two –OH groups, respectively;
  (b) for nitramines, if the –COO–, C–O–C or tertiary amine (NR$_1$R$_2$R$_3$) groups are present, the value of E_{ES}^{+} is 0.75; and
  (c) for both nitroaromatics and nitramines, E_{ES}^{+} is 1.0 in the presence of the amide group (–NH–OC–).
(2) Prediction of E_{ES}^{-}: The value of E_{ES}^{-} equals 1.0 for:
  (a) the attachment of only one CH$_x$– or Ar– to an aromatic ring in the case of nitroaromatic compounds; and
  (b) if the number of methylenenitramine (CH$_2$NNO$_2$) moieties is greater than, or equal to three in cyclic nitramines (except for cage nitramines), that is, 2,4,6,8,10,12-hexanitro-2,4,6,8,10,12-hexaazaisowurtzitane (HNIW).

Example 7.3: The use of eq. (7.4) for both compounds given in previous examples gives:

$$E_{ES} \quad = 5.12 + 2.323\left(\tfrac{a}{d}\right) + 1.513\left(\tfrac{b}{d}\right) + 7.519E_{ES}^{+} - 3.637E_{ES}^{-}$$

TNA:

$$= 5.12 + 2.323\left(\tfrac{7}{7}\right) + 1.513\left(\tfrac{5}{7}\right) + 7.519(2) - 3.637(0)$$
$$= 23.56\,J$$

ORDX:

$$= 5.12 + 2.323\left(\tfrac{4}{6}\right) + 1.513\left(\tfrac{10}{6}\right) + 7.519(0) - 3.637(0)$$
$$= 9.19\,J.$$

Thus, the deviation of the predicted results of TNA and ORDX from the measured values are 5.03 and −1.11 J, respectively. Meanwhile, the use of the complex quantum mechanical method of Wang et al. [486, 487] gives 6.01 J (Dev = 22.58 J) and 9.09 J (Dev = −1.01 J) for TNA and ORDX, respectively.

7.4 Simple prediction of electrostatic spark sensitivity based on the new ESZ KTTV instrument

Since the new instrument of ESZ KTTV gives more reliable experimental data than the old system RDAD, several attempts have been done to introduce the improved correlations based on ESZ KTTV instrument. Moreover, several correlations have also been introduced to extend the output of previous works based on the RDAD instrument to the ESZ KTTV system. These methods are demonstrated in the following sections.

7.4.1 Polynitroarenes based on ESZ KTTV

A suitable correlation has been introduced to estimate the electrostatic sensitivity of polynitroarenes based on ESZ KTTV as follows [499]:

$$E_{ES, PNA}(ESZ\ KTTV) = 220.6 - 7.91d - 71.08f + 191.4E^+_{ES, PNA}, \tag{7.5}$$

where $E_{ES, PNA}(ESZ\ KTTV)$ is the electric spark sensitivity of polynitroarenes based on the ESZ KTTV system in mJ; d and f are equal to the number of moles of oxygen and chlorine atoms, respectively; $E^+_{ES, PNA}$ is a correcting function that increases the predicted results based on d and f. The value of $E^+_{ES, PNA}$ is specified according to the following conditions:

(a) The presence existence of molecular moieties and as well as direct attachment of three alkyl groups or one aromatic ring to another aromatic ring and dinitrobenzene: The value of $E^+_{ES, PNA}$ equals 0.7 for the presence of one of the mentioned molecular moieties.

(b) The presence of three –NH$_2$ groups: The value $E^+_{ES, PNA}$ equals 0.5.

(c) The presence of group: The value of $E^+_{ES, PNA}$ is equal to 0.8.

(d) The presence of : The value of $E^+_{ES, PNA}$ equals 1.5.

Example 7.4: The use of eq. (7.5) and condition (c) for 1,3,7,9-tetranitro-10H-phenothiazine 5,5-dioxide (TNPTD) with the following molecular structure

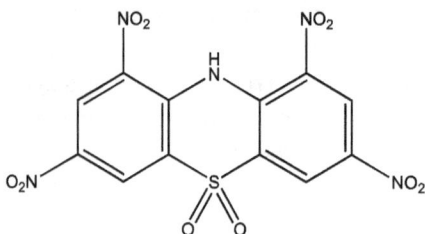

gives:

$$E_{ES, PNA}(ESZ\ KTTV) = 220.6 - 7.91d - 71.08f + 191.4E^{+}_{ES, PNA}$$
$$= 220.6 - 7.91 \times 10 - 71.08 \times 0 + 191.4 \times 0.8 = 294.6\ mJ$$

The measured electrostatic sensitivity of TNPTD using ESZ KTTV instrument is 363.3 mJ [500].

It is possible to correlate the predicted results of the RDAD for the energetic compounds given in Section 7.3 to the ESZ KTTV instrument because most of the available predictive methods [501] and the new software EMDB [71] can estimate the electric spark sensitivity based on RDAD instrument. It was shown that the predicted results of the EMDB [71] can be used to find electric spark sensitivity based on ESZ KTTV instrument as follows [499]:

$$E_{ES, PNA}(ESZ\ KTTV) = 72.5 + 10.40E_{ES, PNA}(RDAD) + 139.4E^{+'}_{ES, PNA} - 62.03E^{-}_{ES, PNA}, \quad (7.6)$$

where $E_{ES, PNA}(RDAD)$ is the electrostatic sensitivity based on the RDAD system for polynitroarenes in mJ; $E^{+'}_{ES, PNA}$ and $E^{-}_{ES, PNA}$ are two correcting functions that can increase and decrease the predicted results based on the RDAD system, respectively. The value of $E^{+'}_{ES, PNA}$ equals 1.0 for polynitroarenes that follow the following conditions:

(a) The presence of and direct attachment of three alkyl groups to the aromatic ring as well as dinitrobenzene.

(b) Direct attachment of picryl group to triazene ring or to another picryl group.

The value of $E^{-}_{ES, PNA}$ is also equal to 1.0 for the presence of –N=N– or –Cl groups.

Example 7.5: The use of eq. (7.6) and the experimental value of $E_{ES,\,PNA}(RDAD) = 13.37\,J$ for (E)-bis(3-methyl-2,4,6-trinitrophenyl)-diazene (DMHNAB) with the following molecular structure

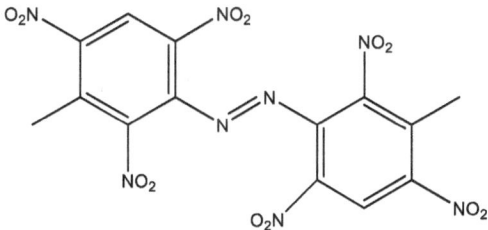

provides:

$$E_{ES,\,PNA}(ESZ\ KTTV) = 72.5 + 10.40 E_{ES,\,PNA}(RDAD) + 139.4 E_{ES,\,PNA}^{+'} - 62.03 E_{ES,\,PNA}^{-}$$
$$= 72.5 + 10.40 \times 13.37 + 139.4 \times 0 - 62.03 \times 1.0 = 149.5\ mJ$$

The reported electrostatic sensitivity for DMHNAB using ESZ KTTV instrument is 118.2 mJ [500].

7.4.2 Nitramines based on ESZ KTTV

A reliable correlation has been introduced for assessment of the electric spark sensitivity of nitramines based on the ESZ KTTV as follows [502]:

$$E_{ES,\,NTA}(ESZ\ KTTV) = 110.7 + 129.4a - 39.8b + 66.2c - 78.7d + 350.0 E_{ES,\,NTA}^{+}, \qquad (7.7)$$

where $E_{ES,\,NTA}(ESZ\ KTTV)$ is the electric spark sensitivity of nitramines based on the ESZ KTTV system in mJ; a, b, c, and d are equal to the number of moles of carbon, hydrogen, nitrogen, and oxygen atoms, respectively; $E_{ES,\,NTA}^{+}$ is a correcting function that increases the predicted results based on elemental composition. The value of $E_{ES,\,NTA}^{+}$ is specified according to the following conditions:
(a) The presence of $-N(NO_2)-CH_2-CH_2-N(NO_2)-$ per cycle in cyclic or acyclic nitramines: The value of $E_{ES,\,NTA}^{+}$ equals 0.5.
(b) The presence of two $-N(NO_2)-CH_2-$ in cyclic nitramines with less than six-membered ring: The value of $E_{ES,\,NTA}^{+}$ is equal to 1.0.

Example 7.6: The use of eq. (7.7) and condition (a) for 1,4,5,8-tetranitrodecahydropyrazino[2,3-b]pyrazine (TNAD) with the following molecular structure:

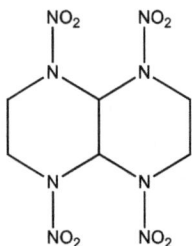

gives:

$$E_{ES,\,NTA}(ESZ\,KTTV) = 110.7 + 129.4a - 39.8b + 66.2c - 78.7d + 350.0E^+_{ES,\,NTA}$$
$$= 110.7 + 129.4 \times 6 - 39.8 \times 10 + 66.2 \times 8 - 78.7 \times 8 + 350.0 \times 0.5$$
$$= 564.1\ mJ.$$

The measured electrostatic sensitivity of TNAD using ESZ KTTV instrument is 520.0 mJ [503].

It is important to have a suitable correlation for conversion of the reported data based on RDAD to the ESZ KTTV, which can be done with the following equation by considering some specific molecular fragments:

$$E_{ES,\,NTA}(ESZ\,KTTV) = 616.8 - 27.80E_{ES,\,NTA}(RDAD) - 296.1E^-_{ES,\,NTA} \qquad (7.8)$$

where $E_{ES,\,NTA}(RDAD)$ is the electric spark sensitivity based on the RDAD system in mJ; $E^-_{ES,\,NTA}$ is a correcting function that decreases the predicted results based on the RDAD system. The value of $E^-_{ES,\,NTA}$ equals 1.0 for cyclic nitramines with the equal number of –CH$_2$– and –N(NO$_2$)– groups in more than five-membered ring as well as cyclic nitramines containing >C(NO$_2$)$_2$ group.

Example 7.7: The use of eq. (7.8) and the measured value of $E_{ES,\,NTA}(RDAD) = 2.96\,J$ for 11,3,5,7,9-pentanitro-1,3,5,7,9-pentazecane (DECAGEN) with the following molecular structure:

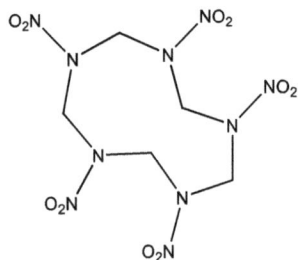

gives:

$$E_{ES, NTA}(ESZ\ KTTV) = 616.8 - 27.80 E_{ES, NTA}(RDAD) - 296.1 E_{ES, NTA}^{-}$$
$$= 616.8 - 27.80 \times 2.96 - 296.1 \times 1.0 = 238.4\ mJ$$

The measured electrostatic sensitivity of DECAGEN using ESZ KTTV instrument is 276.5 mJ [503].

7.4.3 Quaternary ammonium-based energetic ionic liquids or salts based on ESZ KTTV

A simple method has been introduced for the calculation of electrostatic sensitivity of quaternary ammonium-based energetic ionic liquids or salts based on ESZ KTTV as follows [504]:

$$ES_{IL}(mJ) = 635 + 725 a_{cat} - 241 c_{cat} - 409 d_{cat} + 117 a_{ani} + 689 ES_{IL}^{+} - 551 ES_{IL}^{-}, \quad (7.9)$$

where $ES_{IL}(mJ)$ is the sensitivity toward the electrical discharge of a desired energetic ionic compound in mJ; a_{cat}, c_{cat}, and d_{cat} are the number of carbon, nitrogen, and oxygen atoms in cation, respectively; a_{ani} is the number of carbon atoms in the anion. Two parameters ES_{IL}^{+} and ES_{IL}^{-} are two correcting functions. Tables 7.1 and 7.2 show different types of cations and anions for which the values of ES_{IL}^{+} and ES_{IL}^{-} equal 1.0.

Table 7.1: Different types of cations and anions for which the value of ES_{IL}^{+} is 1.0.

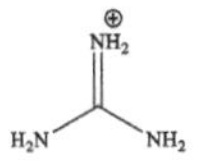

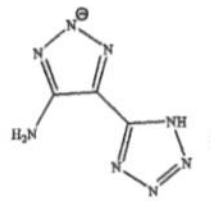

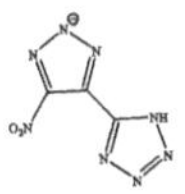

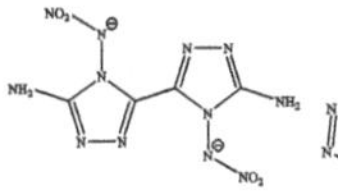

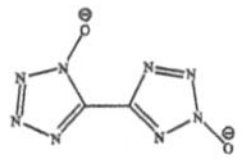

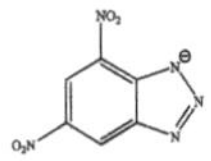

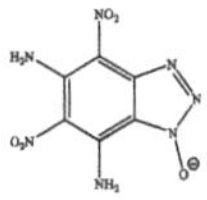

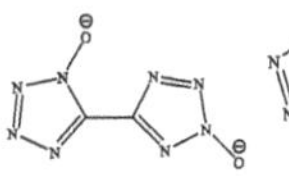

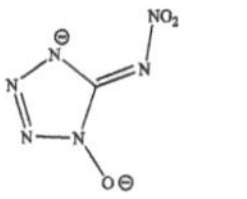

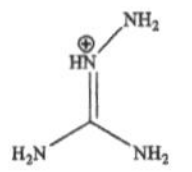

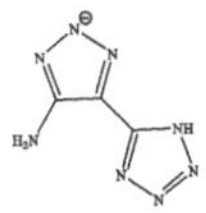

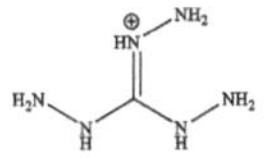

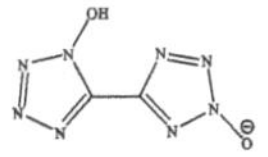

Table 7.2: Different types of cations and anions for which the value of ES_{IL}^- is 1.0.

Cation	Anion

Example 7.8: The use of eq. (7.9) for the following compound

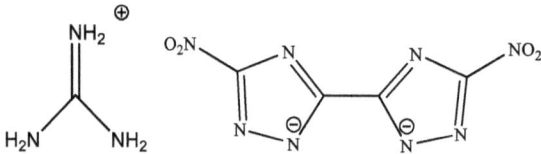

gives:

$$ES_{IL}(mJ) = 635 + 725a_{cat} - 241c_{cat} - 409d_{cat} + 117a_{ani} + 689ES_{IL}^+ - 551ES_{IL}^-$$
$$= 635 + 725 \times 1 \times 2 - 241 \times 3 \times 2 - 409 \times 0 + 117 \times 4 + 689 \times 0 - 551 \times 1.0$$
$$= 556 \; mJ.$$

The measured value of ES_{IL} for this compound is 750 mJ [505].

7.5 Electrostatic discharge (*ESD*) sensitivity of nitrogen-rich heterocyclic energetic compounds (NRHECs)

The *ESD* sensitivity of nitrogen-rich heterocyclic energetic compounds (NRHECs) and their salts is typically measured using specialized *ESD* test apparatuses, as described in the literature [2]. These tests employ spark energies ranging from 0.001 to 20 J [3]. Since *ESD* data are often reported as ranges rather than precise values, a standardized reference scale (per OZM) categorizes sensitivity into six tiers:
- High risk (*ESD* $\leq$ 0.004 J)
- Risky (0.004 < *ESD* $\leq$ 0.04 J)
- Normal sensitivity (0.04 < *ESD* $\leq$ 0.25 J)
- Very low sensitivity (0.25 < *ESD* $\leq$ 0.5 J)
- Low risk (0.5 < *ESD* $\leq$ 1.0 J)
- Very low risk (*ESD* > 1.0 J)

Experimental *ESD* data for 54 NRHECs were obtained from a confidential source [2] for deriving a suitable correlation for predicting *ESD* of NRHECs.

7.5.1 Influence of functional groups on *ESD* sensitivity

The presence of specific functional groups can significantly alter *ESD* sensitivity. For example:
- Amino groups (–NH$_2$ or –NH–) generally reduce sensitivity to impact, electric spark, and shock. In NRHECs, these groups often enhance safety by lowering *ESD* risk. For instance, 1,4-diamino-3,6-dinitropyrazolo[4,3-c]pyrazole and 5,5'-bis(1H-tetrazolyl)hydrazine (Scheme 7.1) exhibit negligible *ESD* sensitivity due to their –NH$_2$/NH– moieties.

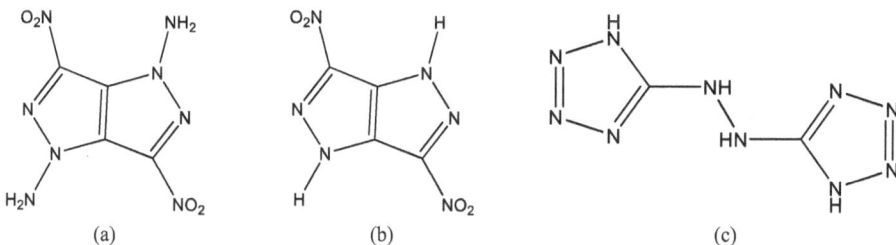

Scheme 7.1: The presence of –NH$_2$ or –NH– groups in these compounds makes them insensitive to electrostatic discharge.

– N=N⁺–O⁻ fragments (e.g., in 6-nitro-7-azido-pyrazol[3,4-d][1,2,3]triazine-2-oxide) also confer insensitivity to ESD (Scheme 7.2).

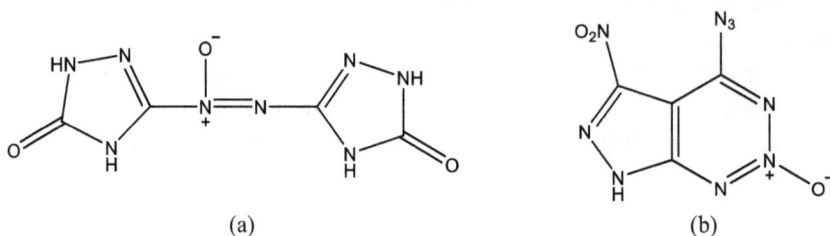

(a)   (b)

Scheme 7.2: The N=N⁺–O⁻ fragment in these compounds renders them insensitive to electrostatic discharge.

– Conversely, –NNO₂ groups (e.g., in bicyclo-HMX) increase *ESD* sensitivity, often resulting in "normal sensitivity" (0.04–0.25 J) (Scheme 7.3).

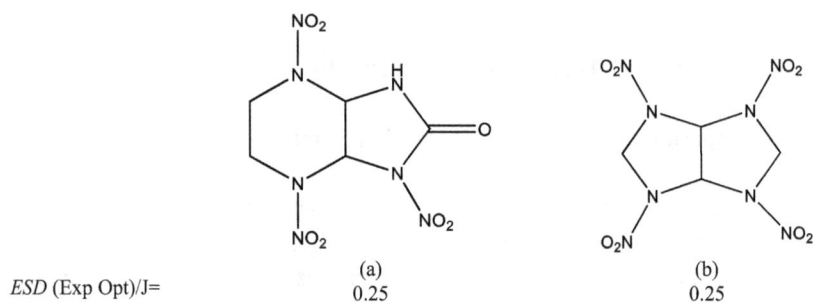

ESD (Exp Opt)/J=

(a)	(b)
0.25	0.25

Scheme 7.3: The –NNO₂ groups in these compounds result in normal electrostatic sensitivity.

7.5.2 Predictive modeling via MLR

The derived correlation is [506]:

$$ESD(J) = 0.36 + 0.731E_{NH_2, NH} + 0.737E_{N^+O^-} - 0.612E_{NNO_2, ONO_2, NO_2}. \tag{7.10}$$

1. $E_{NH_2, NH}$ represents the presence of –NH₂/–NH– groups ($E_{NH_2, NH} = 1.0$, unless negated by ortho-nitro groups in furazan or benzofuroxan derivatives, diazo substituents, =NNH₂ or –N=C(NH₂)₂ or cyclic –NH–C(=O)–NH– (without –NO₂ group) or tetrazine ring or tetrazolium salt for high nitrogen content, the presence of H₂O (or in the form HO–NH₃⁺), and the presence of –C(NO₂)–C(NO₂)– or the ratio of –NNO₂ group is more than two times of –NH– group).

2. $E_{N^+O^-}$ represents the presence of N=N$^+$–O$^-$ fragments ($E_{N^+O^-}$=1.0, except when intramolecular hydrogen bonding nullifies its effect).
3. E_{NNO_2, ONO_2, NO_2} represents the presence of –NNO$_2$/–ONO$_2$/–NO$_2$ groups ($E_{NNO_2, ONO_2,}$ $_{NO_2}$=1.0 for cyclic nitramines without other functional group or nitrotetrazolate salt, but E_{NNO_2, ONO_2, NO_2}= –1.0 for carbonyl-bridged nitramines in form –N(NO$_2$)–C(=O)–N (NO$_2$), attachment of –(CH$_2$)$_n$–ONO$_2$ to bisoxazole, connection of benzene containing nitro groups to furazan or oxadiazole derivatives without other functional groups, and hydrogen-free nitrofurazan derivatives).
4. If the predicted value of *ESD*(J) is negative, the *ESD* value should be taken 0.25 J.

7.5.3 Practical implications

Introducing amino groups is a straightforward strategy to improve thermal stability and reduce sensitivity. However, coexisting functional groups (e.g., –NNO$_2$ or –C(NO$_2$)–C(NO$_2$)–) can counteract this effect.

Example 7.9: Equation (7.10) was applied to tetranitroglycoluril, which has the following molecular structure:

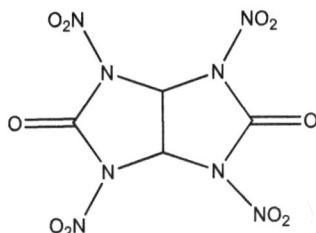

$$ESD(J) = 0.36 + 0.731E_{NH_2, NH} + 0.737E_{N^+O^-} - 0.612E_{NNO_2, ONO_2, NO_2},$$

$$ESD = 0.36 + 0.731(0) + 0.737(0) - 0.612(-1.0) = 0.98 \text{ J}.$$

Since E_{NNO_2, ONO_2, NO_2}= –1.0 for carbonyl-bridged nitramines (–N(NO$_2$)–C(=O)–N(NO$_2$)–), the calculated value closely matches the optimized experimental value, that is, *ESD* (Exp,Opt)/J = 1.1.

Example 7.10: Equation (7.10) was applied to bicyclo-HMX, which has the following molecular structure:

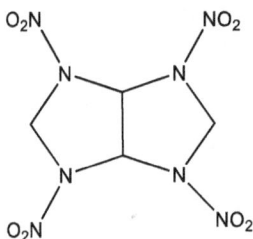

$$ESD(J) = 0.36 + 0.731E_{NH_2, NH} + 0.737E_{N + O^-} - 0.612E_{NNO_2, ONO_2, NO_2}$$

$$ESD = 0.36 + 0.731(0) + 0.737(0) - 0.612(1.0) = -0.25\,J < 0 \rightarrow ESD = 0.25\,J$$

Since $E_{NNO_2, ONO_2, NO_2} = 1.0$ for cyclic nitramines lacking additional functional groups or nitrotetrazolate salts, the calculated value agrees well with the optimized experimental value (ESD (Exp,Opt)/J = 0.25).

7.6 Some aspects of predictive methods

The currently available predictive methods cannot predict electrostatic sensitivity versus the grain size because the electric spark sensitivity depends on the size and shape of crystals. For some isomers, the difference in the sensitivities may be large, and this can be attributed to the different behavior of nitro groups in different positions, for example, electric spark sensitivity for 1,3-dinitrobenzene and 1,4-dinitrobenzene are 3.15 and 18.38 J [481], respectively. Since experimental data of the electric spark sensitivity of different isomers is rare, use of the predictive methods which are available may result in large deviations for some isomers. Fortunately, eqs. (7.2)–(7.4) can predict the electric spark sensitivity which is close to the average value. For example, eq. (7.4) predicts a value of 10.11 J for both 1,3-dinitrobenzene and 1,4-dinitrobenzene and this is close to the average value of the measured values for the two isomers.

7.7 Summary

Some developments for the prediction of the electric spark sensitivity of secondary explosives have been reviewed in this chapter. Since different factors can influence the sensitivity resulting from different stimuli, the main intent in this work was to illustrate the best available simple methods to evaluate the electrostatic sensitivity of energetic compounds. Three simple eqs. (7.2), (7.3), and (7.4) can be can be easily used to theoretically predict the magnitude of the electrostatic sensitivity of based on the RDAD instrument polynitroaromatics, nitramines, and also of both nitroaromatics and nitramines, respectively. Thus, these equations are useful models in terms of their accuracy and simplicity because they require only knowledge of the molecular structure of energetic compounds, which is always known. Due to the different behavior – in terms of the electric spark sensitivity – of nitramines and nitroaromatic compounds, eq. (7.4) has the advantage that it can be applied to nitroaromatic compounds which contain – $N-NO_2$ groups, that is, 1-(methylnitramino)-2,4,6-trinitrobenzene (TETRYL). Since the new instrument of ESZ KTTV gives more reliable experimental data than the old system RDAD, eqs. (7.5) and (7.7) provide two correlations for the prediction of the electrostatic sensitivity based on the ESZ KTTV instrument of polynitroaromatics and nitramines, respectively. Moreover, eqs. (7.6) and (7.8) give a suitable pathway for conversion of the outputs of software EMDB [69] based on the RDAD to those based on the ESZ KTTV for polynitroaromatics and nitramines, respectively. Equation (7.9) can also be used to estimate the electrostatic sensitivity of quaternary ammonium-based energetic ionic liquids or salts based on ESZ KTTV.

ESD sensitivity of nitrogen-rich energetic compounds is classified into six tiers (0.001–20 J range). Amino groups (–NH_2/NH–) reduce sensitivity, while –NNO_2 groups increase it (Schemes 7.1–7.3). A predictive MLR model (eq. (7.10)) was developed using three key parameters: $E_{NH_2,NH}$, $E_{N^+O^-}$, and E_{NNO_2,ONO_2,NO_2}. For carbonyl-bridged nitramines ($E_{NNO_2,ONO_2,NO_2} = -1.0$), predictions match experiments (tetranitroglycolurile: 0.98 J vs 1.1 J). Cyclic nitramines ($E_{NNO_2,ONO_2,NO_2} = 1.0$) default to 0.25 J if predicted negative (bicyclo-HMX: 0.25 J match). Amino groups generally improve safety but nitro groups can counteract this effect.

8 Shock sensitivity

Gap test data is useful to indicate the shock sensitivity of an explosive, and it is nowadays widely used to determine the shock sensitivity of a desired explosive. Different gap tests have been used to qualitatively measure the shock wave amplitude which is required to initiate detonation in explosives, for example, at the Naval Surface Warfare Center (NSWC) and Los Alamos National Laboratory (LANL). Large scale and small scale gap tests are two convenient methods for measuring shock sensitivity [464]. In contrast to the results of impact sensitivity – which are often not reproducible because factors in the impact experiment that might affect the formation and growth of hot spots can strongly affect the measurements – reliable shock sensitivity tests exist. Furthermore, the reported data of impact sensitivities are extremely sensitive to the conditions under which the tests are performed.

There are several reviews in which different methods for predicting the shock sensitivity of different pure and composite explosives have been considered [393, 395, 397]. Price [507] has studied different important factors in shock wave sensitivity tests. For five explosives with closely related structures, that is, TNB, DIPAM, MATB, DATB, and TATB, Storm et al. [464] have shown that there is a linear correlation between the impact and shock sensitivity under specified conditions. Due to the dependence of the results of impact sensitivity tests on the conditions of the experiment, they used the impact sensitivity as measured at LANL and/or NSWC using the Bruceton method, type 12 tools, 2.5 kg weight, 40 mg sample, 5/0 sandpaper, and 25 trials. For seven polynitroaromatics, Owen et al. [415] also found that the measured impact and shock sensitivities can correlate with an approximation of the electronegativity potential at the midpoint of the C–N bond for the longest C–NO_2 bond in each molecule. Tan et al. [508] used quantum mechanical calculations (DFT/BLYP/DNP) to calculate the bond dissociation energies of X–NO_2 (X = C, N, O) and Mulliken charges of nitro groups for 14 examples of nitro compounds.

Yang et al. [509] developed a method to predict how sensitive explosives are to shock. By using reactive molecular dynamics simulations, Yang et al. [509] calculated the critical pressures needed for shock-induced ignition and detonation. Yang et al. [509] found that the number of explosive molecules reacting in a supercell increased linearly as the shock wave traveled through it. The slope of this trend helped determine the explosive's reaction rate under shock loading. Interestingly, different explosives reacted at different rates depending on the pressure, showing two distinct linear stages with different slopes. The transition points between these stages marked the thresholds for shock-induced ignition and detonation, allowing researchers to pinpoint their critical pressures. The results matched well with known data on common explosives like PETN, RDX, and HMX, including their sensitivity differences based on crystal structure. Since this method relies only on crystal structure for accurate predictions, it could be incredibly useful for designing safer explosives and improving handling protocols.

https://doi.org/10.1515/9783112206768-008

Jiang et al. [510] developed amorphous models that mimic locally ordered molecular packing in energetic materials, providing an alternative to crystal structures for predicting shock sensitivity and safety. Using reactive molecular dynamics simulations, Jiang et al. [510] analyzed how these materials decompose under shock, comparing two different simulation methods – the traditional multiscale shock technique (MSST) and a modified quantum-bath-coupled version (QBMSST). The QBMSST method showed faster temperature increases and decomposition rates than MSST. Jiang et al. [510] determined the critical initiation pressure for explosions by tracking when 15% of the key trigger bonds broke in each material. Using QBMSST, these critical pressures were measured at 18.44 GPa for PETN, 24.51 GPa for RDX, 32.00 GPa for TNT, and 40.59 GPa for TATB – all lower than the pressures obtained through MSST. Both methods consistently ranked the materials the most to least shock-sensitive as PETN > RDX > TNT > TATB, matching experimental results. Jiang et al. [510] also revealed that TATB has the highest resistance to compression (bulk modulus), followed by TNT, RDX, and PETN. These findings demonstrate that amorphous models can effectively predict explosive sensitivity, just like crystal models, and may prove valuable for studying more complex mixed or composite explosive systems in the future.

Among the different approaches for the prediction of shock sensitivity, there are two simple methods based on simple structural parameters, which are discussed in this chapter.

8.1 Small-scale gap test

Studies of the shock sensitivity, as measured using the NSWC small-scale gap test, shows that three special structural parameters may affect their values, including:
(1) the distribution of oxygen between carbon and hydrogen;
(2) the existence of nitramine groups or a α-CH linkage in nitroaromatic compounds;
(3) the difference in the number of amino and nitro groups in aminoaromatic (Ar–NH_2) energetic compounds.

Thus, the following general equation can be applied to $C_a H_b N_c O_d$ explosives [511]:

$$P_{90\% \text{ TMD}} = 16.79 + 2.262(a + b/2 - d) - 6.314E^0_{\alpha CH/NNO_2}$$
$$+ 17.72(1.93n_{NH_2} - n_{NO_2})_{\text{pure}}, \tag{8.1}$$

$$P_{95\% \text{ TMD}} = 21.96 + 2.479(a + b/2 - d) - 6.3677E^0_{\alpha CH/NNO_2}$$
$$+ 32.92(1.93n_{NH_2} - n_{NO_2})_{\text{pure}}, \tag{8.2}$$

$$P_{98\% \text{ TMD}} = 25.45 + 2.211(a + b/2 - d) - 4.162E^0_{\alpha\text{CH}/\text{NNO}_2}$$
$$+ 46.39(1.93n_{\text{NH}_2} - n_{\text{NO}_2})_{\text{pure}},$$

(8.3)

where $P_{90\% \text{ TMD}}$, $P_{95\% \text{ TMD}}$, and $P_{98\% \text{ TMD}}$ are the pressures in kbar which are required to initiate material pressed to 90%, 95%, and 98% of theoretical maximum density (TMD); $a + b/2 - d$ is a parameter that shows the distribution of oxygen between carbon and hydrogen; $E^0_{\alpha\text{CH}/\text{NNO}_2}$ is a parameter that indicates the existence of a α-CH linkage in nitroaromatic compounds, or a N–NO$_2$ functional group; $(1.93n_{\text{NH}_2} - n_{\text{NO}_2})_{\text{pure}}$ is the difference in the number of amino and nitro groups in aminoaromatic energetic compounds when $1.93n_{\text{NH}_2} \geq n_{\text{NO}_2}$. The value of $E^0_{\alpha\text{CH}/\text{NNO}_2}$ equals 1.0 for nitramines or for the presence of a α-CH linkage in nitroaromatic compounds, for example, TNT. It is also equal to 1.0 for composite explosives containing more than or equal to 50 % of nitramines or α-C–H linkage in nitroaromatic compounds. The parameter $a + b/2 - d$ may affect the sensitivity of different classes of explosives. Since the presence of the N–NO$_2$ functional group and α-CH linkage in nitroaromatic compounds can increase the sensitivity of these compounds, the coefficient of $E^0_{\alpha\text{CH}/\text{NNO}_2}$ has a minus sign. The attachment of amino groups to an aromatic ring may enhance the stability of an energetic compound, while the addition of electron-withdrawing groups (such as NO$_2$ groups) leads to a reduction in the stabilization of the aromatic ring. Thus, amino groups partially counteract the electron-withdrawing effect of nitro groups, which enhances the stabilization of the aromatic ring through the parameter $(1.93n_{\text{NH}_2} - n_{\text{NO}_2})_{\text{pure}}$.

The deviation of the values obtained from eqs. (8.1) to (8.3) from the measured values becomes large for very fine particle sizes because small particle size can reduce the shock sensitivity at high density.

Example 8.1: The use of eqs. (8.1)–(8.3) to calculate the results of the small-scale gap test of 3,3′-diamino -2,2′,4,4′,6,6′-hexanitrobiphenyl ($C_{12}H_6N_8O_{12}$) gives

$$P_{90\% \text{ TMD}} = 16.79 + 2.262(a + b/2 - d) - 6.314E^0_{\alpha\text{CH}/\text{NNO}_2}$$
$$+ 17.72(1.93n_{\text{NH}_2} - n_{\text{NO}_2})_{\text{pure}}$$
$$= 16.79 + 2.262(12 + 6/2 - 12) - 6.314(0) + 17.72(0)$$
$$= 23.58 \text{ kbar,}$$

$$P_{95\% \text{ TMD}} = 21.96 + 2.479(a + b/2 - d) - 6.3677E^0_{\alpha\text{CH}/\text{NNO}_2}$$
$$+ 32.92(1.93n_{\text{NH}_2} - n_{\text{NO}_2})_{\text{pure}}$$
$$= 21.96 + 2.479(12 + 6/2 - 12) - 6.3677(0) + 32.92(0)$$
$$= 29.40 \text{ kbar,}$$

$$P_{98\% \text{ TMD}} = 25.45 + 2.211(a + b/2 - d) - 4.162E^0_{aCH/NNO_2}$$

$$+ 46.39(1.93n_{NH_2} - n_{NO_2})_{pure}$$

$$= 25.45 + 2.211(12 + 6/2 - 12) - 4.162(0) + 46.39(0)$$

$$= 32.08 \text{ kbar.}$$

The measured values of $P_{90\% \text{ TMD}}$, $P_{95\% \text{ TMD}}$, and $P_{98\% \text{ TMD}}$ are 25.11, 29.71, and 33.04 kbar [464], respectively.

8.2 Large-scale gap test

Different gap tests were used to qualitatively measure the shock wave amplitude required to initiate detonation in an explosive. In the large-scale gap test, a shock pressure of uniform magnitude is produced by a detonating charge of high explosive, which is transmitted to the test explosive through an attenuating inert barrier or gap. Since the thickness of the barrier between the donor and test (acceptor) explosives can be varied, one can determine the barrier thickness required to inhibit detonation in the test explosive half of the time (G_{50}). Two test configurations were used by LANL, in which the diameter of the cylinder acceptor charge in the small-scale test is 12.7 mm and in the large-scale test is 41.3 mm [512]. An explosive in the small-scale test, whose detonation failure diameter is near to or greater than the diameter of the acceptor charge, cannot be tested. The large-scale test has an advantage over the small-scale test; it can be tested in this situation. For the large-scale test, the test method is to fire a few preliminary shots to determine the spacer thickness that allows detonation in the test explosive. When the shots are fired with the spacer thickness being alternately increased and decreased, the spacer thickness that allows a 50 % detonation probability in the acceptor explosive is determined. The dent produced in a witness plate can ascertain detonation of the acceptor charge. Thus, a deep, defined dent in the steel witness plate shows that detonation of the test explosive has occurred.

Large-scale shock sensitivities of various explosives rely on physical and chemical structural parameters, and for various explosives, depend on four main essential parameters:
(1) initial density;
(2) percent void;
(3) distribution of oxygen between carbon and hydrogen; and
(4) structural parameter $C-N(NO_2)-C$ for pure nitramine explosives.

Pure nitramines containing the $C-N(NO_2)-C$ linkage are more sensitive than other pure explosives containing only the $C-NO_2$ linkage. The following general equation can be used for the prediction of the large-scale shock sensitivities of various types of $C_aH_bN_cO_d$ pure and mixed explosives [513]:

$$G_{50} = 171.47 - 69.10\rho_0 - 2.61(a + b/2 - d) - 0.961 Void_{theo}$$
$$+ 12.32(C-N(NO_2)-C)_{pure}, \tag{8.4}$$

where G_{50} is in mm, $Void_{theo}$ is the theoretically calculated percent of voids that can be obtained from

$$\frac{(1/\rho_0 - 1/\rho_{TM})}{1/\rho_0} \times 100,$$

where ρ_{TM} is the theoretical maximum density.

Equation (8.4) can be applied to pure and composite mixtures that are prepared under vacuum cast, cast, hot-pressed and pressed conditions because deviations may be large for creamed, granular, and flake situations.

Example 8.2: PBX-9007 has the composition 90/9.1/0.5/0.4 RDX/Polystyrene/DOP/Rosin ($C_{1.97}H_{3.22}N_{2.43}O_{2.44}$). If the values of ρ_0 and ρ_{TM} are 1.646 and 1.697 g/cm³, respectively, the use of eq. (8.4) gives

$$Void_{theo} = \frac{(1/\rho_0 - 1/\rho_{TM})}{1/\rho_0} \times 100 = \frac{(1/1.646 - 1/1.697)}{1/1.646} \times 100 = 3.005$$
$$G_{50} = 171.47 - 69.10\rho_0 - 2.61(a + b/2 - d)$$
$$- 0.961 Void_{theo} + 12.32(C-N(NO_2)-C)_{pure}$$
$$= 171.47 - 69.10(1.646) - 2.61(1.97 + 3.22/2 - 2.44)$$
$$- 0.961(3.005) + 12.32(0)$$
$$= 51.87 \text{ mm}.$$

The measured values of G_{50} for PBX-9007 are 52.91 mm [512].

8.3 Critical diameter of solid pure and composite high explosives

The critical (failure) diameter is the minimum diameter of a cylindrical charge of a high explosive that sustains a high-order steady-state detonation [514], which depends on confinement, particle size, and the initial temperature of the sample [514, 515]. The critical diameters of primary and high explosives are usually very small, sometimes in the μm and in the range mm to cm region, respectively. Moreover, the experimental values of critical diameters of explosives depend on the conditions in which they are confined, for example, the critical diameters of 15 mm and 7 mm were reported for TNT (cast) with a loading density of about 1.6 g/cm³ encased in 0.2 mm paper at 291 K and 290 K after one temperature cycle to 77 K, respectively [514].

Determination of critical diameters of high explosives with high detonability can be difficult because it is necessary to manufacture high-density charges with diameters less than 1 mm [516]. It is unfeasible to determine the critical diameter of high explosives with low detonability (charges of large diameters >100 mm) under laboratory conditions because it is necessary to investigate them [516]. The charges with higher detonability require a smaller critical diameter [517]. Some attempts have been developed to use suitable predictive methods for reducing the number of experiments of pure and composite explosives. A complex critical-diameter theory based on complex variables as input parameters has been developed for some specific high explosives [516, 518, 519]. For the prediction of the critical diameter of a high-explosive charge under shock-wave compression, Kobylkin [516] has shown that it is necessary to know the shock adiabat, detonation velocity, and the generalized kinetic characteristic of decomposition. The generalized kinetic characteristic of decomposition can be found from the measurement of the shock-wave amplitude on the distance the shock wave that travels during shock-wave initiation of the high-explosive charge. A simple method has been introduced for calculating the critical diameter under unconfined condition as follows [520]:

$$d_c = -3.19c + 5.38d + 21.80C_{Shock}, \tag{8.5}$$

where d_c is the critical diameter in mm; c and d are the numbers of moles of nitrogen and oxygen atoms, respectively; C_{Shock} is a correcting function. For pure and composite explosives, the following situations should be considered for evaluation of C_{Shock}:

(1) *Shock sensitivity and impact sensitivity*: The shock sensitivity of explosives is more important in its safety assessment as compared to the impact sensitivity [464]. Equation (8.3) was used to specify the contribution of the shock sensitivity of both pure and composite explosives. Equations (6.11) and (6.12) are used with the symbol E_{IS} in J unit, that is, $E_{IS} = 0.245H_{50}$, to mention the contribution of impact sensitivity of pure explosives.

(2) *Energetic compounds with restrictions of c and d*: As shown in eq. (8.5), the signs of c and d are negative and positive, respectively. Thus, the ratio of c/d should be less than 1.18.

According to the above situations, different values of C_{Shock} are given as:

(a) *Pure explosives* – Several situations are considered:
　　(i)　The value of C_{Shock} equals −0.40 if $P_{98\% TMD} > 38$ kbar and $E_{IS} > 28$ J.
　　(ii)　For $E_{IS} < 22$ J, the fixed value of critical diameter should be considered, i.e., $d_c = 2.5$.

(b) *Composite explosives* – The assessment of C_{Shock} can be done as follows:
　　(iii)　For composite explosives including PETN, $C_{Shock} = -0.30$.
　　(iv)　The value of C_{Shock} is equal to −0.10 if $P_{98\% TMD} < 24.4$ kbar.

(c) If $c/d > 1.18$: The value of C_{Shock} is 0.80.

The process of casting can affect the value of d_c for the unconfined condition, for example, the values of d_c are $22.0 < d_c < 25.4$ and $12.6 < d_c < 16.6$ mm poured as cloudy

slurry and creamed for TNT, respectively [514]. Moreover, axially oriented TNT crystals give unstable detonation while radially oriented crystals detonate smoothly. The calculated data of eq. (8.5) for cast explosives are more closed to the experimental data than pressed explosives. Two reasons can be introduced for this situation:
(1) The actual density of cast explosives is more closed to its theoretical maximum density than that of pressed explosives because there are fewer voids in cast explosives.
(2) Distributing of stresses in pressed charges exists substantially stronger than that in cast charges.

Example 8.3: z-TACOT (Tetranitro-2,3,5,6-dibenzo-1,3a,4,6a-tetraazapentalene) has the following molecular structure:

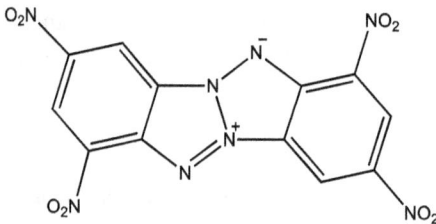

Chemical Formula: $C_{12}H_4N_8O_8$
Molecular Weight: 388.21

The use of eq. (8.3), as well as eqs. (6.11) and (6.12), gives:

$$(\log H_{50})_{core} = -0.584 + 61.62a' + 21.53b' + 27.96c'$$
$$= -0.584 + (61.62 \times 12 + 21.53 \times 4 + 27.96 \times 8)/388.12 = 2.119$$

$$\log H_{50} = (\log H_{50})_{core} + 84.47\frac{F^+}{Mw} - 147.1\frac{F^-}{Mw}$$
$$= 2.119 + 84.47\frac{0}{388.21} - 147.1\frac{0}{388.21} = 2.119$$

$$H_{50} = 131 \text{ cm}$$

$$E_{IS} = 0.245H_{50} = 0.245 \times 131 = 32.2J$$

$$P_{98\% \text{ TMD}} = 25.45 + 2.211(a + b/2 - d) - 4.162E^0_{a\,CH/NNO_2} + 46.39(1.93n_{NH_2} - n_{NO_2})_{pure}$$
$$= 25.45 + 2.211(12 + 4/2 - 12) - 4.162(0) + 46.39(0)$$
$$= 38.72 \text{ kbar.}$$

The value of equals −0.4 because $P_{98\% \text{ TMD}} > 38$ kbar and $E_{IS} > 28$ J. Thus, eq. (8.5) gives:

$$d_c = -3.19c + 5.38d + 21.80C_{Shock}$$
$$= -3.19 \times 8 + 5.38 \times 8 + 21.80(-0.4) = 8.80 \text{ mm.}$$

The measured value of d_c for z-TACOT is 3.0 mm [514].

Example 8.4: Composite explosive Comp A-3 (pressed) consisting of 91% RDX and 9% wax binder has the chemical formula $C_{1.87}H_{3.74}N_{2.46}O_{2.46}$. Since the presence of RDX containing nitramine group $\geq 50\,\%$, $E^0_{\alpha CH/NNO_2} = 1.0$. The use of eq. (8.3) gives:

$$P_{98\%\,TMD} = 25.45 + 2.211(a + b/2 - d) - 4.162E^0_{\alpha CH/NNO_2} + 46.39(1.93n_{NH_2} - n_{NO_2})_{pure}$$

$$= 25.45 + 2.211(1.87 + 3.74/2 - 2.46) - 4.162(1.0) + 46.39(0)$$

$$= 24.12 \text{ kbar.}$$

The value of C_{Shock} is equal to −0.10 because of $P_{98\%\,TMD} < 24.4$ kbar. Thus, eq. (8.5) gives:

$$d_c = -3.19c + 5.38d + 21.80C_{Shock}$$

$$= -3.19 \times 2.46 + 5.38 \times 2.46 + 21.80(-0.1) = 3.21 \text{ mm.}$$

The measured value of d_c for Comp A-3 is 2.2 mm [514].

8.4 Summary

This chapter introduced different approaches for the predicting shock sensitivity of pure and composite explosives using small- and large-scale gap tests. The simple eqs. (8.1)–(8.4) have two major advantages with respect to the impact sensitivity correlations given in Chapter 6, and which are:

(1) since a high percentage of errors are usually attributed to the reported experimental measurements from different sources for impact sensitivities, there is a large uncertainty in the different methods of impact sensitivity predictions as compared to eqs. (8.1)–(8.4) for small- and large-scale gap thickness shock sensitivity;

(2) different correlations for the impact sensitivity can only be applied for pure explosives, but eqs. (8.1)–(8.4) can be used for both pure and composite explosives. Equation (8.5) can be used to find the critical diameter of pure and composite explosives by considering eq. (8.3) as well as eqs. (6.11) and (6.12).

9 Friction sensitivity

Friction is one of the stimuli that can cause explosions and fires in pyrotechnic compositions and explosives [396]. It is important to investigate important aspects of friction sensitivity because it shows the behavior of an energetic compound with respect to friction stimuli. The BAM large friction tester can be used to determine the friction sensitivity of a sample, in which approximately 30 mg of the sample is placed on a porcelain plate [521]. Therefore, the BAM friction tester is widely accepted as a standard friction tester in Europe. The surfaces of both the porcelain plates and pegs have uniform roughness, so that the porcelain pin is lowered onto the sample, and a weight is placed on the arm to produce the desired load. The tester is activated, and the porcelain plate is moved once forward and backward. The results of friction sensitivity are observed as either a reaction in the form of a flash, smoke, and/or audible report, or no reaction.

In contrast to impact, electric spark, and shock sensitivity, friction sensitivity does not attract the attention of theoretical chemists/physicists because of the shortcomings of the influence of the results of friction sensitivity on experimental data. Zeman and Jungová [396] have reviewed several methods that have been used to predict the friction sensitivity of energetic materials. Jungová et al. [522] indicated that there is a relationship between the friction sensitivity of nitramines and their thermal decomposition parameters. Friedl et al. [523] showed the relationship between the friction sensitivity and surface electronic potentials of nitramines. Jungová et al. [524] also compared the friction sensitivity of nitramines with their impact sensitivities and heats of fusion. For some classes of nitramines [525], there is a semi-logarithmic relationship between the impact and friction sensitivities. For the safe handling of energetic compounds, knowledge of the friction sensitivity may be important because friction is frequently encountered during mixing, pouring, sieving, priming, and consolidation operations [526].

Traditional friction sensitivity testing using pendulum tribometers [521] has significant limitations – these large-scale methods are not only dangerous but often yield inconsistent results due to variations in crystal quality and experimental conditions. To overcome these challenges, Wang et al. [527] developed a safer, more precise approach using laser spark spectrometry (LSS) that requires just milligrams of material instead of conventional bulk testing. This innovative technique analyzes the full LSS spectrum to accurately predict sensitivities through physical-parameter-corrected statistical models, with electron density showing a particularly strong correlation with friction sensitivity. The method also reveals other key relationships, including plasma temperature's link to impact sensitivity and oxygen balance's effect on electrostatic sensitivity. By replacing hazardous macroscale detonation tests, this LSS-based approach provides reliable, high-precision sensitivity measurements while dramatically improving safety and reducing costs.

https://doi.org/10.1515/9783112206768-009

Bondarchuk [528] developed a predictive QSPR model to estimate friction sensitivity in nitramine-based energetic materials using genetic function approximation. The model incorporates 10 key molecular descriptors and was trained on 80 diverse nitramine compounds, including both molecular crystals and salts. When tested against 30 additional compounds, the model showed excellent predictive ability with an R^2 of 0.90. Importantly, only two descriptors require quantum-chemical calculations (heat of formation and quadrupole components), while the remaining eight can be quickly determined. By combining this QSPR model with existing methods for rapid detonation property estimation, Bondarchuk [528] successfully modified several energetic salts through strategic ion substitutions. This approach led us to propose 10 new nitramine materials with simultaneously improved friction sensitivity and detonation performance.

Muravyev et al. [529] conducted a comprehensive reliability assessment of published impact and friction sensitivity data by synthesizing and testing over 100 reactive compounds. Their analysis exposed substantial inconsistencies across literature sources, which remained evident even when incorporating their own experimental results. The team identified three primary causes for this variability in impact and friction measurements: (1) non-standardized testing protocols between research groups, (2) inconsistent application of mechanical stress during testing, and (3) differing approaches to accounting for material decomposition effects. To improve reliability, Muravyev et al. [529] developed a validation framework comparing scaled impact and friction sensitivity parameters against established maximum values for compounds with comparable energy content. Their work produced specific guidelines for standardizing mechanical sensitivity reporting, particularly emphasizing impact and friction test methodologies. Through rigorous cross-validation with independent studies, Muravyev et al. [529] established a benchmark dataset for impact and friction sensitivities. The team specifically highlights the need for standardized testing of compounds referenced in their prior work [530] that currently lack comprehensive impact and friction sensitivity data. They recommend that future studies employ their published protocols for both impact and friction testing to ensure data consistency across the field.

Li et al. [531] investigated the complex challenge of predicting friction sensitivity in energetic materials, which had long been complicated by numerous influencing factors. They successfully calculated key electronic and mechanical properties using first-principles methods. Through careful analysis, the team identified several parameters that strongly correlated with friction sensitivity measurements. Their results demonstrated that experimental friction sensitivity values increased with larger band gaps, higher Young's modulus values, and greater effective carrier mass. Conversely, the sensitivity decreased when excess energy was higher. These findings established clear trends: materials with wider band gaps, stiffer mechanical properties, heavier effective carrier mass, and lower excess energy consistently showed reduced friction sensitivity. This work provided researchers with valuable quantitative relationships that could guide the development of safer energetic material formulations.

Two empirical methods are introduced for the assessment of friction sensitivity of nitramines and quaternary ammonium-based energetic ionic liquids.

9.1 Friction sensitivity of nitramines

For nitramines, BAM friction or Julius Peters friction apparatus can be used to determine their friction sensitivity. In these experimental methods, the explosive sample is held between a porcelain plate and a porcelain peg under a given load. Frictional forces are applied by the horizontal movement of the porcelain plate. The relative sensitivity to friction is indicated by the lowest load which leads to ignition, crackling, or explosion. It was shown that the friction sensitivity of nitramines can be related to their molecular structure, as [532]:

$$FS = 600.8 - \frac{2{,}428.6b + 6{,}481.4c + 9{,}560.9d}{Mw} + 54.5P_{FS}^{+} - 77.8P_{FS}^{-}, \tag{9.1}$$

where FS is the friction sensitivity in N; P_{FS}^{+} and P_{FS}^{-} are two parameters that can be predicted on the basis of the molecular structure of the nitramines as follows.

(1) Acyclic nitramines: For those compounds containing more than two repetitive $[-CH_2N(NO_2)-]$ units, the value of P_{FS}^{+} equals 1.0, whilst the value of P_{FS}^{-} is 0.5 for the presence of the $-N(NO_2)-CH_2-CH_2-N(NO_2)-$ molecular fragment. The value of P_{FS}^{+} depends on the number of separate molecular $N(NO_2)-C=$moieties $(n_{N(NO_2)-C=})$. It is equal to $P_{FS}^{+} = 5 - 2n_{N(NO_2)-C=}$ except for $n_{N(NO_2)-C=} \geq 3$ in which $P_{FS}^{+} = 0$.

(2) Cyclic nitramines: For cyclic nitramines containing equal numbers of nitramine and methylene groups, the value equals 1.0. For cyclic nitramines containing five membered rings, the values of P_{FS}^{-} are 0.75 and 1.25 for the presence of two and one $-N(NO_2)-$ groups per ring, respectively.

(3) The attachment of a $-N(NO_2)-$ group to a tetrazole ring: For 5-nitriminotetrazole salts, the values of P_{FS}^{+} and P_{FS}^{-} are 0.5 in the presence of ammonium and hydroxylammonium cations, respectively. The values of P_{FS}^{-} and P_{FS}^{+} equal 1.25 and 0.5 for 5-nitriminotetrazole and its methyl derivatives, respectively.

Example 9.1: The use of eq. (9.1) for the following energetic compound

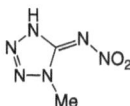

gives its friction sensitivity as

$$FS = 600.8 - \frac{2,428.6b + 6,481.4c + 9,560.9d}{Mw} + 54.5P_{FS}^+ - 77.8P_{FS}^-$$

$$= 600.8 - \frac{2,428.6(4) + 6,481.4(6) + 9,560.9(2)}{144.09} + 54.5(0.5) - 77.8(0)$$

$$= 158.0 \text{ N}.$$

The measured friction sensitivity for this compound is 160 N [533].

9.2 Friction sensitivity of quaternary ammonium-based energetic ionic liquids

A simple correlation has been developed to predict the friction sensitivity of quaternary ammonium-based energetic ionic liquids as follows [534]:

$$FS_{IL}(N) = 224 + 53.5a_{cat} - 25.9b_{cat} + 31.9c_{cat} + 134f_{cat} + 23.3a_{ani} + 162FS_{IL}^+ - 135FS_{IL}^-, \quad (9.2)$$

where $FS_{IL}(N)$ is friction sensitivity in newtons; a_{cat}, b_{cat}, c_{cat}, and f_{cat} are the number of carbon, hydrogen, nitrogen, and chlorine atoms in the cation, respectively; a_{ani} is the number of carbon atoms in the anion; FS_{IL}^+ and FS_{IL}^- are two correcting functions. Tables 9.1 and 9.2 show different types of cations and anions for which the values of FS_{IL}^+ and FS_{IL}^- equal 1.0.

Example 9.2: The use of eq. (9.2) for the following compound

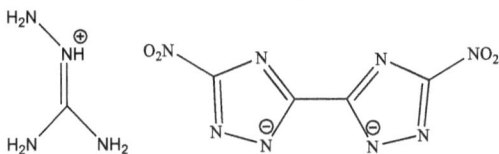

gives:

$$FS_{IL}(N) = 224 + 53.5a_{cat} - 25.9b_{cat} + 31.9c_{cat} + 134f_{cat} + 23.3a_{ani} + 162FS_{IL}^+ - 135FS_{IL}^-$$

$$= 224 + 53.5 \times 1 \times 2 - 25.9 \times 7 \times 2 + 31.9 \times 4 \times 2 + 134 \times 0 + 23.3 \times 4 + 162 \times 0$$

$$- 135 \times 0 = 317 \text{ N}$$

The measured value of FS_{IL} for this compound is 360 N [535].

Table 9.1: Different types of cations and anions for which the value of FS_{IL}^+ is 1.0.

Cation	Anion
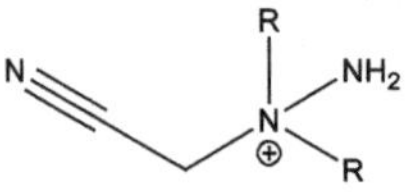 (structure: propargyl ammonium cation with R, R, NH2 groups)	NO_3^-
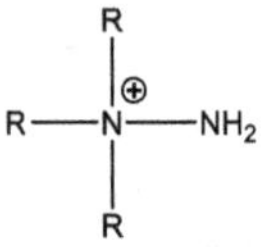 (structure: R—N⁺—NH2 with R, R)	NO_3^- ClO_4^- 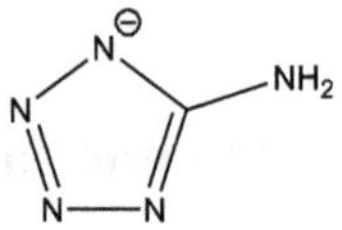 (tetrazole-NH2 structure)
NH_4^+ H_3N^+—NH_2 H_3N^+—OH	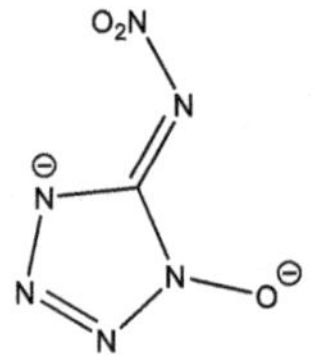 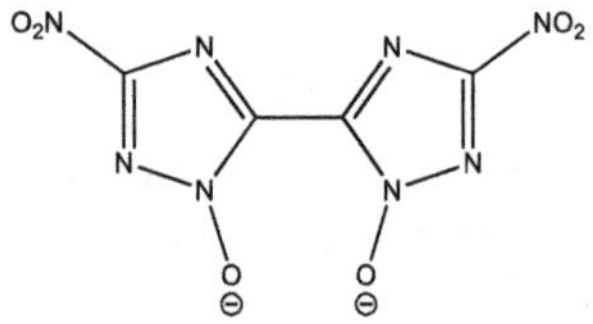 (O_2N-tetrazole structures, O_2N...NO_2 bis-triazole structure)

Table 9.2: Different types of cations and anions for which the value of FS_{IL}^- is 1.0.

Cation	Anion
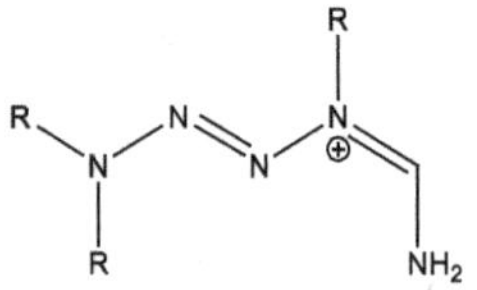 (structure: R, R, R, N=N, N⁺, NH2 chain)	ClO_4^-
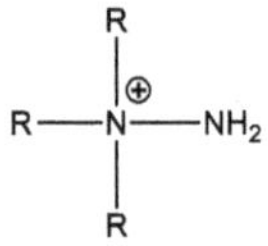 (structure: R—N⁺—NH2 with R, R)	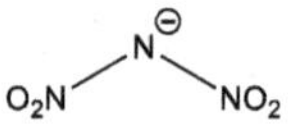 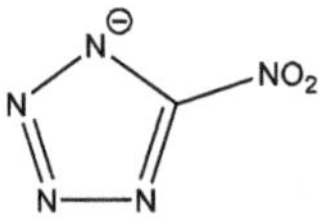 O_2N—N⁻—NO_2 (triazole—NO_2 structure)
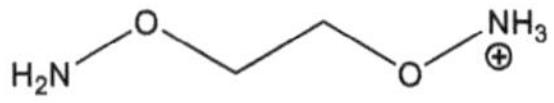 H_2N—O...O—NH_3^+	NO_3^- O_2N—N⁻—NO_2 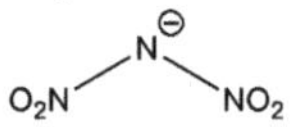

Table 9.2 (continued)

Cation	Anion
NH_4^+ $\overset{\oplus}{H_3N}{-\!\!-}NH_2$ $\overset{\oplus}{H_3N}{-\!\!-}OH$	

9.3 Summary

This chapter introduced a simple but reliable correlation between the friction sensitivity of nitramines and their molecular structures, which may be interesting for chemists and chemical industry. Equation (9.1) assumes that the friction sensitivity of a nitramine with the general formula $C_aH_bN_cO_d$ can be expressed as a function of its elemental composition and structural parameters. Since eq. (9.1) confirms that the initiation reactivity of energetic compounds is intimately related to their chemical character and molecular structures, it can help to elucidate the mechanism of initiation in energetic materials by friction stimuli. Equation (9.2) develops a simple method for prediction of friction sensitivity of quaternary ammonium-based energetic ionic liquids based on the contributions of some atoms in cations and anions, as well as two correcting functions.

10 Heat sensitivity

An ideal energetic compound shows high performance, low sensitivity, and a good shelf life. The heat sensitivity and thermal stability of energetic materials are two important features in their shelf life and safety because knowledge of these properties is important in order to avoid undesirable decomposition or self-initiation during their handling, storage, and application. Prediction of the thermal stability of a desired compound is an important starting point in the evaluation of its stability.

The thermolysis of energetic compounds can be used to estimate their thermal stability [536, 537]. The thermal stability of an energetic compound can be measured by various types of thermal analysis and gasometry, or by a variety of methods based on thermal explosions [538–540]. Among the different experimental methods, differential thermal analysis (DTA), differential scanning calorimetry (DSC), and thermogravimetric analysis methods are the thermoanalytical methods that are most widely used to examine the kinetic parameters of the thermolysis of energetic materials. The Soviet manometric method (SMM) is a suitable isothermal manometric method for energetic compounds, and uses a glass-compensating manometer of the Bourdon type to examine the kinetics of the thermolysis of energetic materials in vacuum. The data obtained by this method can be used to obtain the Arrhenius parameters of the non-autocatalyzed thermal decomposition of energetic materials. Experimental data from DTA and DSC can be converted to SMM data if a relationship such as a calibration curve exists between the results of the DTA and DSC with the results of SMM [541–546]. Since various factors affect the experimental data of the activation energies in the thermolysis of energetic materials, there is no uniform classification of a large majority of results obtained in various laboratories all over the world. In this chapter, some of the different methods for predicting parameters related to the thermal stability and heat sensitivity of energetic compounds – such as the activation energy and deflagration temperature – are reviewed. Some attempts have been done to estimate thermal decomposition kinetic parameters of some classes of energetic compounds using complex methods, e.g., energetic cocrystals by the artificial neural network model [547], and correlations of detonation parameters with activation energy for nitric esters [548]. Some of the simple correlations which are based on the molecular structures of different classes of energetic compounds are also demonstrated.

10.1 Thermal kinetics correlations

Thermoanalytical methods such as DTA, DSC, and thermogravimetry (TGA) can be used to determine Arrhenius parameters. Experimental critical temperatures for explosives of a given size and geometry can be obtained by a variety of tests, such as isothermal cook-off, slow cook-off, one-liter cook-off, and the isothermal time-to-explosion (Henkin test), in which the explosive may be confined or unconfined. Yan and Zeman [395], as

https://doi.org/10.1515/9783112206768-010

well as Zeman and Jungová [396] have reviewed some methods for predicting the kinetic parameters of some classes of explosives. Several approaches are given here.

In the thermolysis of nitramines, homolysis of the $N-NO_2$ bond is a primary step for secondary nitramines, whereas the homolysis of primary nitramines is a bimolecular autoprotolytic reaction [549]. Nitramine groups contribute strongly to the intermolecular potential in the crystalline state because the longest N–N bonds are responsible for the homolytic reactivity of nitramines [550]. There are some linear relationships between the activation energies of nitramine decomposition with the ^{15}N NMR chemical shifts of nitrogen atoms of nitramino groups [392], heats of detonation, or the electronic charges at nitrogen atoms of the nitramines [551, 552]. A Mulliken population analysis of the electron densities using the DFT B3LYP/6–31G** method can be used to calculate the electronic charges at the nitrogen atoms of nitramines [552]. Crystal lattice energies of nitramines do not generally differ from those of polynitroaromatics [454]. The activation energy (which corresponds to the slope in the Kissinger relationship) [553] can be used to evaluate the results of nonisothermal differential thermal analysis. Zeman [555] has used the modified Evans–Polanyi–Semenov (E–P–S) equation to interpret the chemical micromechanism governing the initiation of the detonation of energetic materials. Zeman [554] has used the heat of explosion and activation energy of the low-temperature thermal decomposition to obtain the modified E–P–S equation for energetic materials. There are several simple correlations for the prediction of the activation energy of the thermolysis of several important classes of energetic compounds on the basis of their molecular structures, which are discussed in the following sections.

10.1.1 Nitroparaffins

A suitable relationship has been derived to predict the activation energies for the low-temperature non-autocatalyzed thermolysis (E_a) of nitroparaffins, and the results of SMM are as follows [555]:

$$\ln E_a = \frac{0.4190a + 0.1793b + 1.1914d}{Mw} \times 100. \tag{10.1}$$

Example 10.1. The use of eq. (10.1) for 2,2-dinitropropane ($C_3H_6N_2O_4$) gives:

$$\ln E_a = \frac{0.4190a + 0.1793b + 1.1914d}{Mw} \times 100$$

$$= \frac{0.4190(3) + 0.1793(6) + 1.1914(4)}{134.09} \times 100$$

$$= 5.294$$

$$E_a = 199.1 \text{ kJ/mol.}$$

The measured value for this compound is 198.74 kJ/mol [556].

10.1.2 Nitramines

The elemental composition and structural parameters of nitramines can be used to derive a suitable correlation for predicting E_a as follows [557].

$$\ln E_a = 0.5385 + 0.8951(5.012 - 0.0367a + 0.0255b - 0.0304c + 0.0407d) + 0.1698P_{>5},$$

$$(10.2)$$

where $P_{>5}$ is equal to 1.0 for cyclic nitramines that contain rings which are larger than five-membered rings, as well as nitramine cages.

Example 10.2. 4,10-Dinitro-2,6,8,12-tetraoxa-4,10-diazaisowurtzitane (TEX) has the following molecular structure:

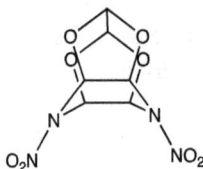

The use of eq. (10.2) for TEX gives

$$\ln E_a = 0.5385 + 0.8951(5.012 - 0.0367a + 0.0255b - 0.0304c + 0.0407d) + 0.1698P_{>5}$$

$$= 0.5385 + 0.8951(5.012 - 0.0367(6) + 0.0255(6) - 0.0304(4) + 0.0407(8)) + 0.1698$$

$$= 5.317$$

$$E_a = 203.8 \text{ kJ/mol}.$$

The experimental value for TEX is 202.5 kJ/mol [549].

10.1.3 Polynitroarenes

It was found that in order to be able to predict the E_a of polynitroarenes, several parameters are important, which include

(1) the contribution of the elemental composition;
(2) the number of $-NH(C=O)-C(=O)NH-$ groups (e.g., N,N'-bis (2,4,6-trinitrophenyl) oxamide) or more than one $-NH_2$ groups (e.g., 2,4,6-trinitrobenzene-1,3-diamine);
(3) the existence of either one α-CH (e.g., 2,4,6-trinitrotoluene), or methoxy group attached to one aromatic ring, or $CH_3O-[C(NO_2)-CH-C(NO_2)]$ – which has the opposite effect with respect to the second parameter.

The number of amino groups can affect the sensitivity and performance, as well as the physicothermal properties of nitroaromatic compounds [285]. Thus, the presence of amino groups can affect the value of the activation energy when more than one amino group is attached to aromatic rings. For polynitroaromatics with α-CH, it was indicated in previous chapters that their sensitivities with respect to different stimuli show different behaviors. It was found that the presence of one α-CH or $CH_3O-[C(NO_2)-CH-C(NO_2)]$ group in the molecular structure of polynitroarenes can influence the value of E_a. The value of E_a is a function of the aforementioned parameters, which can be expressed by the following equation [558]:

$$\log(E_a) = 2.25 + 0.0337 OEC + 0.146 n_{\text{NHCOCONH, NH}_2 > 1}$$
$$+ 0.124 P_{a-\text{CH or } CH_3O-[C(NO_2)-CH-C(NO_2)]}, \tag{10.3}$$

where OEC, $n_{\text{NHCOCONH, NH}_2 > 1}$, and $P_{a-\text{CH or } CH_3O-[C(NO_2)-CH-C(NO_2)]}$ are the contributions of the elemental composition, the second, and the third structural parameters, respectively. The value of $P_{a-\text{CH or } CH_3O-[C(NO_2)-CH-C(NO_2)]}$ is 1.0 for the presence of either one α-CH (e.g., 2,4,6-trinitrotoluene) or a methoxy group attached to one aromatic ring, or a $CH_3O-[C(NO_2)-CH-C(NO_2)]$ group. The value of OEC is the optimized elemental composition of polynitroarenes with the general formula $C_aH_bN_cO_d$, which can be obtained by $OEC = 2.74a - 1.48b - 2.31c - d$.

Example 10.3. 2,2′,2″,4,4′,6,6′,6″-Octanitro-1,1′:3′,1″-terphenyl (ONT) has the following molecular structure:

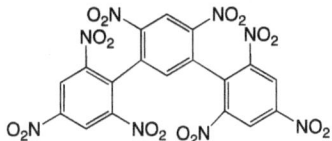

If eq. (10.3) is used, the result is given as

$$OEC = 2.74a - 1.48b - 2.31c - d$$
$$= 2.74(18) - 1.48(6) - 2.31(8) - 16$$
$$= 5.96,$$
$$\log(E_a) = 2.25 + 0.0337 OEC + 0.146 n_{\text{NHCOCONH, NH}_2 > 1}$$
$$+ 0.124 P_{a-\text{CH or } CH_3O-[C(NO_2)-CH-C(NO_2)]}$$
$$= 2.25 + 0.0337(42.92) + 0.146(0) + 0.124(0)$$
$$= 2.451$$
$$E_a = 282.4 \text{ kJ/mol.}$$

The measured value of E_a is 281.58 kJ/mol [554].

10.1.4 Organic energetic compounds

For various organic energetic compounds, the results show that the important factors for predicting E_a can be grouped into additive and nonadditive structural parameters as follows [559]:

$$E_a = 166.36 + 2.85a - 21.2n_{OH} + 31.98E^{+}_{nonadd} - 44.93E^{-}_{nonadd}, \tag{10.4}$$

where n_{OH} represents the number of hydroxyl groups; two functions E^{+}_{nonadd} and E^{-}_{nonadd} show the increasing and decreasing contribution of nonadditive structural parameters, respectively.

10.1.4.1 E^{+}_{nonadd}

The presence of some structural parameters can increase the activation energy and enhance the thermal stability of energetic compounds.

(1) Cyclic nitramines: The value of E^{+}_{nonadd} is 1.0 for cyclic nitramines which corresponds to one of the following conditions:
    (a) ring is larger than a six-membered ring and contains two $-NNO_2$ groups;
    (b) ring larger than a four-membered ring;
    (c) bicyclic ring with only two $-NNO_2$ groups per ring.
    These conditions are consistent with the correlation of activation energy in which the activation energies increase with increasing ring size [557].

(2) Nitroalkanes: For nitroalkanes containing molecular moieties, $R-CH_2NO_2$ and $R-C(NO_2)_2R'$, the values of E^{+}_{nonadd} equal 2.0 and 1.0, respectively.

(3) Nitroaromatics:
    (a) If the amino pyridine derivative, , triazine ring or four adjacent nitrogens in nitroaromatics are present, the value of E^{+}_{nonadd} is 1.0;

    (b) The value of E^{+}_{nonadd} is 2.0 for the presence of the molecular fragment where TNP is 2,4,6-trinitrophenyl.

10.1.4.2 E^{-}_{nonadd}

(1) Acyclic nitramines and cyclic nitramines containing small rings: For acyclic nitramines which contain only one $-NNO_2$ group in the form $Ar(or\ H)-N(NO_2)CH_3$, the value of E^{-}_{nonadd} is 0.75. The value of E^{-}_{nonadd} is equal to 0.4 for cyclic nitramines containing rings which are smaller than five membered rings, or five membered rings with more than one $N-NO_2$ group.

(2) Nitroaromatics:
    (a) If the molecular moieties $-O(or\ S)-R(or\ Ar)$ are present, the value of E^{-}_{nonadd} equals 1.5;

(b) If the [structure: O_2N, H_3C, $H(or\ NH_2)$, NO_2] molecular fragment is present, the value of E_{nonadd}^- equals the number of this molecular moiety;

(c) The values of E_{nonadd}^- equal 1.5, 1.0, and 0.75 for the compound TNP–X where X is –Cl, –NH–, and –N<, respectively;

(d) For energetic compounds of the type TNP–Y–TNP where Y is –N=N–, –CH$_2$–CH$_2$–, and –SO$_2$–, the value of E_{nonadd}^- is 2.0.

(3) Presence of the nitrate group: if the –ONO$_2$ group is present, the value of E_{nonadd}^- is 0.3.

(4) The presence of the nitroso group: The value of E_{nonadd}^- is equal to 0.75.

Example 10.4. If eq. (10.4) is applied to 2,4,6-trinitrobenzene-1,3,5-triol (TNPg) with the following molecular structure

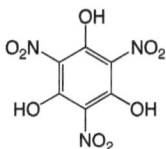

[structure: O_2N, OH, NO_2, HO, OH, NO_2]

it gives

$$E_a = 166.36 + 2.85a - 21.2n_{OH} + 31.98E_{nonadd}^+ - 44.93E_{nonadd}^-$$
$$= 166.36 + 2.85(6) - 21.2(3) + 31.98(0) - 44.93(0)$$
$$= 119.9\ kJ/mol.$$

The measured value of is 114.16 kJ/mol [554].

10.2 Heat of decomposition and temperature of thermal decomposition

Thermal analysis methods can be used to study the heat of decomposition and onset temperature [122, 285, 396, 560]. DSC [1, 3] is a typical example of an experimental screening test, which provides heats of decomposition with uncertainties in the measurement of about 5–10% [560]. The exothermic onset temperature (the temperature at which the first deflection from the baseline is observed), thermal decomposition temperature, and the temperature at which maximum mass loss occurs are three important parameters for assessing the heat sensitivity of different kinds of energetic compounds, and which give better reproducibility than the heats of decomposition.

ML has become a hot topic in energetic materials research, with much attention focused on predicting detonation properties. However, few studies have explored ML models for thermal decomposition temperatures, and those that do often suffer from small datasets, limiting model generalizability. To address this gap, Wu et al. [562] compiled a robust dataset of 1,022 energetic molecules with decomposition temperature values ranging from 38 to 425 °C. After rigorous training, the gradient boosting regressor (GBR) emerged as the top-performing model, achieving an R^2 of 0.65 and a mean absolute error (MAE) of 27.7 on the test set. Key insights came from feature analysis: molecular bond stability (captured by BCUT metrics) and atomic composition (Molecular ID descriptors) proved critical for accurate predictions. Wu et al. [562] also identified outliers, suggesting that adding molecular interaction features could further refine model performance. These findings advance the application of ML in energetic material research – highlighting the importance of dataset quality and strategic feature selection for reliable property prediction.

Zhang et al. [563] employed ML-assisted regression modeling to predict the thermal decomposition temperatures of energetic materials and investigated the key factors governing their thermal stability. Using a dataset of 885 diverse compounds, Zhang et al. [563] evaluated both linear and nonlinear ML algorithms. Among the seven models tested, tree-based methods exhibited particularly strong predictive performance. To uncover the underlying factors influencing thermal decomposition, Zhang et al. [563] applied hierarchical classification to parse the dataset. The results revealed that thermal stability in energetic materials is governed by multiple factors, including molecular composition, electronic distribution, chemical bond characteristics, substituent types, and detonation-related parameters. These insights deepen the understanding of the fundamental mechanisms behind energetic material thermal decomposition, offering valuable guidance for the design and synthesis of new high-energy, stable materials.

Zhang et al. [564] systematically developed five general descriptor sets for 1,091 compounds and combined them with nine machine learning algorithms to build predictive models. These models achieved a MAE between 29 and 41 K – performance on par with state-of-the-art efforts. The findings highlight how multi-level structural interactions critically influence the thermal stability and decomposition of energetic materials. These insights pave the way for better descriptor design, ultimately supporting the development of next-generation energetic compounds.

Pandey and Roy tackled the critical challenge of predicting thermal stability in energetic materials, where balancing performance with heat resistance remains key for developing safer compounds. Their work applies an enhanced quantitative q-RASPR approach, merging QSPR and read-across methods to improve prediction accuracy. Focusing on decomposition temperature ($n = 656$ rigorously curated compounds), their model achieves an R^2 of 0.64. The results identify specific molecular features controlling thermal stability, enabling targeted screening of heat-resistant candidates as ($R^2 = 0.620$):

$$T_{\text{onset}} = 144.45 + 2.684\,C\% - 43.374\,B01[O-O] - 15.109\,B03[N-O] + 8.425\,Hy$$

$$- 8.311\,LOGP99 + 19.520\,nArNO_2 + 16.965(C-005) - 8.233\,B01[N-N] \qquad (10.5)$$

$$+ 0.596\,RA\,function(LK) - 0.870\,SE(LK).$$

The T_{onset} model relies on several key molecular descriptors, each representing different structural or chemical properties. $C\%$ refers to the percentage of carbon atoms in the molecule, while $B01[O-O]$ indicates whether an oxygen-oxygen (O–O) bond exists at a topological distance of 1 (directly bonded). Similarly, $B03[N-O]$ tracks nitrogen–oxygen (N–O) interactions at a distance of 3 (separated by two atoms). Hy , the hydrophilic factor, measures how strongly a molecule interacts with water, and $LOGP99$ (the Wildman–Crippen log P) estimates its hydrophobicity. The descriptor $nArNO_2$ counts the number of aromatic nitro (–NO$_2$) groups, and $C-005$ represents a methyl group (CH$_3$) attached to another atom. Additionally, $B01[N-N]$ checks for nitrogen–nitrogen (N–N) bonds at a distance of 1, $RA\,function(LK)$ is a composite descriptor derived from read-across analysis, and $SE(LK)$ represents the weighted standard error of response values from closed-source compounds. In this model, some descriptors have a positive influence on T_{onset} – meaning higher values of $RA\,function(LK)$, $C\%$, $nArNO_2$, Hy, or $C-005$ lead to an increase in T_{onset}. On the other hand, descriptors like $B01[O-O]$, $B03[N-O]$, $LOGP99$, and $SE(LK)$ have a negative effect, so higher values for these tend to decrease the decomposition temperature. Essentially, compounds with more carbon atoms, nitro groups, or hydrophilic character will resist decomposition at higher temperatures, while those with specific bonds (N–N, O–O, or N–O interactions) or greater hydrophobicity will break down more easily.

Hossain and Roy [565] used QSPR modeling to create a fast, interpretable model for T_{onset} using just 2D molecular descriptors. Their model performed well, balancing accuracy and computational efficiency. It also highlighted structural features that boost stability, like quaternary carbons and conjugated systems.

Designing heat-resistant energetic materials requires a deep understanding of their thermal sensitivity, but predicting this property remains a major challenge. To address this, Zhang et al. [566] used first-principles calculations to develop a theoretical model that evaluates thermal stability. By analyzing key factors like band gap, density of states, and Young's modulus, they derived a new empirical parameter, Ψ, which shows a clear quantitative relationship with thermal decomposition temperature. Zhang et al. [566] tested this model on 10 different energetic materials and found that Ψ strongly correlates with experimental decomposition temperatures – confirming the model's reliability. Essentially, higher Ψ values correspond to higher decomposition temperatures, meaning the material is more thermally stable. This suggests that Ψ can serve as a useful predictor for thermal sensitivity in energetic materials. Zhang et al. [566] performed calculations using first-principles simulations in the Cambridge Serial Total Energy Package (CASTEP) module within Materials Studio (MS). For accuracy, we applied the Perdew–Burke–Ernzerhof (PBE) functionals under the generalized gradient approximation (GGA) method, along with Grimme dispersion correction to account for van der Waals interactions.

In the following sections, some methods for the prediction of these parameters are discussed.

10.2.1 Heat of decomposition of nitroaromatics

Some QSPR studies were undertaken to predict the heats of decomposition of nitroaromatic compounds. For 19 nitrobenzene derivatives, Saraf et al. [567] introduced a correlation based on the number of nitro groups (n_{NO_2}) with a fitting error of 8%. Fayet et al. [568–570] used a series of preliminary multilinear models to derive some correlations from a data set of 22 molecules using some quantum chemical descriptors. Fayet et al. [571, 572] found very robust models by analyzing a more extended data set of 77 nitrobenzene derivatives. A suitable model was introduced for the whole diverse list of structures using a qualitative decision tree with high reliability [572]. For some nitrobenzenes which have no substituent located in the *ortho* position to the nitro group, another complex model was introduced based on quantum mechanical calculations [572]. Three multilinear models were introduced for nitrobenzenes which have no substituents in the *ortho* position relative to the nitro group using a set of complex descriptors [573]. These methods are complex because they require special computer codes and expert users.

A simple correlation on the basis of n_{NO_2} has been introduced to calculate the heat of decomposition by considering inter- and intramolecular interactions rather than using complex molecular descriptors as follows [574]:

$$-\Delta H_{decom} = -53.32 + 362n_{NO_2} - 99.33\Delta H^-_{decom} + 108.9\Delta H^+_{decom}, \tag{10.6}$$

where ΔH_{decom} is the heat of decomposition in kJ/mol; n_{NO_2} is the number of nitro groups; and ΔH^+_{decom} and ΔH^-_{decom} are two correcting functions. The presence of some molecular moieties can influence the values of the heat of decomposition and are referred to as ΔH^+_{decom} and ΔH^-_{decom}, and described in the following sections.

10.2.1.1 ΔH^-_{decom}
(1) Some molecular fragments in nitrobenzene:
   (a) Intramolecular hydrogen bonding: The value of ΔH^-_{decom} is 1.5 for the presence of only –OH or –CH$_2$–COOH groups *ortho* to the nitro group (or another hydrogen bonding group such as –NH$_2$).
   (b) Mono methyl derivatives: The value of ΔH^-_{decom} is 1.5. It was shown that the presence of intramolecular hydrogen bonding or a methyl group may decrease the heats of sublimation in nitroaromatic compounds [76, 98].
(2) The presence of halogens (or –CF$_3$) beside nitro groups in halogenated derivatives of nitrobenzene or polynitrobenzene: The value of ΔH^-_{decom} equals 3.0.
(3) The existence of –COOH beside nitro groups in polynitrobenzene: The value of ΔH^-_{decom} is 4.0.

10.2.1.2 ΔH_{decom}^{+}

(1) Some specific molecular moieties in nitrobenzene: if the (Cl or H)–C=O, –SCN, (NHNH$_2$)–C=O, or –C=C– groups are present, the values of ΔH_{decom}^{+} are 0.8, 2.0, 1.5, and 1.7, respectively.

(2) Some molecular fragments in polynitrobenzene: The values of ΔH_{decom}^{+} are 4.0, 3.0, and 2.0 for the presence of the –C(=O)NH$_2$, halogen, and methyl groups, respectively.

Example 10.5. For 2-amino-4-nitrophenol with the following molecular structure

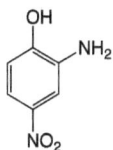

Equation (10.6) gives

$$-\Delta H_{decom} = -53.32 + 362.0 n_{NO_2} - 99.33 \Delta H_{decom}^{-} + 108.9 \Delta H_{decom}^{+}$$

$$= -53.32 + 362.0(1) - 99.33(1.5) + 108.9(0)$$

$$= 159.6 \, kJ/mol.$$

The measured value of $-\Delta H_{decom}$ is 130 kJ/mol [561]. The calculated value by the complex QPRR model of Fayet et al. [572] is 238 kJ/mol (Dev = 108 kJ/mol).

10.2.2 Heats of decomposition of organic peroxides

Several QSPR methods containing complex descriptors were applied to explore:
(1) the relationship between the heat and temperature of decomposition of organic peroxides and their quantum mechanical descriptors [575, 576];
(2) the correlation between the self-accelerating decomposition temperature (SADT) of organic peroxides and their molecular structures [577] or their quantum mechanical properties [578]. These methods require specific computer codes and expert users.

It was found that the molecular structures of organic peroxides can be used to predict the heat of decomposition of these compounds as follows [579]:

$$-\Delta H_{decom}' = 1551.45 - 41.13a + 1014.05 P_{(HO-O \ldots -O-OH)}$$

$$+ 640.13a + 857.68\beta, \tag{10.7}$$

where $\Delta H_{decom}{}'$ is the heat of decomposition in J/g; $P_{(HO-O\ldots-O-OH)}$ is 1.0 for the presence of two hydroperoxy functional groups in the molecular structure. The parameters α and β are nonadditive structural parameters, which can be specified as follows.

Definition of α: if aromatic rings, nonaromatic rings and nonaromatic rings containing methyl substituents are present, the values of α are 0.1, 0.2, and 0.6, respectively.

If the organic peroxide contains the [fragment structure] fragment in its structure, the value of α equals 2.0.

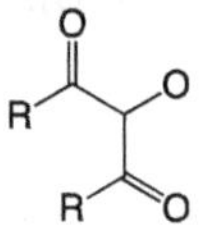

Definition of β: if the [fragment structure] molecular fragment is present in acyclic peroxides, β is 0.5. If two (–O–O–R) or [fragment structure] fragments are present, the value of β is 0.4 and 0.6, respectively.

Example 10.6. Di-*tert*-butyl peroxide has the following molecular structure.

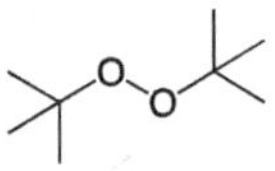

The use of eq. (10.7) gives

$$-\Delta H'_{decom} = 1{,}551.45 - 41.13\alpha + 1{,}014.05 P_{(HO-O\ldots--O-OH)} + 640.13\alpha + 857.68\beta$$

$$= 1{,}551.45 - 41.13(8) + 1{,}014.05(0) + 640.13(0) + 857.68(0)$$

$$= 1{,}222.4 J/g.$$

There are two measured values of $-\Delta H'_{decom}$, i.e., 1,082.5 [580] and 1,175.0 J/g [578].

10.2.3 Onset temperature of polynitroarenes and organic peroxides as well as maximum loss temperature of organic azides

Since experimental data have been reported for the onset and maximum temperature of thermal decomposition for selected classes of energetic compounds, it has been possible to develop different methods to predict these values for other explosives. These methods are discussed in the following sections.

10.2.3.1 Onset temperature of polynitroarenes

For some subgroups of polynitroarenes, there are some relationships between the onset temperature and detonation characteristics [482, 542]. Using the DFT B3LYP/6–3-31G** method, it was found that there is a linear relationship between the onset temperature of thermal decomposition and the electronic charges of nitrogen atoms of selected subgroups of polynitroarenes [551]. For polynitroarenes, it has been shown that the following correlation can be used [581]:

$$T_{\text{onset}} = 571.17 + 30.63a - 21.29b + 32.57c - 43.11d$$
$$+ 15.98\left(n_{\text{NH}_2} - n_{\text{NO}_2}\right) + 50.69\left(|n_{\text{TNB}} - 2| - P_{\text{TNB–CH}_2\text{–TNB}}\right), \tag{10.8}$$

where T_{onset} is the onset temperature in K; $\left(n_{\text{NH}_2} - n_{\text{NO}_2}\right)$ is the difference between the number of amino and nitro groups in energetic compounds containing amino groups; $|n_{\text{TNB}} - 2|$ is the absolute value of the number of 1,3,5-trinitrobenzene rings minus two; $P_{\text{TNB–CH}_2\text{–TNB}}$ is equal to one or zero for the presence or absence of –CH$_2$– between two 1,3,5-trinitrobenzene aromatic rings, respectively.

Example 10.7, The use of eq. (10.8) for the following energetic compound

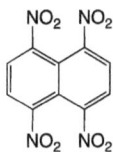

gives:

$$T_{\text{onset}} = 571.17 + 30.63a - 21.29b + 32.57c - 43.11d$$
$$+ 15.98\left(n_{\text{NH}_2} - n_{\text{NO}_2}\right) + 50.69\left(|n_{\text{TNB}} - 2| - P_{\text{TNB–CH}_2\text{–TNB}}\right)$$
$$= 571.17 + 30.63(10) - 21.29(4) + 32.57(4) - 43.11(8)$$
$$+ 15.98(0) + 50.69(0)$$
$$= 577.7 \text{ K.}$$

The measured T_{onset} is 578.8 K [542].

10.2.3.2 Onset temperature of organic peroxides

For organic peroxides, it has been shown that the decomposition onset temperature can be expressed as a function of several structural parameters as [579]:

$$T_{\text{onset}} = 438.96 - 4.92d - 29.11P_{\text{C=O}} - 14.02\lambda_{\text{sym}} - 30.81\lambda', \tag{10.9}$$

where $P_{\text{C=O}}$ is 1.0 for the presence of the carbonyl group; and λ_{sym} is 1.0 for those peroxides that have the same fragments attached to each side of the –O–O– bond, i.e., R–O–O–R' where R=R'. The presence of some molecular fragments may also affect the values of T_{Dec}, which are incorporated as correcting factors. The parameter λ' represents the positive and negative contributions of various structural features, which allow more reliable T_{onset} values to be obtained and can be defined based on the presence of the following molecular fragments:

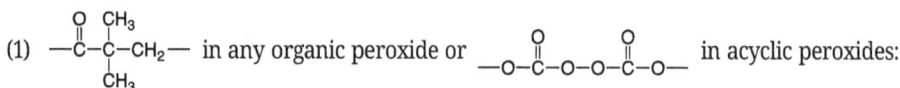

(1) in any organic peroxide or in acyclic peroxides:

The value of λ' is 1.0.

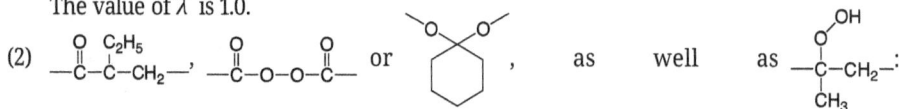

(2) or , as well as

For or , λ' equals 0.7. The value of λ'

for is given as: $\lambda' = 0.7 \times$ the number of , in molecular structure of peroxide.

(3) $-O-C(R)(R')-O-$: For the presence of this molecular fragment, where R and R' can be $-CH_3$ or $-CH_2-CO$, or $-C(CO)_2$ in their molecular structures, $\lambda' = -1.0$.

Example 10.8. If eq. (10.9) is used for the peroxide shown in Example 10.6, it gives

$$T_{onset} = 438.96 - 4.92d - 29.11\lambda_{C\,=\,O} - 14.02\lambda_{sym} - 30.81\lambda'$$

$$= 438.96 - 4.92(2) - 29.11(0) - 14.02(1) - 30.81(0)$$

$$= 415.1\,K.$$

There are two measured values of T_{onset}, i.e., 412.85 [580] and 426.15 K [578].

10.2.3.3 Temperature of maximum mass loss of organic azides

For different types of organic azides, it was shown that the temperatures of maximum mass loss (T_{dmax}) can be given by [582]:

$$T_{dmax} = 405.57 + 1.3959b + 4.3222c + 33.670T_{dmax}^+ - 32.515T_{dmax}^-, \qquad (10.10)$$

where T_{dmax} is in K; T_{dmax}^+ and T_{dmax}^- are two correcting functions which are used to show the increasing and decreasing contributions of nonadditive structural parameters, respectively. The values of T_{dmax}^+ and T_{dmax}^- are given in Tables 10.1 and 10.2 for different molecular fragments. As shown in Tables 10.1 and 10.2, the position of the azide group in an aromatic ring, neighboring groups and the presence of some specific functional groups are important parameters in predicting the values for T_{dmax}^+ and T_{dmax}^-. Since the attachment of Cl–, –CH$_2$OH, and –Ar *ortho* to –N$_3$ group may increase thermal stability, the contribution of T_{dmax}^+ should be considered. The effect of T_{dmax}^+ has also been considered for the presence of the $-C(=O)O-$ group under certain conditions, and also for the attachment of the $-N_3$ group to tertiary carbon. The presence of the $-CH_2-N(cyclic)$ molecular moiety, the existence of Cl *ortho* to the nitro group, and the presence of $-CO-$ *ortho* to the $-N_3$ group can decrease the thermal stability. Therefore, the presence of some molecular fragments or specific groups in an organic azide is responsible for increasing and decreasing the thermal stabilities.

Table 10.1: Summary of the contributions of T_{dmax}^+.

Molecular fragments	Condition	T_{dmax}^+	Examples
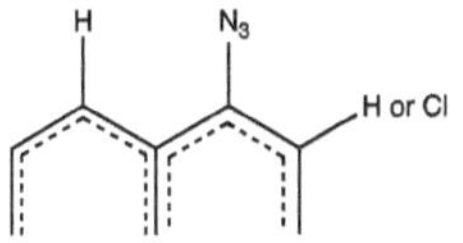		0.7	2-Azido-2H-chromen-2-one
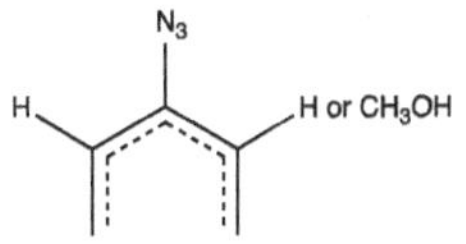		0.9	3-(3-Azidophenyl)-6-chloropyridazin-4-ol
Tertiary carbon – – – – N_3		1.0	3-Azido-7-methyl-phenylquinoline-2,4-(1H,3H)-dione
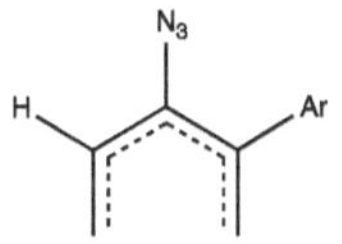		0.7	4-Azido-2-chloro-3-phenylquinoline
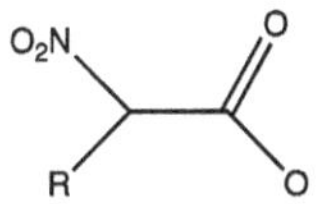		1.0	4-Azido-3-nitro-2H-chromen-2-one
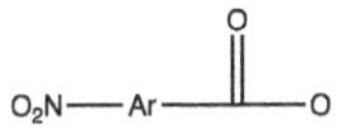		1.5	1,3-Diazidopropan-2-yl 3,5dinitrobenzoate
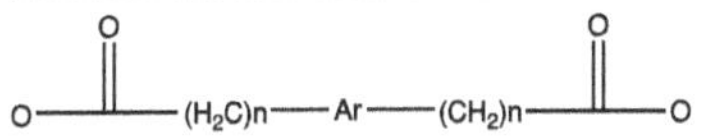	$n \geq 0$	1.0	8-Azidooctyl 2-(2-((8-azidooctyl) oxy)-2-oxoethyl)
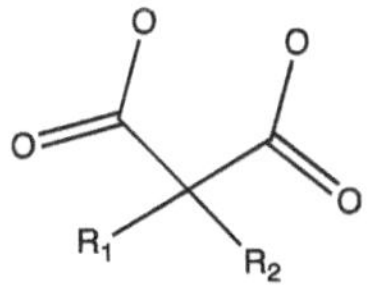		1.2	Bis(1,3-diazidopropan-2-yl)-1H-indene-2,2(3H)-dicarboxylate)
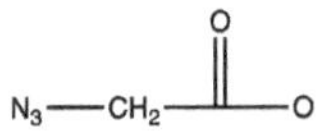		1.5	1,3-Bis(azidoacetoxy)-2-azidoacetoxymethyl-2-ethylpropane

Table 10.2: Summary of the contributions of T_{dmax}^-.

Molecular moieties	T_{dmax}^-	Example
N (cyclic) | CH$_2$ |	1.0	7-Azido-5-oxo-2,3-dihydro-1H,5H-pyrido[3,2,1-*ij*]quinoline-6-carbaldehyde
O$_2$N, CI (structure)	1.0	4-Azido-2-chloro-6-methyl-3-nitropyridine
N$_3$, R or Ar, CO or NO$_2$, H (structure)	0.8	1-(2-Azidophenyl)-1-ethanone

Example 10.9. The use of eq. (10.10) for 4-azido-3-nitro-2H-chromen-2-one with the following molecular structure

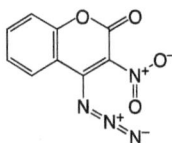

gives

$$T_{dmax} = 405.57 + 1.3959b + 4.3222c + 33.670T_{dmax}^+ - 32.515T_{dmax}^-$$
$$= 405.57 + 1.3959(4) + 4.3222(4) + 33.670(1.0) - 32.515(0)$$
$$= 462.11 \text{ K.}$$

The measured value of T_{dmax} is 463.15 K [583].

10.3 Deflagration temperature

The deflagration temperature is defined as the temperature at which a small sample of an energetic compound gets ignited [584]. This can be determined by heating 0.02 g of the sample in a glass tube in a Wood's metal bath at a heating rate of 5 °C/min. Since various inter- and intramolecular parameters may affect the value of the deflagration temperature, a suitable correlation for estimating the deflagration temperature of energetic compounds containing $-NNO_2$, $-ONO_2$, or $-CNO_2$ groups was established as [48, 585]:

$$DT = 476.6 + 13.08a - 6.21d + 103.7F^+_{nonadd} - 103.1F^-_{nonadd}, \qquad (10.11)$$

where DT is the deflagration temperature in K; two functions F^+_{nonadd} and F^-_{nonadd} show the increasing and decreasing contributions of nonadditive structural parameters, respectively, which are specified in the following sections.

10.3.1 F^+_{nonadd}

(1) Cyclic nitramines containing rings which are larger than six-membered rings or carbocyclic nitroaromatics with

$$\frac{1}{3} \leq \frac{n_{NH_2}}{n_{NO_2}} \leq 1,$$

where n is the number of specified groups: The value of F^+_{nonadd} is 1.0.
(2) Carbocyclic nitroaromatic compounds containing only alkyl substituents: The value of F^+_{nonadd} is 0.5.
(3) The existence of other specific polar groups: The value of F^+_{nonadd} equals 0.5 for the presence of a nitrate salt, $-NHCONH-$ group, or cyclic energetic compounds containing $>N-CO-N<$, two $>N-CO-N(NO_2)-$, or two $-(O_2N)N-CO-N(NO_2)-$ groups. For the presence of a tertiary amine (or $-O^-NH_4^+$), the value of F^+_{nonadd} equals 0.8.

10.3.2 F^-_{nonadd}

(1) The presence of azido, $-N-OH$, cyclic ether groups, as well as the presence of both $>C(NO_2)_2$ and $-OH$: For the presence of the $-N_3$, $-N-OH$ or cyclic ether groups and the presence of both of $>C(NO_2)_2$ and $-OH$ groups, the values of F^-_{nonadd} equal 0.7, 0.5, 0.3 and 1.0, respectively.
(2) Substituents containing $>NNO_2$ or $-NHNO_2$ groups attached to carbocyclic nitroaromatic compounds: For the presence of $>NNO_2$ and $-NHNO_2$ groups, the values of F^-_{nonadd} equal 0.5 and 1.0, respectively.

10.3.3 Energetic compounds containing both F^+_{nonadd} and F^-_{nonadd}

(1) Carbocyclic nitroaromatic compounds containing both nitrate and ether groups in their substituents: For the presence of one nitrate group, the value of F^+_{nonadd} equals 0.5. The value of F^-_{nonadd} equals 0.5 for the presence of more than one nitrate group.

(2) Energetic compounds Ar–NH–Ar': For secondary amines attached to two aromatic rings if Ar and Ar' contain the N–O–N group as well as one nitro group or a tetrazole ring, the value of F_{nonadd}^- is 1.0. Meanwhile, if Ar' contains more than one nitro group, the value of F_{nonadd}^+ is 0.5.

Example 10.10. If eq. (10.11) is used for azido-acetic-acid-3-(2-azido-acetoxy)-2-(2-azido-acetoxymethyl)-2-nitropropylester with the following molecular structure

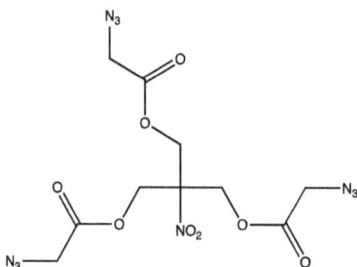

it gives

$$DT = 476.6 + 13.08a - 6.21d + 103.7F_{nonadd}^+ - 103.1F_{nonadd}^-$$
$$= 476.6 + 13.08(10) - 6.21(8) + 103.7(0) - 103.1(0.7)$$
$$= 488.5 \text{ K.}$$

The measured value of T_{dmax} is 487 K [586].

10.4 Thermal stability of selected classes of energetic ionic liquids and salts

Prediction of thermochemical parameters of a new energetic ionic liquid is of the utmost importance for their design to meet specific requirements and industrial applications. Several simple correlations have been developed for prediction activation energy and decomposition temperature (onset) of several classes of ionic liquids and salts, including energetic derivatives, which are discussed here.

10.4.1 Predicting activation energy of thermolysis of some selected ionic liquids

A simple model has been developed for the prediction of the activation energy of thermolysis of imidazolium, pyridinium, and phosphonium-based ionic liquids, mainly based on TGA, through the structure of their anions and cations as follows [587]:

$$E_{a,IL} = 106.12 + 1.1574b_{cat} - 112.06i_{cat} - 6.6481e_{ani} + 23.036h_{ani}$$
$$+ 33.831i_{ani} + 47.681j_{ani} - 58.688E_{a,IL}^- + 56.984E_{a,IL}^+, \quad (10.12)$$

where $E_{a,IL}$ is the activation energy of thermolysis of the desired ionic liquid or salt in kJ/mol; b_{cat}, i_{cat}, and e_{cat} are the number of hydrogen, phosphorous, and fluorine atoms in cation, respectively; h_{ani}, i_{ani}, and j_{ani} are the number of sulfur, phosphorous, and bromine atoms in the anion, respectively; $E_{a,IL}^-$ and $E_{a,IL}^+$ are two correcting functions. Tables 10.3 and 10.4 show the values of $E_{a,IL}^-$ and $E_{a,IL}^+$ equal 1.0 only for some specific imidazolium-based ILS under certain conditions.

Table 10.3: The contribution of $E_{a,IL}^- = 1.0$ in ionic liquids or salts.

Cation	Anion
	Br⁻
R' = Unsaturated alkyl group	
	HO₄S⁻, H₂O₄P⁻
	Cl⁻

Example 10.11. The use of eq. (10.11) for the following energetic ionic liquid

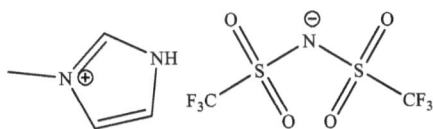

gives:

$$E_{a,\text{IL}} = 106.12 + 1.1574b_{\text{cat}} - 112.06i_{\text{cat}} - 6.6481e_{\text{ani}} + 23.036h_{\text{ani}} + 33.831i_{\text{ani}} + 47.681j_{\text{ani}}$$

$$- 58.688E_{a,\text{IL}}^{-} + 56.984E_{a,\text{IL}}^{+}$$

$$= 106.12 + 1.1574 \times 7 - 112.06 \times 0 - 6.6481 \times 6 + 23.036 \times 2 + 33.831 \times 0$$

$$+ 47.681 \times 0 - 58.688 \times 0 + 56.984 \times 0 = 120.4 \text{ kJ/mol.}$$

The measured value of $E_{a,\text{IL}}$ is 102.7 kJ/mol [588].

Table 10.4: The contribution of $E_{a,\text{IL}}^{+}$ in ionic liquids or salts.

Cation	Anion
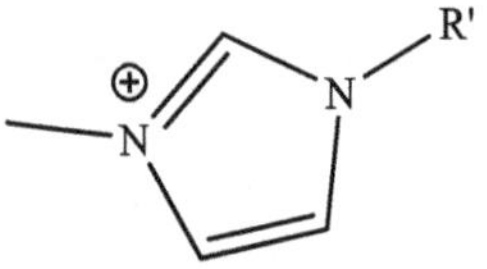 R = Saturated alkyl group with carbon atoms less than 3	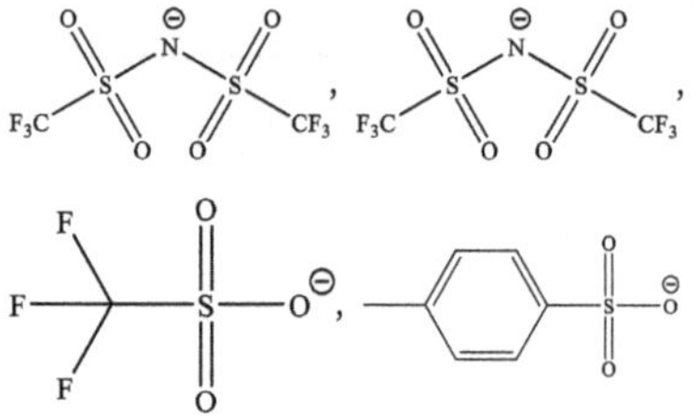
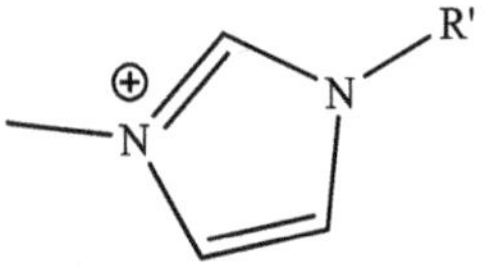 R = Saturated alkyl group with carbon atoms greater than 5	Br^{-}

10.4.2 Decomposition temperature of imidazolium-based energetic ionic liquids or salts

The temperature of decomposition of imidazolium-based energetic ionic liquids or salts can be correlated with their molecular structures as follows [589]:

$$T_{\text{onset,IL}} = \frac{1{,}276b_{\text{cat}} - 1{,}875d_{\text{cat}} + 4{,}439e_{\text{cat}}}{Mw_{\text{cat}}}$$

$$+ \frac{4{,}542a_{\text{ani}} + 5{,}255c_{\text{ani}} + 4{,}958d_{\text{ani}} + 8{,}244e_{\text{ani}} + 20{,}793f_{\text{ani}} + 30{,}766h_{\text{ani}}}{Mw_{\text{ani}}}$$

$$+ \frac{6{,}313i_{\text{ani}} + 6{,}811l_{\text{ani}}}{Mw_{\text{ani}}}$$

$$+ 116.0T_{\text{onset,IL}}^{+} - 126.2T_{\text{onset,IL}}^{-},$$

(10.13)

where $T_{onset,IL}$ is the decomposition temperature in K; b_{cat}, d_{cat}, and e_{cat} are the number of hydrogen, oxygen, and fluorine atoms in imidazolium cation derivatives, respectively; a_{ani}, c_{ani}, d_{ani}, e_{ani}, f_{ani}, h_{ani}, i_{ani}, and l_{ani} are the number of carbon, nitrogen, oxygen, fluorine, chlorine, sulfur, phosphorus, and boron atoms in the anion part of imidazolium ionic liquids, respectively; Mw_{cat} and Mw_{ani} are the molecular weights of the cation and anion, respectively; $T_{onset,IL}^{+}$ and $T_{onset,IL}^{-}$ are correcting functions that can adjust large negative and positive deviations of the first two ratios of eq. (10.13) from the measured data. Tables 10.5 and 10.6 show the values of $T_{onset,IL}^{+}$ and $T_{onset,IL}^{-}$ for the presence of specific cations and anions in some imidazolium-based ionic liquids or salts.

Table 10.5: Specific cation/anion moieties for the estimation of $T_{onset,IL}^{+}$.

Cation	Anion	$T_{onset,IL}^{+}$
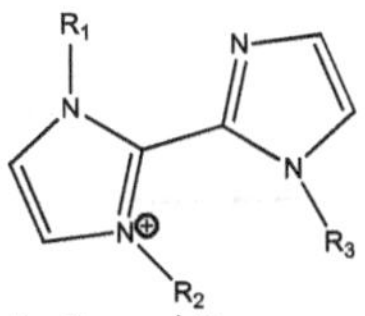 (imidazolium structure with R_1, R_2, R_3) R_1, R_2, and R_3 may contain a fluorine atom	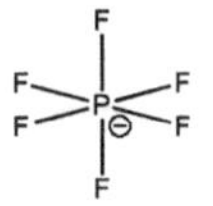 (hexafluorophosphate anion)	0.9
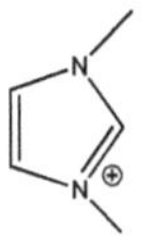 (imidazolium structure)	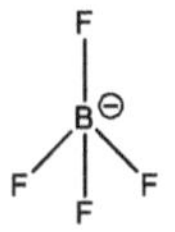 (tetrafluoroborate anion)	0.9
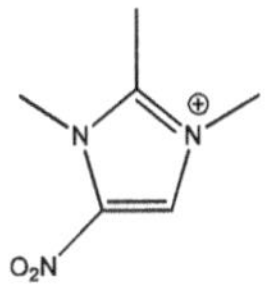 (imidazolium structure with O_2N)	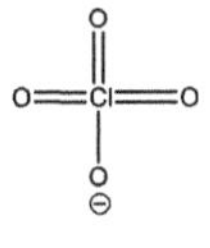 (anion)	0.9
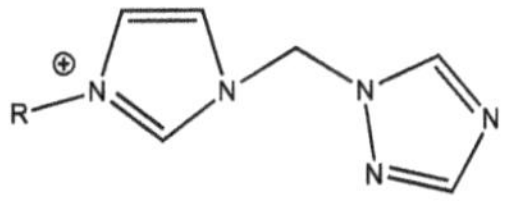 (imidazolium–triazole structure with R)	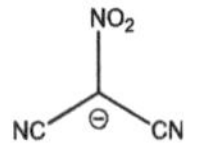 (NO_2, NC, CN anion)	1.0
 (imidazolium structure with R) R is the saturated alkyl group	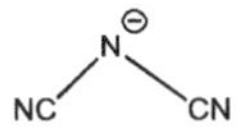 (NC, CN dicyanamide anion)	0.5
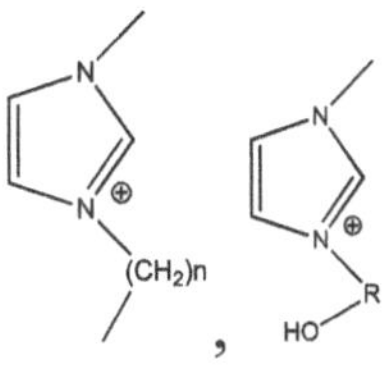 (bis-imidazolium structure with $(CH_2)n$, HO, R; $n<3$)	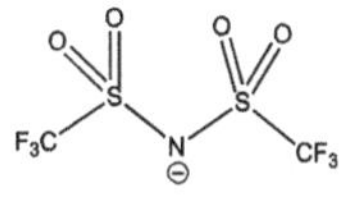 (F_3C, S, N, S, CF_3 anion)	0.6

Table 10.5 (continued)

Cation	Anion	$T^+_{onset,IL}$
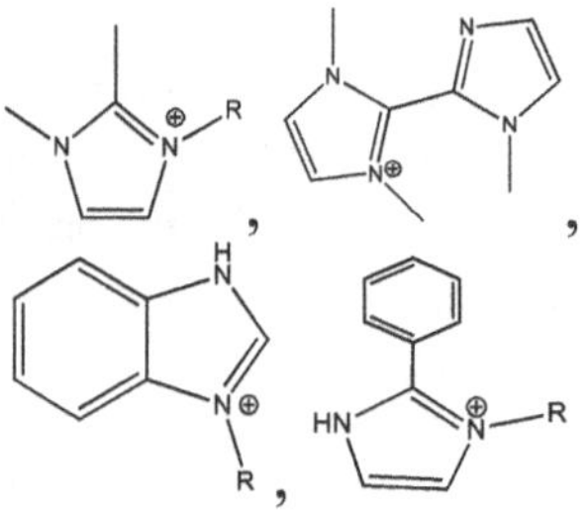		0.7

Table 10.6: Specific cation/anion moieties for estimation of $T^-_{onset,IL}$.

Cation	Anion	$T^-_{onset,IL}$
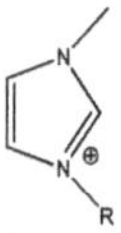 R containing a double bond	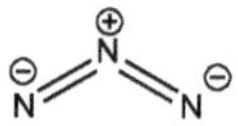	0.4
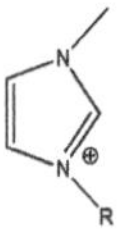 R containing a triple bond		0.7
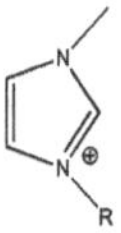	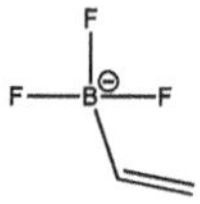	0.5
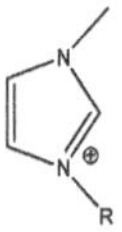 R contains a triple bond	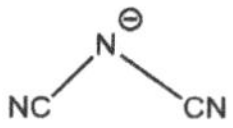	0.5
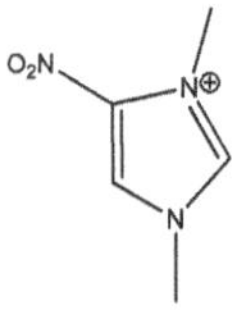	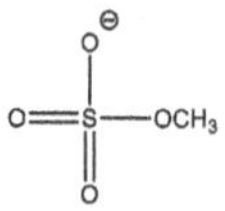	0.9

Table 10.6 (continued)

Cation	Anion	$T^-_{onset,IL}$
R₁ and R₂ contain double bonds	NO_3^-	0.6
R₁ contains fluorine atoms		0.5

Example 10.12. The use of eq. (10.13) for 3-(but-2-ynyl)-1-methyl-1H-imidazol-3-ium azide with the following structures

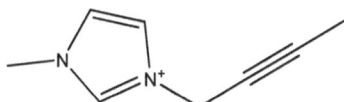

Chemical Formula: $C_8H_{11}N_2^+$
Molecular Weight: 135.19

$^-N{=}N^+{=}N^-$
Chemical Formula: N_3^-
Molecular Weight: 42.02

gives:

$$T_{\text{onset,IL}} = \frac{1{,}276b_{cat} - 1{,}875d_{cat} + 4{,}439e_{cat}}{Mw_{cat}}$$

$$+ \frac{4{,}542a_{ani} + 5{,}255c_{ani} + 4{,}958d_{ani} + 8{,}244e_{ani} + 20{,}793f_{ani} + 30{,}766h_{ani}}{Mw_{ani}}$$

$$+ \frac{6{,}313i_{ani} + 6{,}811l_{ani}}{Mw_{ani}} + 116.0T^{+}_{\text{onset,IL}} - 126.2T^{-}_{\text{onset,IL}}$$

$$= \frac{1{,}276 \times 11 - 1{,}875 \times 0 + 4{,}439 \times 0}{135.19}$$

$$+ \frac{4{,}542 \times 0 + 5{,}255 \times 3 + 4{,}958 \times 0 + 8{,}244 \times 0 + 20{,}793 \times 0 + 30{,}766 \times 0}{42.02}$$

$$+ \frac{6{,}313 \times 0 + 6{,}811 \times 0}{42.02} + 116.0 \times 0 - 126.2 \times 0.7 = 391 \text{ K}.$$

The measured value of $T_{\text{onset,IL}}$ is 388 K [590].

10.5 Decomposition temperature of azole-based energetic compounds

A reliable correlation has been introduced for the prediction of the thermal decomposition temperature of azole-based energetic compounds by their structural parameters, as follows [591]:

$$T_{\text{onset,azole}} = 547.75 + \frac{7{,}777.96a - 2{,}138.50c}{Mw} - 52.40\frac{d}{a} - 118.75\frac{b}{c} - 54.36n_{\text{N}_3}$$

$$- 60.85n_{\text{NH-NO}_2} + 102.62n_{\text{C=NH}} + 51.83T^{+}_{\text{onset,azole}} \qquad (10.14)$$

$$- 65.85T^{-}_{\text{onset,azole}},$$

where $T_{\text{onset,azole}}$ is the decomposition temperature of the desired azole-based energetic compound in K; a, b, c, and d are the number of carbon, hydrogen, nitrogen, and oxygen atoms, respectively; Mw is the molecular weight of the compound; n_{N_3}, $n_{\text{NH--NO}_2}$, and $n_{\text{C=NH}}$ are the number of azido, nitroamino, and imino groups, respectively; $T^{+}_{\text{onset,azole}}$ and $T^{-}_{\text{onset,azole}}$ are nonadditive structural parameters which have been specified in Tables 10.7 and 10.8.

Table 10.7: Specific structural moieties for estimation of $T^+_{onset,azole}$.

Structural moieties	Condition	$T^+_{onset,azole}$
	R_1 and $R_2 = NH_2$	1.0
	$R_1 = H$, $R_2 = NHNO_2$	
	$R_1 = NH_2$, $R_2 = NO_2$	
	The presence of tetrazine in the polycyclic compounds	1.0
	R_1, R_2, and $R_3 = NO_2$ or N_3	0.70
	R_1 and $R_2 = NO_2$ or N_3	2.0
	$R_3 = Cl$	
	$R_1 = CH_3N_3$	0.70
	$R_2 = NO_2$	
	R_1, R_2, and $R_3 = NO_2$ or NH_2 or Cl	0.90
	R_1 and R_2 are nitropyrazole or nitrotriazole	1.0
	$R2 = H$ or NO_2 or NH_2	1.0
	$R1 = H$ or NH_2 or CH_3	
	Situation 1 = –H or –CH$_2$–N(NO$_2$)–CH$_3$	1.0
	Situation 2 = –NH$_2$ or C(NO$_2$)$_3$ or –CO	

Table 10.8: Specific structural moieties for estimation of $T^{-}_{onset,azole}$.

Structural moieties	Condition	$T^{-}_{onset,azole}$
	Repeating of this moiety in the R_1 situation with a C–C bridge or the presence –OH substitution in the R_1 situation, i.e.,: R_1 = or R_1 = –OH	0.50
	R_2 and R_3 = –NO_2 or –N_3	
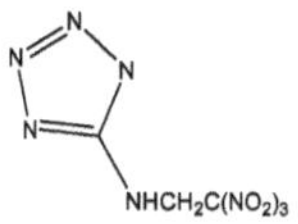	The presence of two or three of this fragment in a molecule which is separated from each other by more than two atoms	0.50
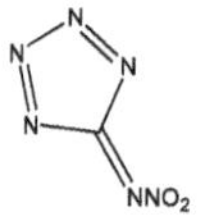	1. In the two-cycle molecules, the presence of one nitroiminotetrazole	0.90
	2. Two nitroiminotetrazole cycles separated by at least four carbon atoms	
	In the one cycle compounds, the presence of nitroimino substitution	0.50
	In the one cycle compounds, the presence of nitroimino, and cyclopentane substitutions	1.5
R_1–N=N–R_2	R_1 and R_2 = triazole or tetrazole ring	1.0
	R_1 and R_2 = tetrazole ring	2.0
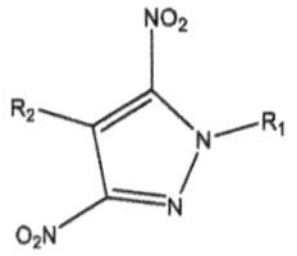	R_1 = –H or –NH_2 or –CH_3	1.0
	R_2 = –H or –N_3	
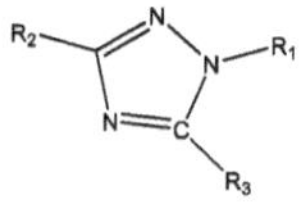	R_1 = –H or –NH_2	1.0
	R_2 = –H or –NO_2 or –CH(NO_2)_2	
	The presence of –CO, –CH_2–ONO_2, and furan fragment in a compound, simultaneously	1.0
	The presence of –CH_2–ONO_2 and a triazole ring in a compound, simultaneously	0.50

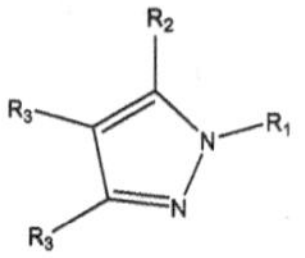

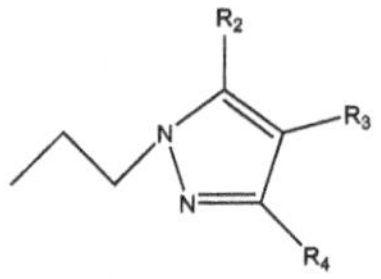

Example 10.13. The use of eq. (10.14) for 1-(2-(5-amino-3-nitro-1H-1,2,4-triazol-1-yl)ethyl)-1H-tetrazol-5-amine with the following structure

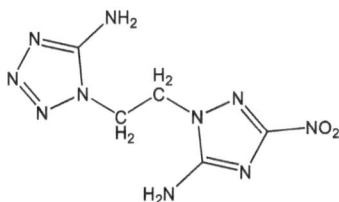

Chemical Formula: $C_5H_8N_{10}O_2$
Molecular Weight: 240.19

gives:

$$T_{onset,azole} = 547.75 + \frac{7{,}777.96a - 2{,}138.50c}{Mw} - 52.40\frac{d}{a} - 118.75\frac{b}{c} - 54.36n_{N_3}$$

$$- 60.85n_{NH-NO_2} + 102.62n_{C=NH} + 51.83T^{+}_{onset,azole} - 65.85T^{-}_{onset,azole}$$

$$= 547.75 + \frac{7{,}777.96 \times 5 - 2{,}138.50 \times 10}{240.19} - 52.40\frac{2}{5} - 118.75\frac{8}{10} - 54.36 \times 0$$

$$- 60.85 \times 0 + 102.62 \times 0 + 51.83 \times 0 - 65.85 \times 0 = 504.7 \text{ K}.$$

The measured value of $T_{onset,azole}$ is 550 K [592].

10.6 Predictive model for thermal decomposition onset in heterocyclic aromatic compounds and salts, incorporating elemental composition and structural features salts

A robust quantitative correlation has been established for predicting T_{onset} across a broad spectrum of heterocyclic aromatic compounds and their corresponding salts. This model encompasses various five- and six-membered nitrogen-containing ring systems, including triazoles, tetrazoles, furazans, triazines, and tetrazines, among others. The developed correlation serves as an efficient predictive tool with demonstrated reliability for rapid estimation of thermal decomposition behavior in both novel and known heterocyclic aromatic systems. The model systematically accounts for three primary factors influencing thermal stability:
(a) Fundamental elemental composition effects
(b) Additive contributions from stabilizing functional groups
(c) Non-additive effects from thermally labile structural motifs.

Through comprehensive analysis, these influences have been quantified according to the following mathematical relationship [593]:

$$T_{\text{onset}} = 485.3 + 34.32 T_{\text{onset, Elem}} + 25.41 T_{\text{onset, Add}} - 45.80 T_{\text{onset, Non - add}}$$

$$+ 62.14 T_{\text{onset}}^{\text{in}} - 62.91 T_{\text{onset}}^{\text{dec}}, \tag{10.15}$$

This approach enables the estimation of thermal stability parameters a priori, facilitating more informed molecular design of thermally robust heterocyclic compounds for applications ranging from pharmaceuticals to energetic materials. The model's simplicity and predictive accuracy make it particularly valuable for preliminary stability assessments during early-stage compound development.

The parameter $T_{\text{onset, Elem}}$ quantifies the elemental composition's contribution through the relationship:

$$T_{\text{onset, Elem}} = a - 0.3339b - 0.2878c - 0.2637d$$

The positive coefficient for carbon (a) and the negative coefficients for other elements indicate that carbon-rich compounds exhibit higher thermal stability, while heteroatoms generally decrease the decomposition temperature. The additive term accounts for specific functional groups and structural features:

$$T_{\text{onset, Add}} = n_{\text{NH}_2 + \text{Cl} - \text{C}\left(\text{NO}_2\right) - \text{C}(\text{NO}_2) + 2\left(\text{CuN} + 1.5\text{HONH}_3^+\right) + 3 \times \text{nitrotetrazolate salt of sodium or silver}}$$

$$+ 0.9779 n_{\text{NH or NHNH}_2} - 0.9122 n_{\text{furazan}}$$

where $n_{\text{NH}_2 + \text{Cl} - \text{C} - \text{C}\left(\text{NO}_2\right)_2 + 2\left(\text{CuN} + 1.5\text{HONH}_3^+\right) + 3 \times \text{nitrotetrazolate salt of sodium or silver}}$ is the sum of NH_2, $\text{Cl} - \text{C}(\text{NO}_2) - \text{C}(\text{NO}_2)$ and two times of CuN bond and three times of hydroxylammonium ions as well as three times of nitrotetrazolate salt of sodium or silver ion; $n_{\text{NH or NHNH}_2}$ is the number of NH or NHNH_2 groups; n_{furazan} is the number of furazan rings. Stabilizing groups with positive coefficients are the number of amino groups (NH_2), chloronitro groups ($\text{Cl}-\text{C}(\text{NO}_2)$), copper–nitrogen bonds (CuN), hydroxylammonium ions (HONH_3^+), metal nitrotetrazolates (Na/Ag salts), and secondary amines (NH/NHNH_2). Destabilizing feature is the number of furazan rings. Thermally labile groups contribute through:

$$T_{\text{onset, Nonadd}} = C_{>\text{C}=\text{NNO}_2 \text{ or } -\text{C}(= \text{NH}) - \text{NH}\left(\text{NO}_2\right)} + 2.20 C_{\text{N}-\text{N}(\text{NH})-\text{C}\left(\text{NO}_2\right) \text{ or } \text{N} = \text{N}-\text{NH}-\text{C}\left(\text{NO}_2\right)}$$

$$+ 1.171 C_{\text{Azide}} + 1.037 C_{\text{ONO}_2 \text{ or } \text{CH}(\text{NO}_2)_2 \text{ or dinitroamide ion}}$$

where $C_{>\text{C}=\text{NNO}_2 \text{ or } -\text{C}(= \text{NH}) - \text{NH}\left(\text{NO}_2\right)}$, $C_{\text{N}-\text{N}(\text{NH}) - \text{C}\left(\text{NO}_2\right) \text{ or } \text{N} = \text{N}-\text{NH} - \text{C}\left(\text{NO}_2\right)}$, C_{Azide} and $C_{\text{ONO}_2 \text{ or } \text{CH}(\text{NO}_2)_2 \text{ or dinitroamide ion}}$ show the presence of the specified groups or molecular fragments. These presences of the mentioned groups reduce thermal stability: weak N-NO$_2$ bonds in nitroimines ($>\text{C}=\text{NNO}_2$), acidic protons adjacent to nitro groups (N–NH–C(NO$_2$)), explosive decomposition of azides (N$_3$), and low bond dissociation energies in nitroesters (ONO$_2$). The model incorporates two adjustment factors:

Table 10.9: The optimized values for T^{in}_{onset} and T^{dec}_{onset}.

Type of compounds	T^{in}_{onset}	T^{dec}_{onset}	Examples
	0.5	0	

(continued)

Table 10.9 (continued)

Type of compounds	T_{onset}^{in}	T_{onset}^{dec}	Examples
	0.7	0	

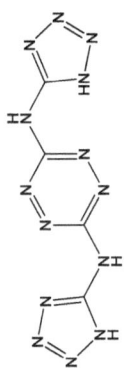

1.0   0   0

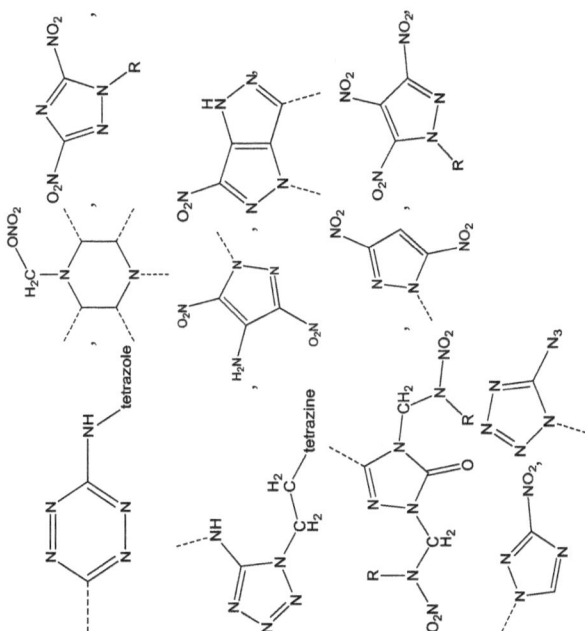

(continued)

Table 10.9 (continued)

Type of compounds	T_{onset}^{in}	T_{onset}^{dec}	Examples
	0	0.5	
	0	0.7	

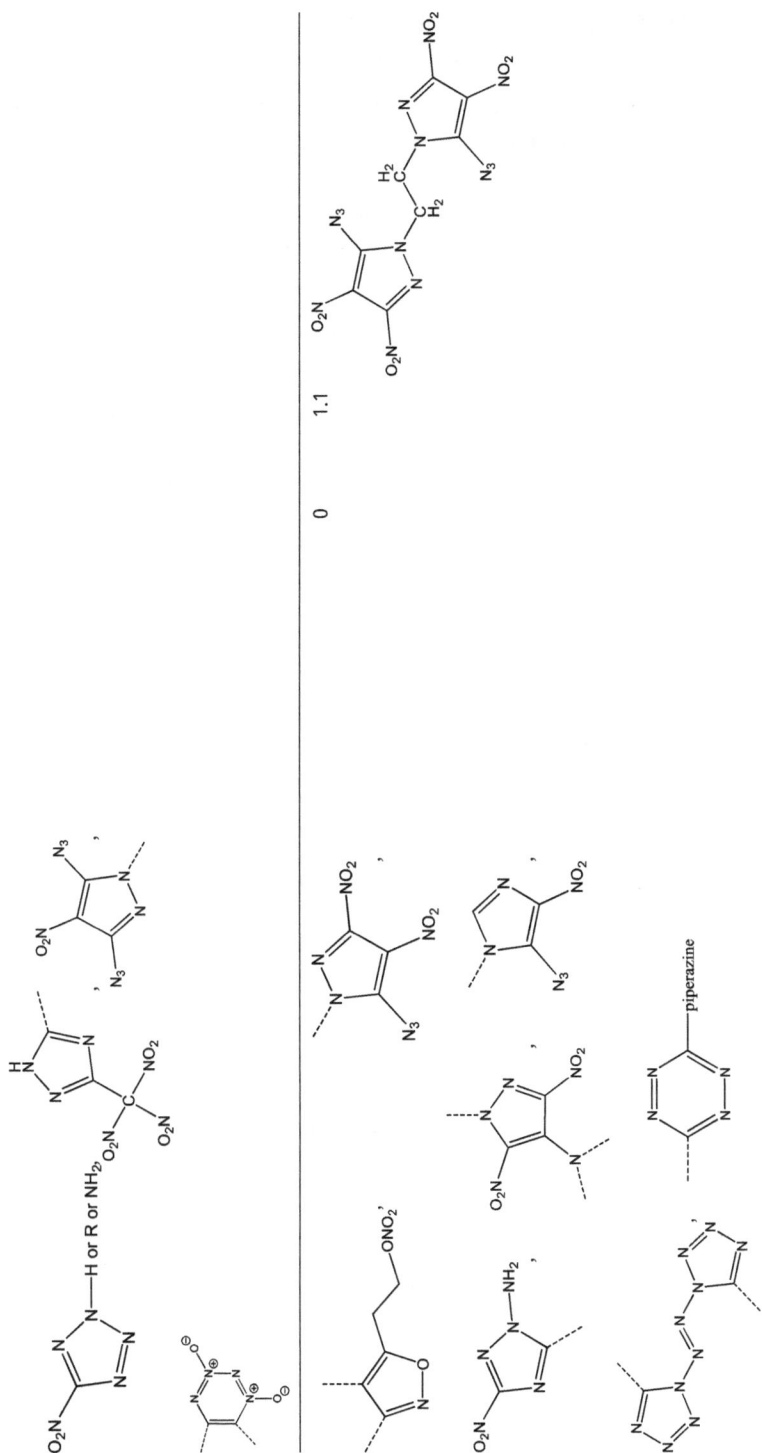

piperazine

T_{onset}^{in}: Positive corrections for synergistic stabilization

T_{onset}^{dec}: Negative corrections for destabilizing interactions

Reference values for these parameters are provided in Table 10.9. This quantitative framework enables precise prediction and optimization of thermal stability in heterocyclic compound design.

Example 10.14. The use of eq. (10.15) for bis([1,2,5]oxadiazolo)[3,4-b:3′,4′-e]pyrazine-4,8-diylbis(methylene) dinitrate with the following structure

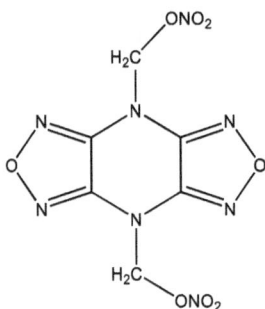

Chemical Formula: $C_6H_4N_8O_8$

gives:

$$T_{onset, Elem} = a - 0.3339b - 0.2878c - 0.2637d = 6 - 0.3339(4) - 0.2878(8) - 0.2637(8) = 0.2524,$$

$$T_{onset, Add} = n_{NH_2 + Cl - C(NO_2) - C(NO_2) + 2(CuN + 1.5HONH_3^+) + 3 \times nitrotetrazolate\ salt\ of\ sodium\ or\ silver}$$

$$+ 0.9779 n_{NH\ or\ NHNH_2} - 0.9122 n_{furazan} = 0 + 0.9779(0) - 0.9122(2) = -1.8244,$$

$$T_{onset, Nonadd} = C_{>C=NNO_2\ or\ -C(=NH)-NH(NO_2)} + 2.20 C_{N-N(NH)-C(NO_2)\ or\ N=N-NH-C(NO_2)}$$

$$+ 1.171 C_{Azide} + 1.037 C_{ONO_2\ or\ CH(NO_2)_2\ or\ dinitroamide\ ion} = 0 + 2.20(0) + 1.71(0) + 1.037(1) = 1.037.$$

Due to presence of molecular fragment 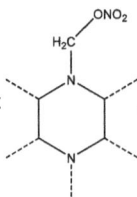, T_{onset}^{in} is 1.0. Thus,

$$T_{onset} = 485.3 + 34.32 T_{onset, Elem} + 25.41 T_{onset, Add} - 45.80 T_{onset, Non-add} + 62.14 T_{onset}^{in} - 62.91 T_{onset}^{dec}$$
$$= 485.3 + 34.32(0.2524) + 25.41(-1.8244) - 45.80(1.037) + 62.14(1.0) - 62.91(0) = 399.0\ K$$

The reported T_{onset} for this compound is 469 K [593].

Example 10.15. The use of eq. (10.15) for 5,7-diamino-4,6-dinitrobenzofuroxan dihydrate, sodium salt with the following structure

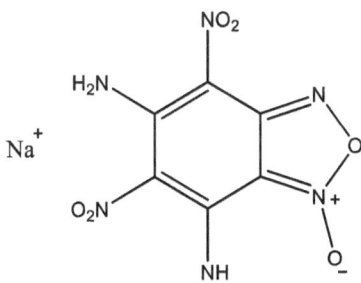

Chemical Formula: $NaC_6H_3N_6O_6$

gives:

$$T_{onset,Elem} = a - 0.3339b - 0.2878c - 0.2637d = 6 - 0.3339(3) - 0.2878(6) - 0.2637(6) = -0.1737,$$

$$T_{onset,Add} = n_{NH_2+Cl-C(NO_2)-C(NO_2)+2(CuN+1.5HONH_3^+)+3 \times nitrotetrazolate\ salt\ of\ sodium\ or\ silver}$$

$$+ 0.9779 n_{NH\ or\ NHNH_2} - 0.9122 n_{furazan} = 1 + 0.9779(1) - 0.9122(0) = 1.9779,$$

$$T_{onset,Nonadd} = C_{>C=NNO_2\ or\ -C(=NH)-NH(NO_2)} + 2.20 C_{N-N(NH)-C(NO_2)\ or\ N=N-NH-C(NO_2)}$$

$$+ 1.171 C_{Azide} + 1.037 C_{ONO_2\ or\ CH(NO_2)_2\ or\ dinitroamide\ ion} = 0 + 2.20(0) + 1.71(0) + 1.037(0) = 0.$$

Thus,

$$T_{onset} = 485.3 + 34.32 T_{onset,Elem} + 25.41 T_{onset,Add} - 45.80 T_{onset,Nonadd} + 62.14 T_{onset}^{in} - 62.91 T_{onset}^{dec}$$

$$= 485.3 + 34.32(-0.1737) + 25.41(1.9779) - 45.80(0) + 62.14(0) - 62.91(0) = 530\ K.$$

The reported T_{onset} for this compound is 542 K [593].

10.7 Summary

This chapter introduced different methods for predicting: the activation energies of low-temperature non-autocatalyzed thermolysis, heats of decomposition, exothermic onset temperatures, thermal decomposition temperatures of polynitroarenes, and the temperature at which the maximum loss of mass occurs, as well as the deflagration temperature. Equations (10.1)–(10.4) were used to predict the activation energies of low-temperature, non-autocatalyzed thermolysis of nitroparaffins, nitramines, and organic energetic compounds, respectively. Equation (10.4) is a general correlation, which can be applied to a wide range of energetic compounds including those energetic compounds that satisfy eqs. (10.1)–(10.3), but it is more complex. Equations (10.6) and (10.7) are used for the prediction of the decomposition temperatures of polynitroarenes and organic peroxides, respectively. Equations (10.8) and (10.9) can be applied for the prediction of the onset temperature of polynitroarenes and organic peroxides, respectively. Equation (10.10) is used to estimate the temperatures at which maximum mass loss occurs for organic azides. Equation (10.11) is a simple, reliable correlation for the prediction of the deflagration temperature of organic energetic compounds containing the – NNO_2, $–ONO_2$, or $–CNO_2$ functional groups. Equation (10.12) provides a simple method for the calculation of the activation energy of thermolysis of imidazolium, pyridinium, and phosphonium-based ionic liquids. Equation (10.13) gives the temperature of decomposition of imidazolium-based energetic ionic liquids or salts. Equation (10.14) introduces a reliable correlation for the prediction of the thermal decomposition temperature of azole-based energetic compounds. Equation (10.15) provides a predictive model for thermal decomposition onset in heterocyclic aromatic compounds and salts incorporating elemental composition and structural features.

11 Relationships between different sensitivities

In previous chapters, different methods for the prediction of the impact, shock, electric spark, friction, and heat sensitivities have been discussed. Among the different sensitivity tests, the impact sensitivity test is extremely easy to implement. Since the results of the impact sensitivity test greatly depend on the conditions under which the tests are performed, the experimental data from impact sensitivity tests are not often reproducible. Hot spots in an energetic compound may contribute to initiation in the drop-weight impact test. Due to the large number of impact sensitivity data which has been reported, some efforts have been undertaken to correlate the impact sensitivity of selected classes of energetic compounds to other sensitivities [403, 492, 594–598]. Wu and Huang [599] used a complex micro-mechanics model to describe hot spot formation in the energetic crystal powders of the two well-known explosives, HMX and PETN, subjected to drop-weight impact. Many attempts have been undertaken to illustrate the mechanism of initiation of an energetic material by impact stimulus, but this feature is not yet fully understood. In this chapter, several simple relationships between different sensitivities are demonstrated.

11.1 Relationship between impact sensitivity of energetic compounds and activation energies of thermal decomposition

It was indicated that the initiation of the decomposition of an energetic compound containing $-NNO_2$, $-ONO_2$, and $R-NO_2$ (or $Ar-NO_2$) groups through impact and heat stimuli can be related as follows [403]:

$$E_{IS} = 18.07 - 0.1130E_a + 14.68(b/d) + 22.65E_{IS}^{++} - 11.30E_{IS}^{--} \qquad (11.1)$$

where E_a is the activation energy of low-temperature non-autocatalyzed thermolysis in $kJ\,mol^{-1}$; E_{IS} is the impact sensitivity in J; two correcting functions E_{IS}^{++} and E_{IS}^{--} show the increasing and decreasing contributions of nonadditive structural parameters, respectively. The presence of some molecular moieties can increase or decrease the impact sensitivity values of different classes of energetic compounds through E_{IS}^{++} and E_{IS}^{--}, which are described in the following sections.

11.1.1 Nitroaromatics

(1) $-NH_2$ group: If per aromatic ring, then the values of E_{IS}^{++} are 0.9, 3.2 and 4.8 for $n_{NH_2} = 1$, 2, and 3 per aromatic ring, respectively.

https://doi.org/10.1515/9783112206768-011

(2) One 2,4,6-trinitrophenyl (TNP) in the form TNP–X: The values of E_{IS}^{++} equal 0.5, 1.0, and 1.5 for X = –NH–R (or –OH), –R, and –OR, respectively.

(3) Two aromatic rings in the form Ar–X–Ar or Ar–Ar: The value of E_{IS}^{--} equals 0.5 for X = –S– and –CH$_2$–R– except for 2,2′,4,4′,6,6′-hexanitro-1,1′-biphenyl (HNB).

(4) Polynitrobenzene containing more than one alkyl group: The value of E_{IS}^{--} is 1.0.

(5) Polynitronaphthalene: The value of E_{IS}^{++} is 0.5.

11.1.2 Nitramines

(1) Cyclic nitramines: For rings containing up to six ring atoms where, the value of E_{IS}^{++} is 0.5. For cyclic nitramines which have rings larger than six-membered ring and in which

$$0.75 \leq \frac{n_{NNO_2}}{n_{CH_2}} < 1 \quad \text{and} \quad \frac{n_{NNO_2}}{n_{CH_2}} < 0.75,$$

the values of E_{IS}^{--} equal 1.0 and 2.0, respectively. The value of E_{IS}^{--} is equal to 0.5 for cyclic nitramines, which are smaller than six-membered rings and in which

$$\frac{n_{NNO_2}}{n_{CH_2}} \geq 1.$$

(2) Acyclic nitramines: For acyclic nitramines in which the number of –N(NO$_2$)–CH$_2$–N(NO$_2$)– groups >1 and ≤1, the values of E_{IS}^{--} and E_{IS}^{++} equal 0.75 and 1.75, respectively. The value of E_{IS}^{--} equals 1.0 for the presence of the –N(NO$_2$)–CH$_2$–N(NO$_2$)– molecular fragment.

Example 11.1: 1,5-Endomethylene-3,7-dinitro-1,3,5-tetraazacyclooctane (DPT) has the following molecular structure:

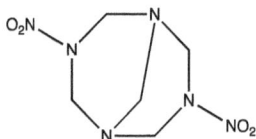

The measured E_a is 192.3 kJ/mol [554]. The use of eq. (11.1) gives

$$E_{IS} = 18.07 - 0.1130E_a + 14.68(b/d) + 22.65E_{IS}^{++} - 11.30E_{IS}^{--}$$
$$= 18.07 - 0.1130(192.3) + 14.68(10/4) + 22.65(0.5) - 11.30(0)$$
$$= 10.44 \, J.$$

The measured E_{IS} is 10.20 J [594].

11.2 Relationship between electric spark sensitivity and impact sensitivity of nitroaromatics

For nitroaromatics, it was shown that the electric spark and impact sensitivities can be correlated as [492]:

$$E_{ES} = 6.17 + 0.0797E_{IS} + 10.1E_{cor}^{+} - 3.21E_{cor}^{-}, \tag{11.2}$$

where E_{ES} is the electric spark sensitivity in J; E_{cor}^{+} and E_{cor}^{-} are two correcting functions that have been used to adjust large deviations of E_{ES} and E_{I}, and which can be specified as follows.

E_{cor}^{+}: The values of E_{cor}^{+} are 1.8, 0.5, and 0.9 for the presence of the –OR group, two –OH groups and the attachment of more than one nitroaromatic ring to another nitroaromatic ring, respectively.

E_{cor}^{-}: The value of E_{cor}^{-} equals 1.0 for the attachment of only one CH_x– or Ar– group to the aromatic ring in the case of CHNO nitroaromatics.

Example 11.2: If eqs. (6.11) and (6.12) are used to predict the impact sensitivity of 2-methyl-4-[(3-methyl-2,4,6-trinitrophenyl)thio]-1,3,5-trinitrobenzene (DIMEDIPS) with the following molecular structure:

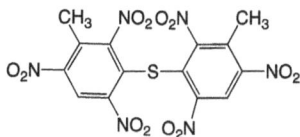

$E_{IS} = 26.05$ J is obtained. The use of this value in eq. (11.2) gives

$$E_{ES} = 6.17 + 0.0797E_{IS} + 10.1E_{cor}^{+} - 3.21E_{cor}^{-}$$
$$= 6.17 + 0.0797(26.05) + 10.1(0) - 3.21(0)$$
$$= 8.25 \text{ J}.$$

The measured E_{ES} is 8.57 J [481].

11.3 A general correlation between electric spark sensitivity and impact sensitivity of nitroaromatics and nitramines

It was shown that electric spark sensitivity, based on the RDAD instrument, can be correlated with the impact sensitivity of nitramines [600]. Examination of various molecular moieties of nitroaromatics and nitramines has shown that the following general equation can be used to correlate their electric spark sensitivity, based on the RDAD instrument, and impact sensitivity as [601]:

$$E_{ES} = 6.16 + 0.0843E_{IS} + 8.16E^+_{ES, NAr, NiA} - 3.43E^-_{ES, NAr, NiA} + 9.48E_{N\,excess}, \qquad (11.3)$$

where $E^+_{ES, NAr, NiA}$, $E^-_{ES, NAr, NiA}$, and $E_{N\,excess}$ are correcting functions. The values of $E^+_{ES, NAr, NiA}$ and $E^-_{ES, NAr, NiA}$ are given in Table 11.1.

Table 11.1: Contribution of $E^+_{ES, NAr, NiA}$ and $E^-_{ES, NAr, NiA}$.

Organic compounds containing energetic groups	$E^+_{ES, NAr, NiA}$	$E^-_{ES, NAr, NiA}$	Condition
Cyclic nitramine	0	1.0	Equal numbers of $>NNO_2$ and $>CH_2$ or 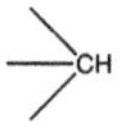 $\diagup\!\!\!\diagdown CH$ groups in form $-NNO_2-CH_x-NNO_2-$ with more than five-membered rings
	0.8	0	Equal numbers of $>NNO_2$ and $-CH_2-CH_2-$
	0.6	0	The presence of $-O-$ group
Acyclic nitramine	0.8	0	The presence of $-C(=O)-O-$ group
Nitroaromatics	0	1.0	(a) Direct attachment of two nitroaromatic rings or the presence of a specific group (X) between nitroaromatic rings (Ar), where X is $>CH_2$, $>NH$, $-CH_2=CH_2-$, and $-CH_2-CH_2-$
			(b) The attachment of both alkyl and hydroxyl groups to 2,4,6-trinitrobenzene
	2.2	0	The attachment of $-OR$ group to carbocyclic nitroaromatics
	0.6	0	The attachment of two $-OH$ groups to one nitroaromatic ring
	1.2	0	Direct attachment of three nitroaromatic rings
	1.0	0	The presence of $-NH-C(=O)-C(=O)-NH-$
			The attachment of $-OR$ group *ortho* to the nitrogen atom ring
	0.5	0	The presence of $-SO_2-$ group

The value of $E_{N\,excess}$ is for the presence of excess nitrogen atoms in cyclic nitramines containing excess nitrogen atoms besides $-NNO_2$ and $-NO_2$.

Example 11.3: The reported E_{IS} of 1,4-dinitrotetrahydroimidazo[4,5-d]imidazole-2,5(1H,3H)-dione (DINGU), with the following molecular structure, is 5.55 J [605].

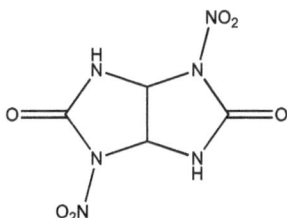

The use of this value in eq. (11.3) gives:

$$E_{ES} = 6.16 + 0.0843E_{IS} + 8.16E^+_{ES,\,NAr,\,NiA} - 3.43E^-_{ES,\,NAr,\,NiA} + 9.48E_{N\,excess}$$

$$= 6.16 + 0.0843(5.55) + 8.16(0) - 3.43(0) + 9.48(1) = 16.11 J$$

The measured E_{ES} is 15.19 J [498].

11.4 Relationship between electric spark sensitivity and activation energy of the thermal decomposition of nitramines

For various nitramines, it has been shown that the mechanism of spark energy transfer can be related to the activation energy of low-temperature nonautocatalyzed thermolysis, as [597]:

$$E_{ES} = 9.826 - 0.047E_a + 7.432(a/d) + 7.680E^+_{ES,\,corr}, \tag{11.4}$$

where $E^+_{ES,\,corr}$ is an increasing factor. The value of $E^+_{ES,\,corr}$ equals 1.0 for the presence of the following molecular fragment or condition:
(1) cyclic –O–, >N– and –C(=O)–N–;
(2) acyclic –C(=O)–O–;
(3) acyclic nitramines with more than one N–NO$_2$ functional group, in which

$$\frac{n_{NNO_2}}{n_{CH_2}} \geq 2;$$

(4) for cyclic nitramines with more than one N–NO$_2$ group, in which

$$\frac{n_{NNO_2}}{n_{NCH_2CH_2N}} \leq 1,$$

where $n_{NCH_2CH_2N}$ is the number of –NCH$_2$CH$_2$N– groups.

Example 11.4: The molecular structure of 2,4,6-trinitro-2,4,6-triazaheptane (ORDX) is given as

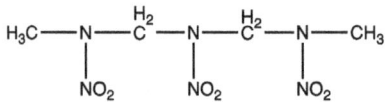

$$H_3C-N-\underset{H_2}{C}-N-\underset{H_2}{C}-N-CH_3$$

with NO_2 substituents on each nitrogen.

If the measured E_a of ORDX is 179.2 kJ/mol [549], the use of eq. (11.4) gives

$$E_{ES} = 9.826 - 0.047E_a + 7.432(a/d) + 7.680E^+_{ES,\,corr}$$
$$= 9.826 - 0.047(179.2) + 7.432(4/6) + 7.680$$
$$= 6.36\,J.$$

The measured E_{ES} is 8.08 J [498].

11.5 Correlation of the electrostatic sensitivity and activation energies for the thermal decomposition of nitroaromatics

It was shown that the following equation can be introduced as a suitable correlation for predicting the relationship between the electric spark sensitivity and the activation energy of low-temperature non-autocatalyzed thermolysis [595]:

$$E_{ES} = -9.72 + 0.034E_a + 7.41\left(\tfrac{c+d}{a}\right) + 3.49E^+_{ES,\,Ar} - 2.50E^-_{ES,\,Ar}, \tag{11.5}$$

where the functions $E^+_{ES,\,Ar}$ and $E^-_{ES,\,Ar}$ show increasing and decreasing contribution of structural parameters, respectively, which depend on the attachment of some groups to TNP which increase and decrease the electrostatic sensitivity. They are defined as follows.

(1) One TNP: For compounds with the general structure TNP(X)$_2$ or TNP(X)$_3$, where X is $-CH_3$, $-NH_2$, or $-OH$, the values of $E^+_{ES,\,Ar}$ equal 1.75, 0.75, and 1.25, respectively. If $-Cl$ is present, the value of $E^-_{ES,\,Ar}$ equals 2.0.

(2) Two TNP: For energetic compounds with general formula TNP–Y–TNP, where Y is $-SO_2-$,

$$-HN-\underset{\|}{\overset{O}{C}}-\underset{\|}{\overset{O}{C}}-NH-$$

, or TNP, the value of $E^+_{ES,\,Ar}$ equals 1.40. If the $-NH-$ or $-S-$ groups are present, or if two TNP groups are directly attached together, the value of $E^-_{ES,\,Ar}$ equals 1.25.

Example 11.5: 1-Methyl-3-hydroxy-2,4,6-trinitrobenzene (TNCr) has the following molecular structure:

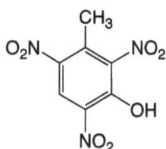

The measured E_a is 192.46 kJ/mol [364]. The use of eq. (11.5) gives

$$E_{ES} = -9.72 + 0.034E_a + 7.41\left(\frac{c+d}{a}\right) + 3.49E^+_{ES,\,Ar} - 2.50E^-_{ES,\,Ar}$$

$$= -!9.72 + 0.034(192.46) + 7.41\left(\frac{3+7}{7}\right) + 3.49(0) - 2.50(0)$$

$$= 7.41 \text{ J}.$$

The measured E_{ES} is 5.21 J [498].

11.6 Relationship between the activation energy of thermolysis and friction sensitivity of cyclic and acyclic nitramines

The elemental composition and some structural parameters can be used to correlate friction sensitivity with the activation energy of thermolysis for cyclic and acyclic nitramines as [596]

$$FS = 212.0 + 32.67a - 10.21c - 14.50d - 85.07E_a/Mw$$
$$+ 81.92FS^+ - 48.19FS^-,$$

(11.6)

where functions FS^+ and FS^- show the presence of some structural parameters that can increase and decrease the predicted friction sensitivity values on the basis of the elemental composition and E_a/Mw. For cyclic and cage nitramines, as well as acyclic nitramines, the correcting functions FS^+ and FS^- are defined as follows.

(1) Cyclic and cage nitramines: If the ratio of the number of $-NNO_2$ groups to the number of $-CH_2-$ groups, i.e., n_{NNO_2}/n_{CH_2}, in cyclic nitramines equals 1.0, the value of FS^+ is 0.8. The value of FS^- equals 0.7 for those compounds in which the ratios of n_{NNO_2}/n_{CH_2} in cyclic nitramines and n_{NNO_2}/n_{CH} in cage or unsaturated nitramines are less than 1.0, where n_{CH} is the number of $>CH-$ or $=CH-$ groups.

(2) Acyclic nitramines:
(a) For compounds with the general formula $TNP-N(NO_2)-R$, $FS^+ = 1.4$.

   (b) For compounds containing the neutral tetrazole molecular moieties, the values of FS^- and are equal to 0.8 and 1.0 for the absence and presence of the R–N< group attached to the tetrazole ring, respectively.

   (c) For compounds with the general formula $-RN-(CH_2)_n-NR-$, $FS^- = 1.4$.

Example 11.6: 1-Methyl-5-nitriminotetrazole has the following molecular structure:

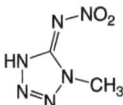

The calculated E_a is 189.16 kJ/mol [559]. The use of eq. (11.6) gives

$$FS = 212.0 + 32.67a - 10.21c - 14.50d - 85.07E_a/Mw$$
$$+ 81.92FS^+ - 48.19FS^-$$
$$= 212.0 + 32.67(2) - 10.21(6) - 14.50(2) - 85.07(189.16/144.09)$$
$$+ 81.92(1.0) - 48.19(0)$$
$$= 157.3 \text{ N}.$$

The measured FS using the BAM friction tester is 160 N [602].

11.7 Relationship between shock sensitivity of nitramine energetic compounds based on small-scale gap test and their electric spark sensitivity

An attempt has been done to correlate shock sensitivity to electric spark sensitivity based on the RDAD instrument of nitramine compounds [603]. Tan et al. [604] applied an Mn–Cu manometer to measure the output pressures of shock waves. They improved a set of small-scale gap tests to pass through aluminum gaps with different thicknesses for drawing a standard curve. For the tested explosive pillars, they measured the thickness of aluminum gaps with the calibrated set in terms of "go" or "no go". Since the reported data of Tan et al. [457] and Storm et al. [336] are different from each other, there is a linear correlation between $P_{90\%\text{ TMD}}$, which is calculated by eq. (8.1), and the improved method of Tan et al. [604] as follows [605]:

$$(x\,\text{Gap})_{90\%\text{ TMD}} = 12.50 - 0.1220P_{90\%\text{ TMD}}, \tag{11.7}$$

where $(x\,\text{Gap})_{90\%\text{ TMD}}$ is the thickness of aluminum in mm at 90% of theoretical maximum density (TMD). Thus, eq. (11.7) can convert the calculated shock sensitivity based on NSWC data using the Navy small-scale gap test method of Storm et al. [464] to the improved method of Tan et al. [604].

It was shown that shock sensitivities at 90%, 95%, and 98% of TMD and electric spark sensitivities based on the measurements of RDAD can be correlated for nitroaromatics and nitramines as:

$$P_{90\% \text{ TMD}} = 8.62 + 0.4970 E_{ES} \tag{11.8}$$

$$P_{95\% \text{ TMD}} = 12.29 + 0.7583 E_{ES} - 9.113 C_{NH_2} \tag{11.9}$$

$$P_{98\% \text{ TMD}} = 15.43 + 0.9723 E_{ES} - 16.17 C_{NH_2}, \tag{11.10}$$

where $P_{90\% \text{ TMD}}$, $P_{95\% \text{ TMD}}$, and $P_{98\% \text{ TMD}}$ are the pressures in kbar required to initiate material pressed to 90%, 95%, and 98% of TMD, respectively; C_{NH_2} shows the contribution of amino groups in nitroaromatics. The values of C_{NH_2} are 1.0, 2.0, and −2.0 for the presence of one, two, and three amino groups, respectively.

Example 11.7: 1,3,5-Trinitro-1,3,5-triazinane (RDX) has the following molecular structure:

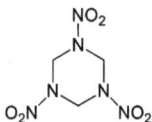

The use of eq. (8.1) gives:

$$P_{90\% \text{ TMD}} = 16.79 + 2.262(a + b/2 - d) - 6.314 E^0_{a\,CH/NNO_2} + 17.72(1.93 n_{NH_2} - n_{NO_2})_{\text{pure}}$$

$$= 16.79 + 2.262(3 + 6/2 - 6) - 6.314(1) + 17.72(0)$$

$$= 10.48 \text{ kbar}.$$

The use of eq. (11.7) gives:

$$(x\,\text{Gap})_{90\% \text{ TMD}} = 12.50 - 0.1220 P_{90\% \text{ TMD}} = 12.50 - 0.1220(10.47) = 11.22 \text{ mm}.$$

The measured $(x\,\text{Gap})_{90\% \text{ TMD}}$ is 11.68 mm [604].

The use of eq. (7.3) provides:

$$E_{ES} = 3.460 + 6.504(a/d) - 4.059 C_{CH_2NNO_2 > 2,\, C(=O)(O\,or\,NH)}$$

$$= 3.460 + 6.504(3/6) - 4.059(1)$$

$$= 2.65 \text{ J}.$$

The use of eqs. (11.8), (11.9), and (11.10) gives:

$$P_{90\% \text{ TMD}} = 8.62 + 0.4970E_{ES} = 8.62 + 0.4970(2.65) = 9.94 \text{ kbar},$$

$$P_{95\% \text{ TMD}} = 12.29 + 0.7583E_{ES} - 9.113C_{NH_2} = 12.29 + 0.7583(2.65) - 9.113(0)$$

$$= 14.30 \text{ kbar},$$

$$P_{98\% \text{ TMD}} = 15.43 + 0.9723E_{ES} - 16.17C_{NH_2} = 15.43 + 0.9723(2.65) - 16.17(0)$$

$$= 18.01 \text{ kbar}.$$

The reported values of $P_{90\% \text{ TMD}}$, , and $P_{98\% \text{ TMD}}$ for RDX are 10.97, 15.77, and 20.35 kbar, respectively [464].

11.8 Summary

The relationships between the impact, electric spark, friction sensitivities and activation energies of low-temperature non-autocatalyzed thermolysis were discussed in this chapter. Equation (11.1) correlates the impact sensitivity of an energetic compound containing $-NNO_2$, $-ONO_2$, or $R-NO_2$ (or $Ar-NO_2$) groups with its activation energy of low-temperature non-autocatalyzed thermolysis. Equation (11.2) gives another correlation between the electric spark sensitivity, based on the RDAD instrument, and the impact sensitivity of nitroaromatics. Equation (11.3) provides a general correlation between spark sensitivity, based on the RDAD instrument, and the impact sensitivity of nitroaromatics and nitramines. Equations (11.4) and (11.5) introduce suitable relationships between the electric spark sensitivity, based on the RDAD instrument, and the activation energy of low-temperature non-autocatalyzed thermolysis of nitramines and nitroaromatics, respectively. Equation (11.6) relates the friction sensitivity of nitramines with the activation energies of low-temperature non-autocatalyzed thermolysis. Equation (11.7) gives a linear correlation between $(x\,\text{Gap})_{90\% \text{ TMD}}$ and $P_{90\% \text{ TMD}}$. Equations (11.8), (11.9), and (11.10) show the relationships between, $P_{95\% \text{ TMD}}$, and $P_{98\% \text{ TMD}}$, and the electric spark sensitivity, based on the RDAD instrument.

12 Estimation of the properties of metal-containing energetic complexes and energetic metal-organic frameworks (MOFs)

HEDMs have captivated researchers for decades due to their vital roles in explosives, propellants, and pyrotechnics [606, 607]. The ideal HEDM must balance competing traits: high detonation power, thermal stability, and low sensitivity [608]. Meanwhile, materials with outstanding explosive performance often suffer from instability or dangerous sensitivity [609]. To solve this, scientists are now designing innovative HEDMs that maximize energy output while minimizing risks [610–612]. Two leading approaches have emerged: metal-containing energetic complexes and energetic metal-organic frameworks (EMOFs). While both use metals paired with energetic components [613, 614], their structures and behaviors differ dramatically. Energetic complexes pack a powerful punch – these dense molecular structures feature metals tightly bound to explosive ligands, delivering impressive energy output. However, this comes with a tradeoff: they tend to be notoriously sensitive compared to conventional organic explosives [615]. What makes them unique is how they break down. Unlike traditional explosives, the metal-ligand bonds in these complexes release less energy during decomposition, and the presence of metals alters their reaction pathways [615]. There's an interesting paradox with heavy-metal-based explosives – while they boast high density, this actually translates to lower heat output per unit mass during detonation [616]. This has driven significant research efforts in recent years, particularly in developing and testing energetic transition metal coordination compounds [617]. The secret to their power lies in their nitrogen-rich azole-based ligands [618]. But before these complexes fully decompose, they undergo several preliminary changes – you might see them melt, lose water molecules, or shed some of their coordinating ligands first [619]. In contrast to energetic complexes, EMOFs form extended crystalline networks where metal nodes connect via organic linkers, enabling precise tuning of properties like density and stability [620]. The flexibility of EMOFs makes them especially promising. By selecting nitrogen-rich ligands (like tetrazoles or pyrazoles), researchers can exploit energetic C–N and N≡N bonds to boost performance while maintaining structural integrity [621, 622]. These ligands also offer multiple binding sites for metals, further enhancing design control [623]. However, some high-energy groups (e.g., azides) can undermine thermal stability – a challenge mitigated by using rigid 3D frameworks to reinforce the structure [624–626]. Real-world examples highlight these tradeoffs. Traditional energetic complexes, such as nickel-hydrazine perchlorate, pack tremendous energy but are notoriously unstable [627]. Meanwhile, advanced EMOFs like copper-triazole frameworks achieve densities near 2.0 g/cm^3 and withstand temperatures over 320 °C without decomposing [628]. This combination of power and safety positions EMOFs as frontrunners for next-generation applications. To accelerate progress, this chapter introduces three predictive models for key metal-containing energetic com-

https://doi.org/10.1515/9783112206768-012

plexes and EMOFs properties: density, heat of formation, and thermal stability. These tools aim to streamline the development of safer, more powerful materials.

12.1 A molecular weight-dependent model for predicting the density of EMOFs

Dalirandeh et al. [629, 630] proposed two simple models for predicting the density of EMOFs based on their molecular structures. Among these, the model relying more on experimental data and molecular weight – rather than experimental radius values – is presented here for EMOFs with the general formula $C_aH_bN_cO_dK_hNa_iLi_jAg_kPb_l$ [630]. The model is expressed as [630]:

$$\rho = 3.13 + \frac{-22.35a - 12.74(b+c+d) - 31.78(h+i+j) + 79.72k + 351.66l}{Mw} + 0.98\rho_{EMOF}^+ - 0.27\rho_{EMOF}^-,$$

(12.1)

where h, i, j, k, and l are the number of potassium, sodium, lithium, silver, and lead atoms of an EMOF, respectively. The terms ρ_{EMOF}^+ and ρ_{EMOF}^- are correction functions, defined in Tables 12.1 and 12.2, respectively.

Table 12.1: Different values of ρ_{EMOF}^+.

Structural parameter*	ρ_{EMOF}^+	Example**
H_2Tztr ligand (bridging mode)	0.2	
Existence of an oxygen atom (as a bridge to link a metallic ion with a ligand)	0.08	
Existence of a hydroxyl group or an oxygen atom (as a bridge to link two adjacent node atoms)	0.18	

Table 12.1 (continued)

Structural parameter*	ρ_{EMOF}^{+}	Example**
Existence of H$_2$TZEG ligand	0.12	
Existence of ClO$_4$ (coordinated to metal centers)	0.39	
Existence of NO$_3$ (coordinated to metal centers)	0.08	
Number of uncoordinated water =[H$_2$O]$_{uncoordinated\ water}$	$0.08[H_2O]_{uncoordinated\ water}$ if $0 < [H_2O]_{uncoordinated\ water} \leq 5$	
	$0.1[H_2O]_{uncoordinated\ water}$ if $6 \leq [H_2O]_{uncoordinated\ water} \leq 9$	
	$0.46[H_2O]_{uncoordinated\ water}$ if $10 \leq [H_2O]_{uncoordinated\ water}$	
Number of IO$_3^-$ =$n_{IO_3^-}$	$0.3n_{IO_3^-}$	

*Abbreviations of different ligands are given in Appendix D.
**M is a metal node.

Table 12.2: Different values of $\rho_{\overline{EMOF}}$.

Structural parameter*	$\rho_{\overline{EMOF}}$	Example**
Direct attachment of –O– between two furazan rings that conjugates with the tetrazole ring	2.55	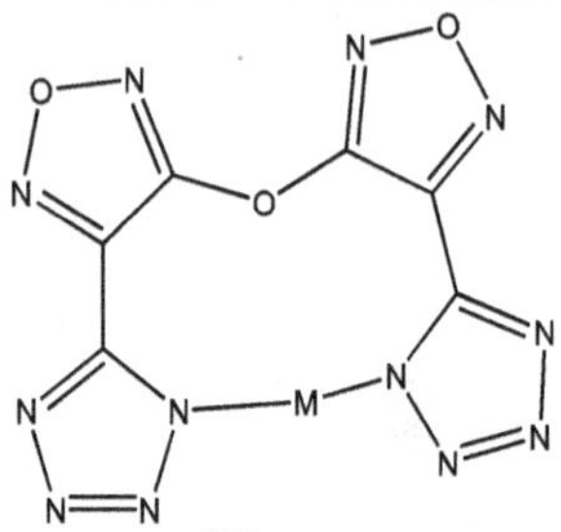
Number of melamine ligands = $n_{melamine}$	$n_{melamine}$	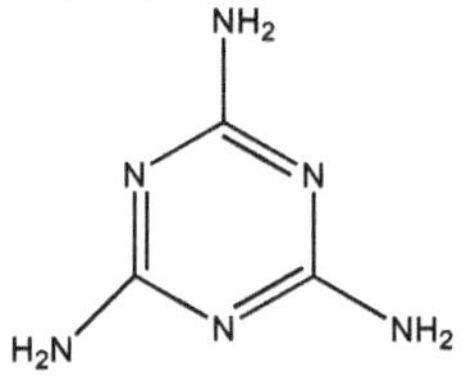
Number of cyclo-N_5^- ligands $=n_{cyclo-N_5^-}$		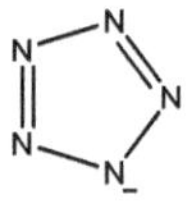
Direct attachment of –N=N– between two triazole rings	0.5	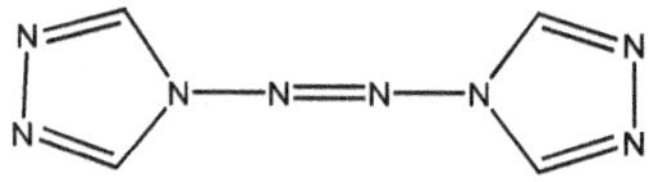
Existence of –N=N– between two triazole rings, which contain carbonyl groups	1.1	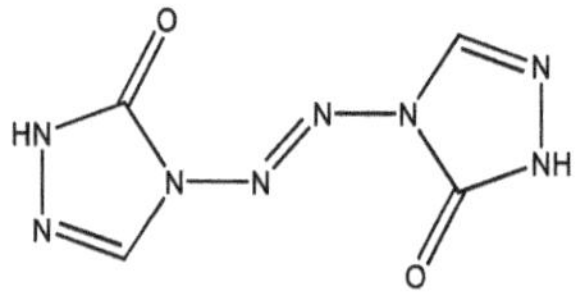
H$_3$DTTZ ligand	1	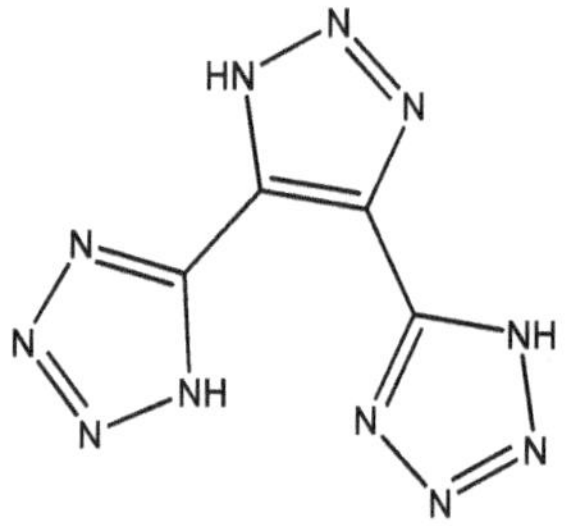

Table 12.2 (continued)

Structural parameter*	$\bar{\rho}_{EMOF}$	Example**
H$_3$BTT ligand	1	
H$_2$Tztr ligand (chelating mode)	2.2	
H$_2$Tztr ligand (bridging with chelating modes)	0.8	
Number of coordinated water = [H$_2$O]$_{coordinated\ water}$	0 if $0 < [H_2O]_{coordinated\ water} \leq 5$	
	$0.2[H_2O]_{coordinated\ water}$ if $6 \leq [H_2O]_{coordinated\ water} \leq 9$	
	$1.6[H_2O]_{coordinated\ water}$ if $10 \leq [H_2O]_{coordinated\ water}$	
Existence of ClO$_4$ or NO$_3$ (as free counterion)	0.12 for ClO$_4$	
	0.6 for NO$_3$	
Existence of N$_3$ group	0.18	

*Abbreviations of different ligands are given in Appendix D.
**M is metal node.

Example 12.1: [Cu(ntz)(N$_3$)(H$_2$O)]$_n$ has the following molecular structure:

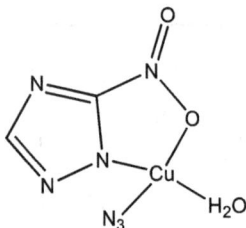

Chemical Formula: C$_2$H$_3$CuN$_7$O$_3$
Molecular Weight: 236.64

The use of eq. (12.1) and the presence of N$_3$ group (Table 12.2) give:

$$\rho = 3.13 + \frac{-22.35a - 12.74(b+c+d) - 31.78(h+i+j) + 79.72k + 351.66l}{Mw} + 0.98\rho_{EMOF}^+ - 0.27\rho_{EMOF}^-,$$

$$\rho = 3.13 + \frac{-22.35(2) - 12.74(3+7+3) - 31.78(0+0+0) + 79.72(0) + 351.66(0)}{236.64} + 0.98(0) - 0.27(0.18) = 2.19 \text{ g/cm}_3.$$

Since the experimental value of the density of [Cu(ntz)(N$_3$)(H$_2$O)]$_n$ is 2.26 g/cm^3 [631], the calculated density is close to the experimental value.

12.2 Predictive model for condensed-phase heat of formation in EMOFs

Dalirandeh et al. [632] developed a simple model to predict the condensed-phase heat of formation of EMOFs using experimental data from 63 diverse compounds. Their dataset included EMOFs with both aromatic and non-aromatic linkers coordinated to various cations, such as transition metals (Cu^{2+}, Zn^{2+}, Cd^{2+}, Ag$^+$, Co^{2+}, Mn^{2+}), alkali metals (Li$^+$, Na$^+$, K$^+$, Cs$^+$), alkaline earth metals (Ba^{2+}), and Pb^{2+}. The model correlates the condensed-phase heat of formation values with easily obtainable structural features, including specific chemical bonds, elemental composition, and framework descriptors that systematically enhance or reduce stability. By relying solely on molecular structure-derived parameters, this approach eliminates the need for additional computational tools while maintaining predictive accuracy [632]:

$$\Delta_f H^\theta(c) = -198.79n_{N-H} + 529.68n_{C=O} + 214.98n_{C=N} + 725.25 \sum_i \Delta_f H^\theta_{Inc,i}$$

$$- 91.16 \sum_j \Delta_f H^\theta_{Dec,j}, \tag{12.2}$$

where $\Delta_f H^\theta(c)$ is in kJ/mol; n_{N-H}, $n_{C=O}$, and $n_{C=N}$ are the number of N–H, C=O, and C=N bonds; and the $\sum_i \Delta_f H^\theta_{Inc,i}$ and $\sum_j \Delta_f H^\theta_{Dec,j}$ represent the sum of the standard enthalpies of formation of certain molecular fragments, as provided in Tables 12.3 and 12.4.

Table 12.3: The values of $\sum_i \Delta_f H^\theta_{Inc,i}$.

Structural parameters	$\sum_i \Delta_f H^\theta_{Inc,i}$ [a]
Metal nodes:	
Mn	0.6
Zn	0.95
Ag or Cu	1.1
Li	1.4
Free counter ions:	
$N(NO_2)_x$	0.54*n
$C(NO_2)_x$	0.34*n
Attachment of $N(NO_2)_x$ to triazoles	3.45*n
Coordination of dimethylformamide to nodes	1.85*n
Cyclo N_5	0.18*n
Azido	0.77*n
Azo between two furazans	0.56*n
NH between two tetrazoles	1
Attachment of tetrazole with furazan	3.35*n
Attachment of tetrazole with triazole	1.6

[a] n means the number of the specified fragment in the molecule.

Table 12.4: The values of $\sum_j \Delta_f H^\theta_{Dec,j}$.

Structural parameters	$\sum_j \Delta_f H^\theta_{Dec,j}$ [a]	Examples
Pb	7.8	
Free counter ions:		
Aminoguanidinium	3.25*n	
Triaminoguanidinium	13.62*n	
CH_3CN	6.72*n	
NO_3	5.55*n	

Table 12.4 (continued)

Structural parameters	$\sum_j \Delta_f H^\theta{}_{Dec,\,j}$ [a]	Examples
Coordination to nodes:		
NO_3	6.65^*n	
CH_3CN	8.91^*n	
Attachment to triazole:		
$C(NO_2)_2$	11.7^*n	
$C(NO_2)_3$	14.3^*n	
Attachment to tetrazole:		
Methyl	9.6	
Methylhydrazinyl	3.7	
Rings:		
Benzene	8.06^*n	
Pyridyl	14.64^*n	
Attachment of COOH to benzene	11.04^*n	
Existence of aliphatic ligand (e.g., 1,2-diaminopropane)	34.9	
Attachment of two triazoles to each other	15.55^*n	
Number of coordinated waters:		
$1 < x < 5$	1.9	
≥ 5	24.41	
Uncoordinated waters	5	
H_2TZEG ligand	14.7	
O atoms between benzene and nodes	1.93^*n	
Attachment of acetic acid to tetrazole without any further substituents	5.95	
Bridging of OH or O between two nodes	4.5^*n	
Attachment of nitro to triazole while coordinated to the node via an O atom	25.2^*n	

[a] n means the number of the specified fragment in the molecule.

Example 12.2: [Cu(atrz)(IO$_3$)$_2$]$_n$ has the following molecular structure:

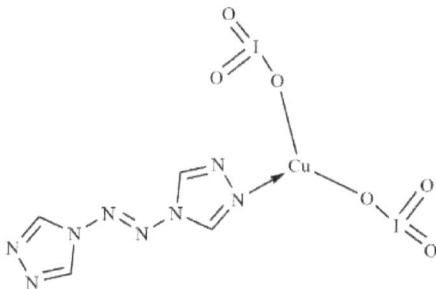

The use of eq. (12.2) and the presence of the Cu node (Table 13.3) give:

$$\Delta_f H^\theta(c) = -198.79 n_{N-H} + 529.68 n_{C=O} + 214.98 n_{C=N} + 725.25 \sum_i \Delta_f H^\theta_{\text{Inc, }i} - 91.16 \sum_j \Delta_f H^\theta_{\text{Dec, }j},$$

$$\Delta_f H^\theta(c) = -198.79(0) + 529.68(0) + 214.98(4) + 725.25(1.1) - 91.16\ (0) = 1658 \text{ kJ/mol}.$$

The experimental value of $\Delta_f H^\theta(c)$ is 1181 kJ/mol [633].

12.3 A reliable predictive model for the onset temperature of energetic complexes

A reliable method was developed to predict the onset temperature T_{onset} of energetic complexes, which is crucial for evaluating their safety and chemical stability [634]. The model was built using the largest available experimental dataset, consisting of 157 different energetic complexes. It incorporates four key structural features: the number of azide groups, sulfur atoms, electron-donating groups attached to aromatic rings, and the total number of carbocyclic and heterocyclic aromatic rings (excluding tetrazole rings). To improve accuracy, the model also includes two adjustable parameters ($T^{\text{Inc}}_{\text{onset}}$ and $T^{\text{Dec}}_{\text{onset}}$) that correct large deviations between predicted and experimental values. The model was validated using 134 compounds for training and 23 for testing, resulting in the following final equation [634]:

$$T_{\text{onset}} = 241.3 - 8.735 n_{\text{Don}} - 12.59 n_{\text{N}_3} + 15.50 n_{\text{Ar + Het}} + 79.97 n_{\text{S}} + 122.7 T^{\text{Inc}}_{\text{onset}} - 97.17\ T^{\text{Dec}}_{\text{onset}},$$

$$(12.3)$$

where n_{N_3} and n_{S} are the numbers of azide groups and sulfur atoms, respectively; n_{Don} is the number of electron-donating groups attached to the aromatic ring; $n_{\text{Ar + Het}}$ is the sum of carbocyclic and heterocyclic aromatic rings except the tetrazole ring. Table 12.5 provides different values of $T^{\text{Inc}}_{\text{onset}}$ and $T^{\text{Dec}}_{\text{onset}}$.

Table 12.5: The values of T_{onset}^{Inc} and T_{onset}^{Dec}.

No.	Ligand	Cation	Condition	T_{onset}^{Inc}	T_{onset}^{Dec}	Example
1		Ba$_{2+}$	–	1.0	0	
2			One ligand and not more than two H$_2$O ligands			
3		Pb^{2+}	More than one H$_2$O ligand	3.0	0	
			One H$_2$O ligand	1.0		

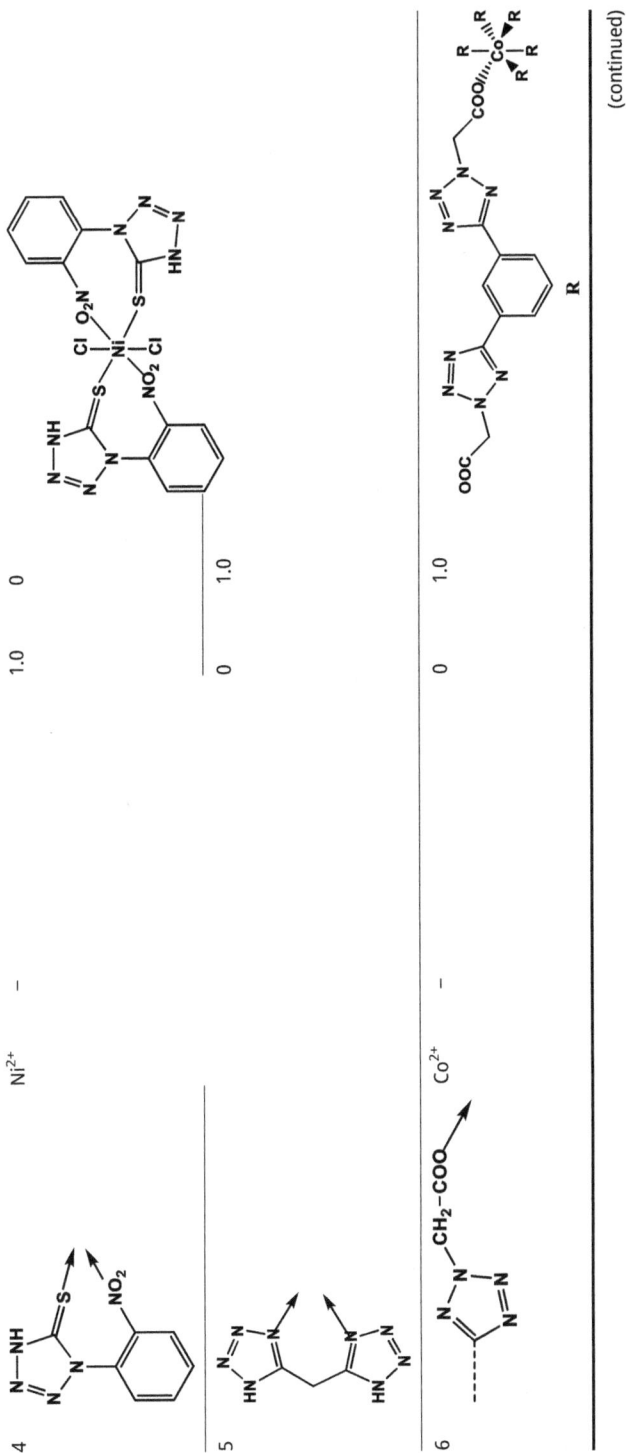

(continued)

Table 12.5 (continued)

No.	Ligand	Cation	Condition	T_{onset}^{Inc}	T_{onset}^{Dec}	Example
7		Cu^{2+}, Ag^{+}	Except the presence of ligands $O_3Cl\!-\!O\!\longrightarrow$ or $Cl\longrightarrow$	0	1.0	
8		Mn^{2+}, Cd^{2+}, Zn^{2+}	More than three	1.0	0	

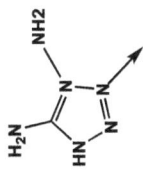

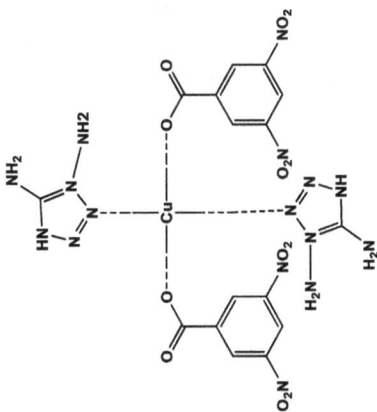

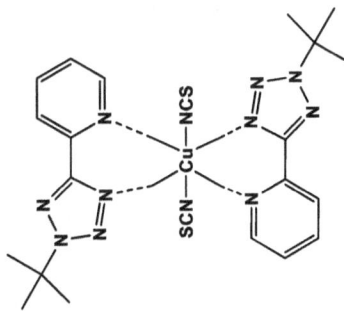

Equation (12.3) can estimate T_{onset} of energetic complexes containing structurally different ligands with the biggest types of cations, containing alkali metals (Li^+, Na^+, K^+, Rb^+, and Cs^+), and alkaline earth cations (Ca^{2+} and Ba^{2+}), and other cations (Cu^{2+}, Fe^{2+}, Ag^+, Co^{2+}, Ni^{2+}, Mn^{2+}, Zn^{2+}, Cd^{2+}, Sr^{2+}, Pb^{2+}, Bi^{3+}, and Ir^{3+}).

Example 12.3: Consider the following energetic complexes:

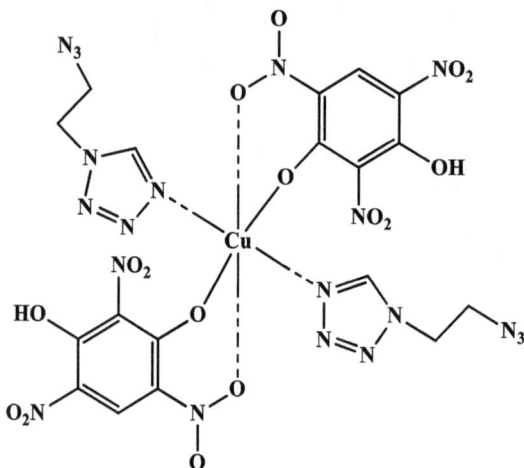

The use of eq. (12.3) gives:

$$T_{onset} = 241.3 - 8.735n_{Don} - 12.59n_{N_3} + 15.50n_{Ar + Het} + 79.97n_S + 122.7T_{onset}^{Inc}$$
$$- 97.17\, T_{onset}^{Dec},$$

$$T_{onset} = 241.3 - 8.735(6) - 12.59(2) + 15.50(2) + 79.97(0) + 122.7(0)$$
$$- 97.17(0) = 195\ K.$$

The measured value is 121 [635].

Example 12.4: Consider the following energetic complexes:

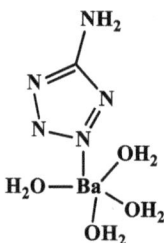

The use of eq. (12.3) and Table 12.5 gives:

$$T_{onset} = 241.3 - 8.735n_{Don} - 12.59n_{N_3} + 15.50n_{Ar + Het} + 79.97n_S + 122.7T_{onset}^{Inc}$$
$$- 97.17\, T_{onset}^{Dec},$$
$$T_{onset} = 241.3 - 8.735(1) - 12.59(0) + 15.50(0) + 79.97(0) + 122.7(1)$$
$$- 97.17(0) = 355 \text{ K}.$$

The measured value is 360 [636].

12.4 Summary

The development of high-energy-density materials represents a significant challenge in fields ranging from defense to aerospace engineering. These materials must carefully balance explosive power with thermal stability – a combination that has proven difficult to achieve. Recent research has focused on two primary approaches: conventional energetic metal complexes and the more novel EMOFs. While metal complexes deliver impressive energy density, their sensitivity remains problematic. EMOFs, with their tunable crystalline structures, offer greater stability but require careful design to maintain competitive performance. Three new predictive models are helping address these challenges by estimating critical material properties from molecular structure alone. Equation (12.1) calculates EMOF density with surprising accuracy, incorporating factors like metal type and crystalline water content. Equation (12.2) estimates the heat of formation based on chemical bond types and structural features, though refinements are ongoing after initial tests showed some variance from measured values. Equation (12.3) predicts decomposition temperatures for traditional energetic complexes using straightforward molecular parameters like azide group counts and aromatic ring structures. These approaches stand out for their practical simplicity. Rather than requiring complex computational simulations, they rely on easily identifiable structural features and correction factors. For example, simply counting sulfur atoms or electron-donating groups allows for reasonably accurate stability assessments. While not infallible, these models significantly streamline the materials development process by helping researchers identify promising candidates before synthesis.

13 Estimating the properties of energetic polymers

Energetic materials, which release energy through chemical or nuclear reactions, are classified into explosives, propellants, and pyrotechnics based on their sensitivity and power. Propellants, particularly solid ones, are vital in applications ranging from mining to space exploration. They generate thrust in rockets and missiles through steady combustion and are composed of a polymer binder, metallic fuel, and oxidizer [637]. Solid propellants can be categorized into double-base propellants, which are homogeneous and primarily consist of nitrocellulose (NC) and nitroglycerine (NG), and composite propellants, which are heterogeneous types typically including ammonium perchlorate (AP) and a polymer binder like hydroxy-terminated polybutadiene (HTPB). While double-base propellants are among the oldest types, they can decompose over time, requiring stabilizers. HTPB is widely used but is considered inert, meaning it does not enhance energy release during combustion. This can lead to inefficiencies, as it consumes effective oxygen and diminishes performance. Moreover, traditional binders often compromise the specific impulse, a critical metric of fuel efficiency [638]. Emerging energetic polymers are being developed to replace HTPB and enhance performance. These binders contain high-energy functional groups, such as azido and nitro groups, which increase energy output and improve combustion characteristics. The key benefits of energetic binders include higher energy output, smokeless combustion, and improved safety, as enhanced stability reduces the risk of accidental detonation. Research indicates that even a modest increase of 5% in specific impulse can significantly enhance the range of intercontinental ballistic missiles [639]. Advancements in binder materials are crucial for achieving such improvements. Binders are essential for maintaining structural integrity and mechanical properties. However, traditional inert binders like HTPB do not contribute positively to energy release and can hinder performance. Emerging energetic binders show promise for replacing HTPB to achieve higher specific impulse and better combustion characteristics. The oxygen balance concept quantifies the relative amount of oxygen present in a mixture compared to what is needed for complete combustion. A balanced oxygen content is crucial for ensuring efficient energy release during combustion. For example, the oxygen balance of 3,3-bis(azidomethyl)oxetane (BAMO) (−123.7) indicates better energetic properties compared to 3-azido-methyl-3-methyl oxetane (AMMO) (−169.9), which suggests that BAMO has superior performance characteristics [640]. The development of innovative energetic polymers as binders represents a significant advancement in solid propellant technology. These materials not only enhance performance and safety but also address environmental concerns associated with traditional propellants. As research continues, the integration of computational methods and novel synthesis techniques will pave the way for next-generation energetic materials that meet the demands of modern aerospace applications. This chapter discusses some important properties of energetic polymers and also demonstrates predictive models.

https://doi.org/10.1515/9783112206768-013

13.1 Predicting solubility of energetic polymers: computational models, group contribution methods, and experimental validation

The solubility parameter (δ), proposed by Hildebrand [641], quantifies a solvent's dissolving capability based on cohesive energy density, representing the energy required to separate molecules. Materials with similar δ values typically exhibit miscibility or swelling behavior [642]. While Hildebrand's approach works well for nonpolar substances, Hansen extended this concept by introducing three-component parameters (δ_D, δ_P, δ_H) to account for dispersion, polar, and hydrogen-bonding interactions [643]. These Hansen parameters enable more accurate predictions of solvent-polymer compatibility through a three-dimensional Hansen space, where solvents with parameters closer to the polymer's values demonstrate better dissolution [644]. The Relative Energy Difference (RED) serves as a practical indicator, with RED < 1 signifying a "good" solvent and RED > 1 indicating a "bad" solvent [644]. This framework has found broad applications across coatings, pharmaceuticals, and food science [645, 646].

Three primary methods exist for determining solubility parameters: experimental measurements using vaporization energy for solvents [647], indirect techniques like DSC or ultraviolet-visible (UV) spectroscopy for polymers [648, 649], and correlations with physical properties such as surface tension or refractive index [650]. Key factors influencing polymer solubility include molecular weight (higher values reduce solubility), crystallinity (more crystalline polymers are less soluble), and crosslinking (which inhibits dissolution) [650]. For solid materials, semi-empirical approaches like fluorescence spectroscopy are commonly employed [651].

Prediction methods fall into two categories: group additivity methods and QSPR approaches. Albahri's group additivity model predicts δ for organic solvents using structural group contributions [652], while Hansen group additivity methods estimate δ_D, δ_P, and δ_H for complex molecules [653]. QSPR models leverage computational descriptors for enhanced accuracy [654–656], with Koç and Koç's genetic programming-based QSPR model showing superior performance over linear regression methods [657].

A simple approach is introduced for the rapid calculation of δ for various polymers with the structural parameters of repeating unit structures $-(C^1H_2-C^2R^3R^4)-$ and general formula $C_aH_bN_cO_dF_eCl_fS_hBr_iSi_j$, including energetic polymers, based on their elemental composition and repeating unit structures as follows [658]:

$$\delta = 22.96 + \frac{-45.16b + 300.90c + 51.06d - 243.63e + 129.21h - 287.71j}{Mw_{unit}} + 3.75\delta^+ - 2.04\delta^-,$$

(13.1)

where δ is in MPa$^{1/2}$; Mw_{unit} is the molecular weight of the repeating unit structure in g/mol; h and j are the number of moles of sulfur and silicon atoms per mole of repeating unit structures, respectively; δ^+ and δ^- are two correcting functions for the adjustment of the underestimating and overestimating results. Different values of δ^+ and δ^- were derived as:

(1) *Description of hydrogen bonding and polar groups under definite circumstances*: Great deviations of the experimental data based on hydrogen bonding and polar groups in repeating unit structures. Since the effects of hydrogen bonding and polar groups depend on the kind of groups and environmental conditions around each group, the effective structural parameters of polymers are stated for great deviations. The existence of δ^+ corresponds to hydrogen bonding groups and polar groups, where their existences can enhance the attraction between polymers and polar solvents. Meanwhile, the contribution of δ^- displays the polar groups, where their presence can decrease the predicted results.

(2) *Optimization of δ^+ and δ^-*: The best values of δ^+ and δ^- were gained by regression of various amounts for these variables to provide the maximum and minimum values of r^2 and RMSE, respectively. Thus, the optimum values of δ^+ and δ^- are given as follows:

(i) **The values of δ^+ for hydrogen bonding groups**: Direct attachment of hydrogen bonding groups to the carbon atom of the main chain can provide stronger attractions with polar solvents than those attached to the main chain through a benzene ring. The long-distance between the hydrogen bonding group and the main chain can decrease the value of solubility. Thus, the maximum and minimum values of δ^+ are considered for direct attachment and the attachment of the hydrogen bonding group at least through a benzene ring.

1) –OH group-The maximum-minimum values of δ^+ $(\delta^+_{Max} - \delta^+_{Min})$ for –OH are in the range 3.0–0.0. The values of δ^+ for the presence of more than one –OH, one –OH (without the existence of more than one polar group), and phenolic hydroxyl group are 3.0, 2.0, and 1.25, respectively. The value of δ^+ is 0.0 for the existence of more than one polar group. For example, the values of δ^+ for cellulose, poly(vinyl alcohol), and cellulose diacetate are 3.0, 2.0, and 0.0, respectively.

2) –COOH or –CONH$_2$: For the existence of –COOH or –CONH$_2$, the δ^+ value is 1.0.

3) Alkoxy group: For the attachment of the alkoxy group to the main chain containing an alkyl group substituent, the value of δ^+ is 0.25.

4) –S(O)$_2$–NH$_2$ group: The presence of –S(O)$_2$–NH$_2$ provides the highest contribution δ^+, i.e., $\delta^+ = 5.0$. Since the contribution of δ^+ for the other hydrogen bonding groups can be neglected, the value of δ^+ equals 0.0.

(ii) **The value δ^+ for polar groups**: As can be expected, the kind and position of polar groups as the part of the main chain or substituent are important for the contribution of δ^+ because they can adjust the underestimated results of eq. (13.1).

5) –NH–(CH$_2$)$_n$–C(=O)O–, –NH–(CH$_2$)$_n$–NH–, –CH$_2$–CF$_2$–, and –CF$_2$–S–: For the presence of –NH–(CH$_2$)$_n$–C(=O)O–, –NH–(CH$_2$)$_n$–NH–, –CH$_2$–CF$_2$–, and –CF$_2$–S– in the main chain, the value of δ^+ is 1.0.

6) Furan ring, $-C(=O)CH_2-$ and $-CN$ (without the alkyl group): The value of δ^+ is 2.0 and 0.8 for the existence of the furan ring in the main chain and the attachment of $-C(=O)CH_2-$ to the main chain, respectively. For the attachment of $-CN$ to the main chain without the alkyl group, the value of δ^+ is 1.0.

(ii) **The δ^- values for polar groups:** In contrast to the presence of strong hydrogen bonding groups, the existence of some polar groups in a specific position can provide large positive deviations. Thus, correcting function δ^- can reduce the overestimating results.

7) CF_3-CF_2-, $(CH_3)_3-O(C=O)-$, and $-CH_2-O(C=O)-CH_3$ as well as $-CN$ (containing the alkyl group and without any polar group): For the existence of CF_3-CF_2-, $(CH_3)_3-O(C=O)-$, and $-CH_2-O(C=O)-CH_3$ as well as the attachment of $-CN$ to the repeating unit structure containing the alkyl group and without any polar group, δ^- value equals 1.0.

8) $-CH_2-S-$, phenyl, and $-NHC(=O)NH-$: The value of is 1.5 for the existence of $-CH_2-S-$ in the main chain. For the attachment of phenyl group, which may contain electron-withdrawing group, to the repeating unit structure without other substituents, the value of δ^- is 0.6. The existence of $-NHC(=O)NH-$ in the main chain has the largest value of δ^-, i.e., $\delta^- = 2.0$.

Polymers containing energetic groups such as glycidyl azide polymer (GAP), poly(3-azidomethyl-3-methyl oxetane) (poly-AMMO or PAMMO), poly-NIMMO or polyNIMMO or PNIMMO, where NIMMO is 3-nitratomethyl-3-methyloxetane, and poly-glycidyl nitrate (poly-GLYN) can be used as energetic binders in the polymer-bonded explosives.

Example 13.1: Poly-AMMO has the following repeating unit structure:

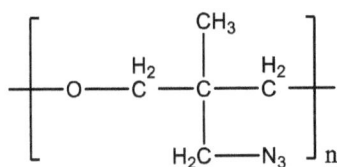

The use of eq. (13.1) gives:

$$\delta = 22.96 + \frac{-45.16b + 300.90c + 51.06d - 243.63e + 129.21h - 287.71j}{Mw_{unit}} + 3.75\delta^+ - 2.04\delta^-$$

$$\delta = 22.96 + \frac{-45.16(9) + 300.90(3) + 51.06(1) - 243.63(0) + 129.21(0) - 287.71(0)}{127} + 3.75(0) - 2.04(0)$$

$$= 27.27 \text{ MPa}^{1/2}.$$

Thus, suitable solvents within ± 2 MPa$^{1/2}$ of the polymer poly-AMMO may be selected.

13.2 Characterization and prediction of intrinsic viscosity in energetic polymer solutions

The intrinsic viscosity ($[\eta]$) of dilute polymer solutions is a fundamental property that reflects molecular weight, chain conformation, and polymer-solvent interactions [659]. It is determined by extrapolating reduced viscosity (η_{sp}/C_{poly}), where η_{sp} is the specific viscosity as the relative change in viscosity (η_{rel}) that occurs with the addition of the polymer and C_{poly} is concentration of polymer, to zero concentration using capillary viscometry in which flow times of solvent and solution are compared to calculate relative viscosity [660, 661]. For nonionic polymers, the Huggins and Kraemer equations ($\eta_{sp}/C_{poly} = [\eta] + k_h[\eta]^2 C_{poly}$ and $\ln(\eta_{rel})/C_{poly} = [\eta] + k_k[\eta]^2 C_{poly}$) are employed, with the Huggins constant (k_h) indicating solvent quality (0.3–0.5 for good solvents and 0.5–1.0 for poor solvents) [662–664]. Polyelectrolytes exhibit unique behavior due to charge repulsion, often following the Fuoss law ($\eta_{sp}/C_{poly} \propto C_{poly}^{-1/2}$) at low concentrations, though deviations occur in salt-free solutions [665–669]. Predictive models include empirical approaches like van Krevelen's solubility parameter-based equation ($R^2 = 0.324$) [659] and more accurate QSPR methods, such as Afantitis et al.'s 8-descriptor model ($R^2 = 0.759$) [670], Gharagheizi's SVM/radial-based function neural network (RBFNN) models [671], and Wang et al.'s DFT-based SVM model ($R^2 = 0.90$) [672]. Challenges include capillary adsorption artifacts for high-molecular-weight polymers [673, 674] and shear/thinning effects in polyelectrolytes [675]. Intrinsic viscosity is widely applied in material design (e.g., drug delivery) and biological system analysis [672].

A simple method has been introduced to predict the intrinsic viscosity of polymer-solvent combinations based on the largest available experimental data. It is a function of δ, b, and c per Mw_{unit} as follows [676]:

$$[\eta] = -42.14 + \frac{271.34\delta + 455.57b + 10853.04c}{Mw_{unit}} + 69.53\eta^+ - 65.44\eta^-, \tag{13.2}$$

where $[\eta]$, δ, and Mw_{unit} are in cm³/g, MPa$^{1/2}$, and g/mol, respectively. Two correcting functions η^+ and η^- can adjust large deviations of the reported intrinsic viscosities of polymer-solvent combinations. They are based on the interactions of specific polymers with aromatic or cycloalkane solvents and solvents containing δ_D, δ_P, and δ_H in certain ranges as follows:

1. **Repeating unit without double bonds and only alkyl substituents**
   - *Solvent:* Cycloalkanes
   - *Values:* $\eta^+ = 1.0$, $\eta^- = 0$
2. **Repeating unit containing polar groups**
   - *Solvent criteria:* $\delta_H > 5$, $1 < \delta_P < 12$, $\delta_D > 16$
   - *Values:* $\eta^+ = 1.0$, $\eta^- = 0$

3. **Repeating unit with etheric functional group in the main chain (no substituents)**
   - *Solvent criteria:* δ_P and $\delta_H > 11$
   - *Values:* $\eta^+ = 1.0$, $\eta^- = 0$
4. **Repeating unit with unsaturated C=C bonds**
   - *Solvent:* Aromatic compounds
   - *Values:* $\eta^+ = 1.0$, $\eta^- = 0$
5. **Repeating unit with etheric functional group in main chain + alkyl substituent**
   - *Solvent:* -
   - *Values:* $\eta^+ = 0.5$, $\eta^- = 0$
6. **Repeating unit without double bonds + multiple alkyl substituents**
   - *Solvent:* Aromatic compounds
   - *Values:* $\eta^+ = 0$, $\eta^- = 1.0$
7. **Repeating unit with etheric functional group in the main chain**
   - *Solvent criteria:* $1 < \delta_P < 11$, $\delta_D < 16$
   - *Values:* $\eta^+ = 0$, $\eta^- = 1.0$
8. **Repeating unit with polar groups**
   - *Solvent criteria:* $6 < \delta_H < 9$, $\delta_D > 19$
   - *Values:* $\eta^+ = 0$, $\eta^- = 1.0$
9. **Repeating unit without double bonds + one alkyl substituent**
   - *Solvent:* Aromatic compounds
   - *Values:* $\eta^+ = 0$, $\eta^- = 0.5$
10. **Repeating unit with polar groups**
    - *Solvent criteria:* $\delta_H > 12$
    - *Values:* $\eta^+ = 0$, $\eta^- = 0.5$

Example 13.2: The values of δ_D, δ_P, and δ_H for the solvent dimethyl sulfoxide (DMSO) are 18.4, 16.4, and 10.2 MPa$^{1/2}$, respectively [659]. The calculated δ for dimethyl sulfoxide is 26.68 MPa$^{1/2}$ ($= \sqrt{18.4^2 + 16.4^2 + 10.2^2}$). According to the answer in Example 13.1, the δ value for poly-AMMO is 27.27 MPa$^{1/2}$. Since suitable solvents are those with δ values ± 2 MPa$^{1/2}$ of the polymer, dimethyl sulfoxide may be a suitable solvent for poly-AMMO. Because dimethyl sulfoxide does not satisfy the conditions for η^+ and η^-, these values are zero. Applying eq. (3.12) yields:

$$[\eta] = -42.14 + \frac{271.34\delta + 455.57b + 10,853.04c}{Mw_{unit}} + 69.53\eta^+ - 65.44\eta^-,$$

$$[\eta] = -42.14 + \frac{271.34(27.27) + 455.57(9) + 10,853.04(3)}{127} + 69.53(0) - 65.44(0)$$

$$= 304.8 \text{ cm}^3/\text{g}.$$

13.3 Glass transition in energetic polymers: experimental characterization, predictive modeling, and performance implications

Polymers are indispensable in modern applications due to their tunable properties and cost-effectiveness, with the glass transition temperature (T_g) serving as a critical parameter that determines their mechanical behavior and suitability for specific uses [677–680]. The T_g marks the transition between rigid, glassy, and flexible rubbery states, profoundly affecting properties like thermal expansion, modulus, and diffusion rates [44, 45]. Experimental determination of T_g typically employs DSC, dynamic mechanical analysis (DMA), or thermomechanical analysis (TMA), though measured values can vary significantly depending on heating rates, measurement techniques, and sample history [677–680]. Standardized methods using DMA (5 °C/min, 1 rad/s) or DSC (10 °C/min) help enable reliable comparisons, with extensive T_g data available in databases like PoLyInfo [681] and NIST's Polymer Property Database [682].

Various approaches exist for predicting T_g, including group contribution methods pioneered by Van Krevelen and Te Nijenhuis [659] and later refined by Camacho-Zuniga and Ruiz-Trevino [46], who developed a more flexible scheme using 66 group contributions that account for both main and side-chain effects. However, these methods are limited to polymers containing previously characterized structural groups [659, 683]. More sophisticated QSPR and machine learning models have achieved greater accuracy, with ANN and SVM approaches demonstrating R^2 values exceeding 0.9 for diverse polymer sets [684–686]. Recent advances incorporate active learning frameworks to efficiently identify high-T_g polymers [687] and combine molecular dynamics simulations with machine learning for conjugated polymers [688]. Polymer informatics approaches, particularly for biopolymers like polyhydroxyalkanoate (PHA)-based polymers, have enabled rapid prediction of structure–property relationships [689], while Monte Carlo optimization methods have improved correlation accuracy [690].

Despite these advances, challenges remain in developing universally applicable models due to the nonlinear nature of T_g -structure relationships [691] and the need for extensive, high-quality training data [692]. Current research focuses on developing better structural descriptors and hybrid experimental-computational approaches to overcome these limitations [687, 693]. The field continues to evolve through the integration of advanced computational techniques with experimental validation, offering promising avenues for accelerated polymer design and discovery [694, 695].

A suitable model has been introduced for the prediction of T_g based on elemental composition and structural factors as follows [695]:

$$logT_g = 2.472 + 0.01480 \left(\log T_g\right)_{\text{Elem}} - 0.09797 \left(\log T_g\right)_{\text{NonAr, Ar}} + 0.08124 \left(\log T_g\right)_{\text{PG}}$$
$$+ 0.09910 IT_g - 0.1185 DT_g,$$

(13.3)

where $(\log T_g)_{\text{Elem}}$, $(\log T_g)_{\text{NonAr, Ar}}$, and $(\log T_g)_{\text{PG}}$ represent the contributions of elemental composition, unsaturated double bonds, and polar groups to T_g, respectively. These parameters are defined as follows:

$$(\log T_g)_{\text{Elem}} = a - 0.7783b - 0.5837e + 2.444f + 8.775j, \tag{13.4}$$

$$(\log T_g)_{\text{NonAr, Ar}} = E_{\text{NonAr}} - 0.6262E_{\text{Ar}}, \tag{13.5}$$

$$(\log T_g)_{\text{PG}} = E_{\text{OH}} - 0.5776E_{\text{Ether}} + 1.162E_{\text{NH}} - 3.391E_{\text{Si-O}}. \tag{13.6}$$

The parameters in these equations are defined as:
- E_{NonAr} and E_{Ar} in eq. (13.5): Presence of non-aromatic and aromatic carbon–carbon double bonds in repeating unit structures, respectively
- E_{OH}, E_{Ether}, E_{NH}, and $E_{\text{Si-O}}$ in eq. (13.6): Presence of hydroxyl, ether, –NH–, and –Si–O– groups in repeating unit structures, respectively

DT_g and IT_g are structural correction parameters (provided in Table 13.1) that adjust for overestimated and underestimated outputs. As shown in Table 13.1, both parameters contain five classes with specific conditions.

Table 13.1: The values of DT_g and IT_g.

Class	Structural parameters	Conditions	DT_g	Examples
1		R = Alkyl group X = R or Ar (aromatic group)	1.0	
2		U = Methyl group or H atom or Cl atom W = Methyl group or H atom or Cl atom	1.0	
3		Z = Small alkyl group with less than three carbon atoms or H atom	0.6	

Table 13.1 (continued)

Class	Structural parameters	Conditions	DT_g	Examples
4	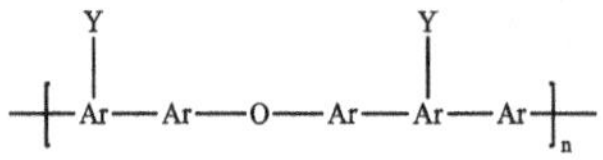	Y = Ar or H atom	0.75	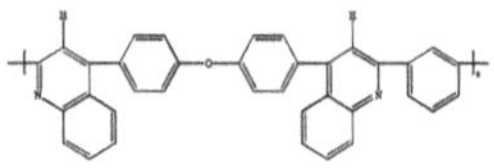
5	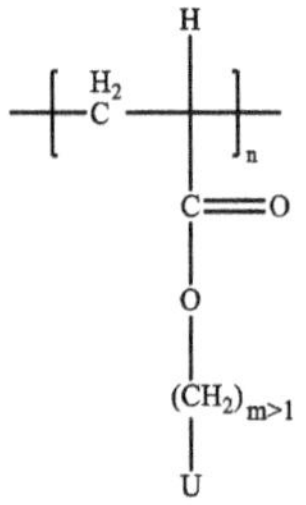	U = R or cyanide	0.5	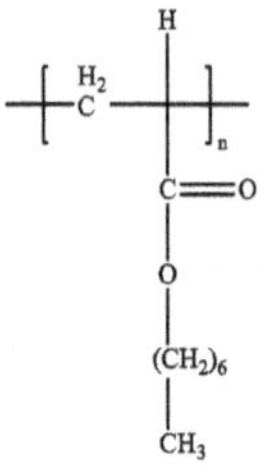

Class	Structural parameters	Condition	IT_g	Example
1	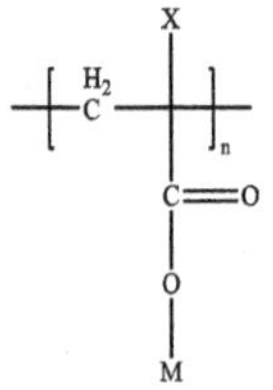	X = R or cyanide M = R or cyclic hydrocarbon	0.5	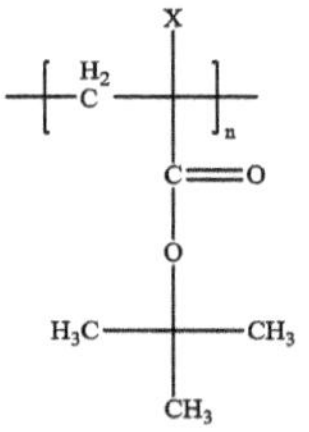
2	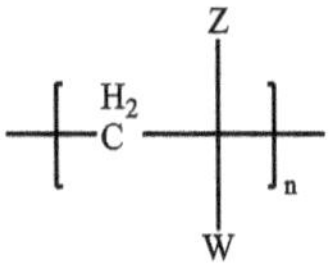	Z = R or H atom W = Cyclic hydrocarbon or cyanide	1.2	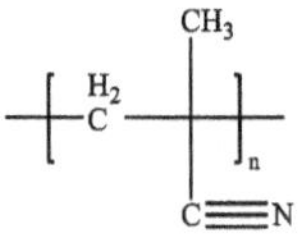
3	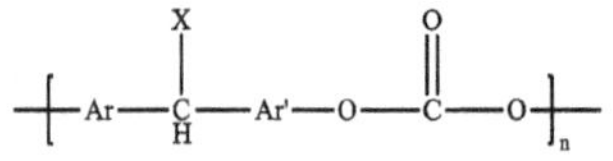	X = Cyclic hydrocarbon or R	1.0	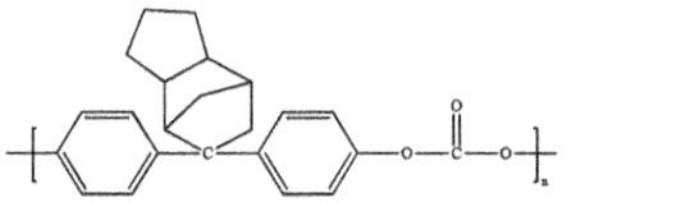
4	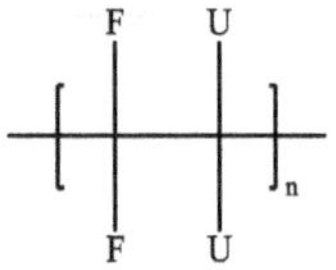	U = H atom or F atom	0.1	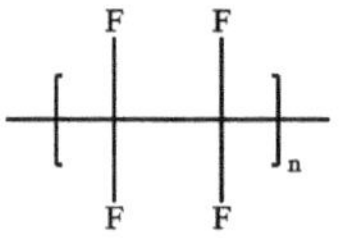
5	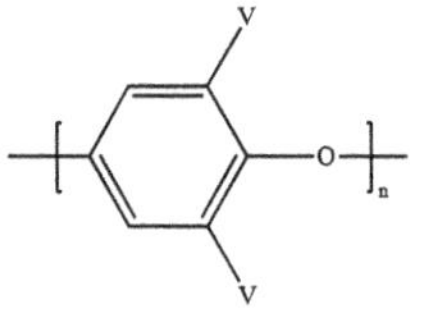	V = R or Ar	0.9	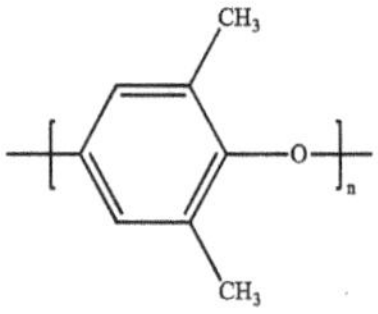

Example 13.3: Poly-GLYN has the following repeating unit structure:

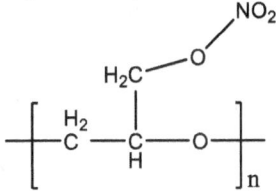

The use of eqs. (13.3) to (13.6) gives:

$$(\log T_g)_{Elem} = a - 0.7783b - 0.5837e + 2.444f + 8.775j$$

$$= 3 - 0.7783(5) - 0.5837(0) + 2.444(0) + 8.775(0) = -0.8915,$$

$$(\log T_g)_{NonAr, Ar} = E_{NonAr} - 0.6262E_{Ar} = 0 - 0.6262(0) = 0,$$

$$(\log T_g)_{PG} = E_{OH} - 0.5776E_{Ether} + 1.162E_{NH} - 3.391E_{Si-O}$$

$$= 0 - 0.5776(1) + 1.162(0) - 3.391(0) = -0.5776.$$

The values of DT_g and IT_g are zero because there are no molecular fragments given in Table 13.1.

$$\log T_g = 2.472 + 0.01480(\log T_g)_{Elem} - 0.09797(\log T_g)_{NonAr, Ar} + 0.08124(\log T_g)_{PG} + 0.09910IT_g - 0.1185DT_g$$

$$= 2.472 + 0.01480(-0.8915) - 0.09797(0) + 0.08124(-0.5776) + 0.09910(0) - 0.1185DT_g(0) = 2.4119.$$

The calculated T_g is 258 K.

13.4 Energetic polymers with engineered refractive index for multifunctional applications

The refractive index (RI) of polymers, defined as the ratio of light speed in a vacuum to that in the material ($RI = \dfrac{c_{vacuum}}{c_{mater}} = \sqrt{\varepsilon_r}$), is a fundamental property for optical applications [696]. Here, c_{vacuum} represents the speed of light in a vacuum, c_{mater} is the speed of light in the material, and ε_r is the relative permittivity (dielectric constant). According to the Lorentz–Lorenz equation, RI depends on molecular polarizability (a) and number density (N) through the relation $RI = \sqrt{\dfrac{1 + 2aN/3\varepsilon_0}{1 - aN/3\varepsilon_0}}$, where ε_0 is dielectric permittivity of free space. High-refractive-index polymers (HRIPs) are particularly valuable for optoelectronics, antireflective coatings, light-emitting diode (LED) encapsulants, complementary metal-oxide-semiconductor (CMOS) image sensors, and lithography [697–699].

RI measurement techniques include refractometry (ASTM D542 standard) [700], wavelength-dependent analysis [701], and nondestructive scanning-angle Raman spectroscopy [702]. Recent advances in QSPR modeling have enabled *RI* prediction using molecular descriptors. For instance, Khan et al. [703] developed four predictive models ($R^2 > 0.89$) based on 2D computational descriptors, including mean first ionization potential (*Mi*), molecular linear free energy relations (*MLFER_E*), mean atomic polarizability (*Mp*), and binary indicators for O-Si bonds at a topological distance of 1 (*B01 [O-Si]*). Similarly, Erickson et al. [704] achieved $R^2 = 0.90$ using a four-descriptor model incorporating Geary autocorrelation weighted by polarizability (*GATS1p*) and spectral mean absolute deviation from the adjacency matrix (*SpMAD_A*). Machine learning approaches, including interactive machine learning (IML) that integrates expert knowledge, have also shown promise (Lightstone et al., $R^2 = 0.88$) [705, 706].

A key challenge in *RI* prediction is the limited experimental data relative to the vast chemical space of possible polymers [706]. Additionally, many machine learning models operate as black boxes, lacking clear physicochemical interpretations [707]. Despite these limitations, HRIPs remain essential for advanced applications in telecommunications, medical devices, and optoelectronic components [708, 709]. Future research should focus on expanding experimental datasets, improving model interpretability, and extending predictive capabilities to copolymer systems.

Due to the complex structural parameters of polymers, directly evaluating descriptors from polymer molecules is challenging. To address this, a simple model has been developed to predict the *RI* of polymers with the general formula $C_aH_bN_cO_dF_eCl_fS_hBr_iSi_j$ based on their elemental composition. The model also accounts for increases (*IRI*) and decreases (*DRI*) in *RI* caused by specific structural parameters, as described [710]:

$$\log RI = 0.1701 + 0.004133a - 0.001950b + 0.008969c - 0.002089d - 0.006760e + 0.002875f$$

$$+ 0.02000h + 0.007175i - 0.009816j + 0.03319 IRI - 0.01928 DRI,$$

$$(13.7)$$

Table 13.2 gives different values of *IRI* and *DRI* under specific structural conditions.

Table 13.2: Different values of *IRI* and *DRI*.

No.	Structural parameters	Condition	*IRI*	Example
1		$m > 5$	0.5	
2		R = Alkyl group or hydrogen atom, X = Containing aromatic group	0.5	
3		R = Alkyl group or hydrogen atom, Y = Containing aromatic or cyclic group	1.0	
4		R = Alkyl group or hydrogen atom, Z = Aromatic or cyclic group without $-CF_3$ or isopropyl substituent	0.5	
5		Without $-S(=O)_2-$ or $-C(CH_3)_2-$ group in repeating unit	0.5	
6		W = Cyclic ether containing the hydroxyl group	1.0	

Table 13.2 (continued)

No.	Structural parameters	Condition	*IRI*	Example
7	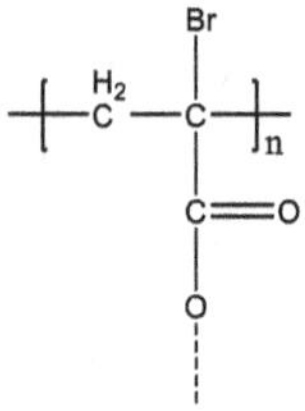		0.25	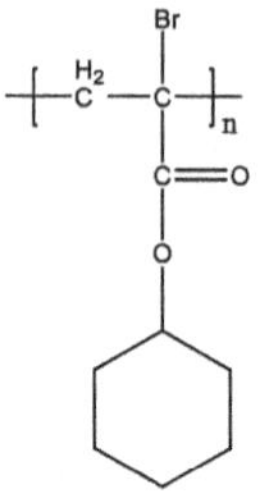

No.	Structural isomers	Condition	*DRI*	Example
1	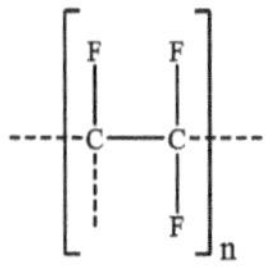		1.0	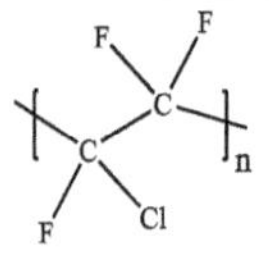
2	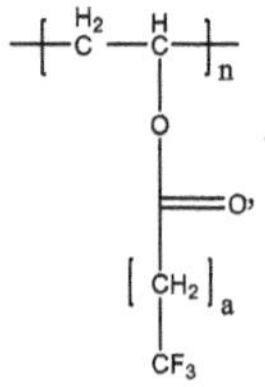 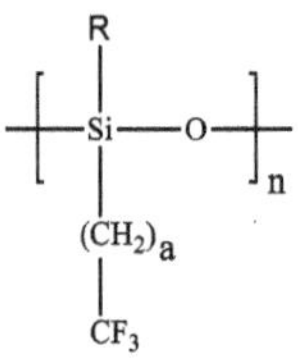	$a \geq 0$ R = Alkyl group	1.0	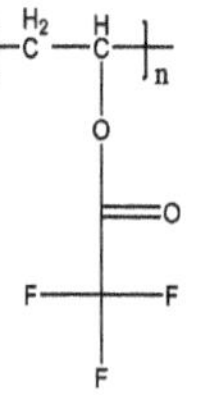

Example 13.3: The use of eq. (13.7) for poly-GLYN gives:

$$\log RI = 0.1701 + 0.004133a - 0.001950b + 0.008969c - 0.002089d - 0.006760e + 0.002875f + 0.02000h$$
$$+ 0.007175i - 0.009816j + 0.03319IRI - 0.01928DRI,$$

$$\log RI = 0.1701 + 0.004133(3) - 0.001950(5) + 0.008969(1) - 0.002089(4) - 0.006760(0) + 0.002875(0)$$
$$+ 0.02000(0) + 0.007175(0) - 0.009816(0) + 0.03319(0) - 0.01928(0) = 0.173.$$

The calculated RI is 1.491.

13.5 Summary

This chapter develops computational methods to predict key properties of energetic polymers for propellants, explosives, and pyrotechnics applications. Equation (13.1) calculates the δ from elemental composition and structural corrections for hydrogen bonding and polar groups, enabling efficient solvent screening. Building on this, equation (13.2) predicts $[\eta]$ by relating it to the solubility parameter and elemental composition while accounting for specific polymer-solvent interactions. For thermal properties, eqs. (13.3) through (13.6) estimate T_g through a logarithmic model incorporating elemental contributions, unsaturated bonds, polar group effects, and structural corrections. Finally, eq. (13.7) models RI based on elemental composition and structural increments for optical property optimization. These predictive tools address critical material design challenges by establishing quantitative structure–property relationships while balancing computational efficiency with accuracy, though they require expanded experimental validation and broader training sets for novel energetic groups. The framework supports the rational design of energetic polymers with tailored solvation, processing, thermal, and optical characteristics, advancing performance and safety in aerospace and defense applications while providing a foundation for future multi-scale simulation approaches.

14 Prediction of the properties of energetic materials

The development of novel energetic materials (explosives and propellants) has historically been constrained by the inherent challenges of traditional synthesis methods – high costs, prolonged timelines, and significant safety risks [3]. Each experimental failure not only delays progress but also exposes researchers to potential hazards, underscoring the need for more efficient discovery approaches. In recent decades, computational chemistry has emerged as a transformative tool, enabling scientists to predict key properties and performance metrics before committing to labor-intensive synthesis.

This chapter examines the evolution and application of computational methods in energetic materials research, from empirical correlations to sophisticated quantum mechanical models. Early empirical approaches, such as those implemented in EMDB 2.1 [127–129, 711] and RoseBoom [712–714], provided rapid screening capabilities but were limited by their reliance on simplified assumptions. However, significant challenges persist, especially for ionic compounds. The prediction of lattice energies for salts remains problematic, with widely used methods like the Jenkins equation demonstrating poor accuracy for non-1:1 stoichiometries [715, 716]. As the field progresses, integrating multiscale modeling with high-throughput screening may overcome current bottlenecks. This chapter not only reviews these technological advances but also emphasizes their practical implications – how virtual testing accelerates discovery while reducing laboratory risks. By critically evaluating successes and shortcomings, we aim to guide future research toward more reliable predictive frameworks for next-generation energetic materials.

14.1 Computational prediction of energetic materials: methods, challenges, and applications

The laboratory synthesis of new energetic materials (explosives and propellants) has many disadvantages:
- it is time-consuming;
- it is expensive; and
- it can be potentially hazardous.

Therefore, the prediction of the properties of new and hitherto unknown energetic compounds is important for focusing research on the most promising candidates. Numerous empirical and thermodynamic codes have been developed over the past 60 years (Figure 14.1).

While the empirical codes like EMDB 2.1 [127–129, 711] and RoseBoom [712–714] do not rely on physical input data, such as the density or the enthalpy of formation, they

https://doi.org/10.1515/9783112206768-014

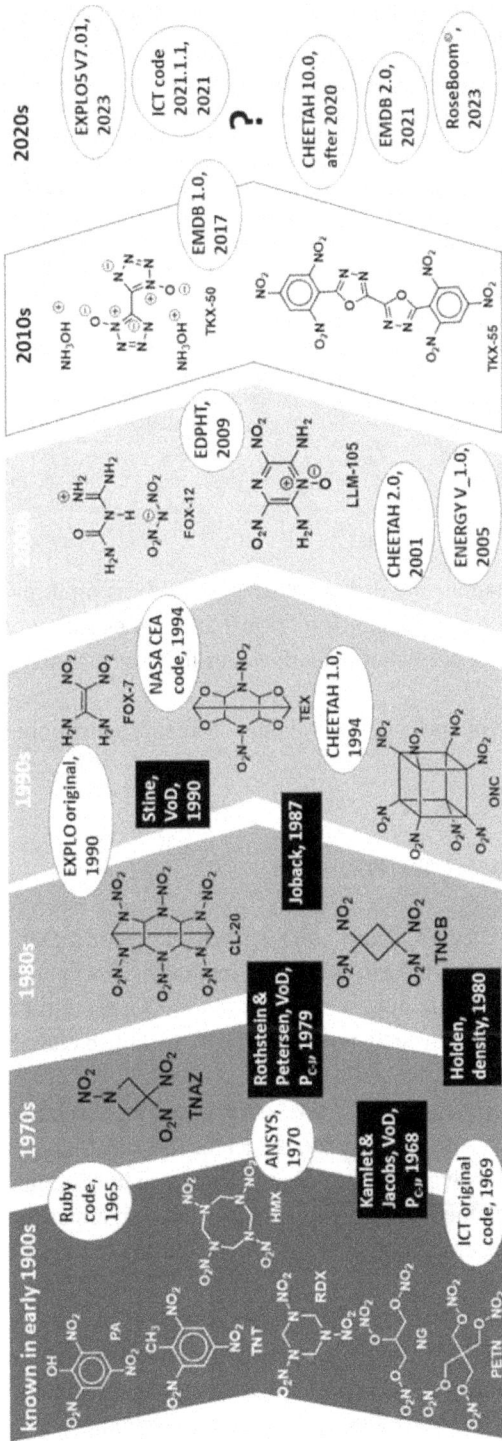

Figure 14.1: An extremely basic overview of developments in energetic materials research showing some selected landmark compounds as well as computational advances [Figure taken from: T. M. Klapötke, *Chemistry of High-Energy Materials*, 7th edn., de Gruyter, **2025**] [3].

are less accurate than thermodynamic codes like the ICT code [717, 718], CHEETAH [369, 719], or EXPLO5 [126, 719, 720].

The most desirable values for the prediction of (secondary) explosives and propellants are:

Explosives: Detonation velocity (VoD, D)
              Detonation pressure (p_{C-J})
              Heat of detonation (Q_{ex})
              Detonation temperature (T_{ex})
              Volume of detonation gases (V_0)
              TNT equivalents

Propellants: Heat of combustion (Q_{comb})
              Combustion temperature (T)
              Specific impulse (I_{sp})

On the other hand, for the thermodynamic equilibrium codes one needs as input parameters the density of the explosive (or propellant), ρ (g/cm^3) and the enthalpy of formation, $\Delta H°_f$ (kJ mol^{-1}) or $\Delta_f H^\theta(c)$. The prediction of both values for experimentally unknown compounds is not trivial.

The **density** can be estimated on the basis of several empirical codes, including the methods by
- Keshavarz (EMDB 2.1) [127–129, 711]
- Holden [711]
- RoseBoom (machine learning method) [712–714].

While relatively standard methods are used by almost all synthetic energetic materials groups world-wide for the calculation of the **enthalpies of formation** for neutral compounds and salts, it has recently become apparent that there are severe limitations and errors that result from using Jenkins equation to estimate the lattice enthalpies of salts [722].

For neutral compounds and ions in the *gas*-phase the best way to obtain reliable enthalpies of formation is still based on composite methods such as CBS-4M, CBS-QB3 or CBS-APNO [723]. These increase in the given order in accuracy but also in cpu time. However, the accuracy gained from one method to the next higher one can almost be neglected in comparison with the inaccuracy that is made by calculating the lattice energies for salts.

Therefore, for **neutral compounds**, it is recommended to calculate the enthalpy of formation in the gas-phase at the CBS-QB3 level of theory, then convert this value into the standard state (solid or occasionally liquid) using the enthalpies of sublimation or vaporization calculated based on ML methods using the RoseBoom© code

[388]. The CBS-4M has been modified by the inclusion of diffuse functions in the geometry optimization step to give CBS-QB3. The five-step CBS-QB3 series of calculations starts with a geometry optimization at the B3LYP level, followed by a frequency calculation to obtain thermal corrections, zero-point vibrational energy, and entropic information. The next three computations are single-point calculations (SPCs) at the CCSD (T), MP4SDQ, and MP2 levels. The CBS extrapolation then computes the final energies.

For **salts**, the situation is more complicated, and it appears that the problem lies in the estimation of lattice energies/enthalpies of salts. While it has been shown recently that the Jenkins equation is not good enough to estimate the lattice energies and enthalpies for 1:1 salts, the improved equation by Gutowski (which is a reparametrization of the Jenkins equation for 1:1 salts which is more suitable for energetic compounds) [724], may be used in combination with the CBS-QB3 calculated values for the enthalpy of formation for the individual ions in the gas phase:

$$\Delta U_L = |z_+||z_-|v \left[\frac{\alpha}{\sqrt[3]{V_M}} + \beta \right]$$

with the constants: $\alpha = 19.9$ (kcal nm)/mol and $\beta = 37.6$ kcal/mol.

The conversion from density (ρ) to molecular volume (V_M) of a substance is straightforward and is carried out using the following equations, depending on whether the molecular volume is required in $Å^3$ or nm^3:

$$V_M\left[nm^3\right] = \frac{M\left[\frac{g}{mol}\right] 10^{21}\left[\frac{nm}{cm^3}\right]}{\rho\left[\frac{g}{cm^3}\right] N_L} \qquad V_M\left[Å^3\right] = \frac{M\left[\frac{g}{mol}\right] 10^{24}\left[\frac{Å}{cm^3}\right]}{\rho\left[\frac{g}{cm^3}\right] N_L}.$$

For 1:2 and 2:1 salts, the situation is more complicated since there is no re-parametrized version of the Jenkins equation available, and for salts of the type A_2B such as TKX-50, the Jenkins equation provides a very unsatisfactory value for the lattice enthalpy [715, 716].

At the moment, one option to estimate the enthalpy of formation is 1:2 (A_2B) and 2:1 (AB_2) salts which avoids the necessity of estimating the lattice enthalpy is to undertake the following steps:

1. The enthalpy for the "salt reaction" between the neutral base in its standard state and the nitric acid in its standard state is calculated. The value of $\Delta H_{salt\ reaction}$ is specific to every individual base (Table 4.9a). It is obtained according to Sinditskii's method [725] by subtracting from the experimentally determined value of the enthalpy of formation of the Base-$H^+NO_3^-$ (s) salt, the enthalpy of formation of the base and nitric acid in their standard states. It is then assumed that this value for the $\Delta H_{salt\ reaction}$ is the same for a specific base, regardless of which acid is involved in protonating the base. In order to calculate the enthalpy of formation of the ionic salt, the enthalpies of formation of the neutral base, and acids in their

standard states as well as the $\Delta H_{\text{salt reaction}}$ for the given base are added together. An example is given below using TKX-50:

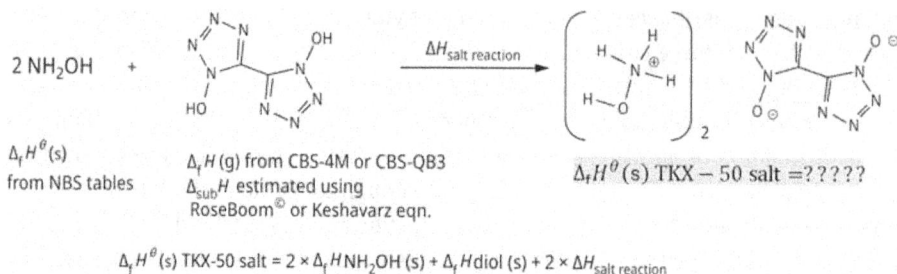

2 NH$_2$OH   +

$\Delta_f H^{\theta}(s)$
from NBS tables

$\Delta_f H$ (g) from CBS-4M or CBS-QB3
$\Delta_{\text{sub}} H$ estimated using
RoseBoom$^{\odot}$ or Keshavarz eqn.

$\Delta_f H^{\theta}(s)$ TKX − 50 salt =?????

$\Delta_f H^{\theta}(s)$ TKX-50 salt = 2 × $\Delta_f H$NH$_2$OH (s) + $\Delta_f H$diol (s) + 2 × $\Delta H_{\text{salt reaction}}$

2. In the second step, the enthalpy of the formation of a (hypothetical) neutral adduct between the acid and base(s) in the gas-phase is calculated at, for example, the CBS-QB3 level of theory. The enthalpies of sublimation/vaporization for the component acid and base are then subtracted from the gas-phase enthalpy of formation of the adduct. An example is given below using TKX-50:

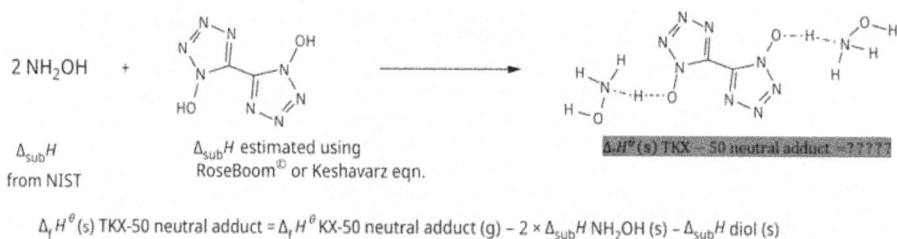

2 NH$_2$OH   +

$\Delta_{\text{sub}} H$
from NIST

$\Delta_{\text{sub}} H$ estimated using
RoseBoom$^{\odot}$ or Keshavarz eqn.

$\Delta_f H^{\theta}(s)$ TKX − 50 neutral adduct =?????

$\Delta_f H^{\theta}(s)$ TKX-50 neutral adduct = $\Delta_f H^{\theta}$ KX-50 neutral adduct (g) − 2 × $\Delta_{\text{sub}} H$ NH$_2$OH (s) − $\Delta_{\text{sub}} H$ diol (s)

3. In the final step, the values from steps 1 and 2 are weighted 80:20 to obtain an estimated value for the enthalpy of formation for the salt:

$\Delta_f H^{\theta}(s)$ TKX-50 (est.) = $0.8 \times \Delta_f H^{\theta}(s)$ TKX − 50 salt + $0.2 \times \Delta_f H^{\theta}(s)$ TKX − 50 neutral adduct

Table 14.1 gives a summary of the estimated values for the $\Delta H_{\text{salt reaction}}$ for a range of neutral bases. The experimentally determined values for the enthalpy of formation ($\Delta_f H_m^{\circ}(s)$) of the corresponding base-H$^+$NO$_3^-$ salts were used, and not of other salts such as the corresponding perchlorate salt.

Table 14.2 shows a summary of the calculated and measured enthalpies of formation for various 2:1 salts using the Jenkins equation and the 80:20 method as described above.

Table 14.1: Estimated values for the $\Delta H_{salt\ reaction}$ for a range of bases, estimated using the values of $\Delta_f H_m^\circ$ (s) of only the corresponding nitrate salts.

Salt	$\Delta_f H_m^\circ$ (salt, s)/ kJ/mol	$\Delta_f H_m^\circ$(base)/ kJ/mol	$\Delta_f H_m^\circ$ (acid)/ kJ/mol	$\Delta H_{salt\ reaction}$/ kJ/mol
$NH_3OH^+NO_3^-$	−366.5	$NH_2OH(s) = -114.2$	$HNO_3(l) =$ −174.10	−78.2
$NH_4^+NO_3^-$	−365.56	$NH_3(g) = -46.11$	$HNO_3(l) =$ −174.10	−145.35
$C(NH_2)_3^+NO_3^-$	−386.94	guanidine(s) = −56.1	$HNO_3(l) =$ −174.10	−156.8
$AG^+NO_3^-$	−278.7	AG(s) = +58.5	$HNO_3(l) =$ −174.10	−163.1
$DAG^+NO_3^-$	−157.3	DAG(s) = +167.4	$HNO_3(l) =$ −174.10	−150.6
$TAG^+NO_3^-$	−50.2	TAG(s) = +287.7	$HNO_3(l) =$ −174.10	−163.8
$5\text{-}AT^+NO_3^-$	−27.6[	5-AT(s) = +207.9	$HNO_3(l) =$ −174.10	−61.4
$N_2H_5^+NO_3^-$	−251.58	$N_2H_4(l) = +50.63$	$HNO_3\ (l) =$ −174.10	−128.1
$(H_2N)_2COH^+NO_3^-$	−564.0	Urea(cr) = −333.51	$HNO_3(l) =$ −174.10	−56.39
3-Amino-1,2,4-triazolium $^+NO_3^-$	−171.1	3-Amino-1,2,4-triazole (s) = +77.0[16a]	$HNO_3(l) =$ −174.10	−74
Anilinium$^+NO_3^-$	−182.	Aniline (l) = +31.3	$HNO_3(l) =$ −174.10	−40
$EtNH_3^+NO_3^-$	−366.9	$EtNH_2(l) = -74.1$	$HNO_3(l) =$ −174.10	−118.7
$Me_2NH_2^+NO_3^-$	−352.0	$Me_2NH(l) = -43.9$	$HNO_3(l) =$ −174.10	−134.0
$Me_3NH^+NO_3^-$	−343.9	$Me_3N(l) = -46.0$	$HNO_3(l) =$ −174.10	−123.8
$MeNH_3^+NO_3^-$	−354.4	$MeNH_2(l) = -47.3$		−132.6

Table 14.1 (continued)

Salt	$\Delta_f H_m^\circ$ (salt, s)/ kJ/mol	$\Delta_f H_m^\circ$ (base)/ kJ/mol	$\Delta_f H_m^\circ$ (acid)/ kJ/mol	$\Delta H_{salt\ reaction}$/ kJ/mol
$Et_2NH_2^+NO_3^-$	−418.8	$Et_2NH(l) = -103.3$	$HNO_3(l) = -174.10$ $HNO_3(l) = -174.10$	−141.7
$Et_3NH^+NO_3^-$	−447.7	$Et_3N(l) = -169.0$	$HNO_3(l) = -174.10$	−104.6
NH_3^+ ... NO_3^- 4-Amino-1,2,4-triazolium nitrate	+2	NH_2 ... 4-Amino-1,2,4-triazole(s) = +223.13	$HNO_3(l) = -174.10$	−47.03

Example 14.1: A nice example of the prediction of a new energetic molecule has recently been published by Jean'ne Shreeve et al. [726], a master in the art of the synthesis of new energetic materials: UIX (Figure 14.2):

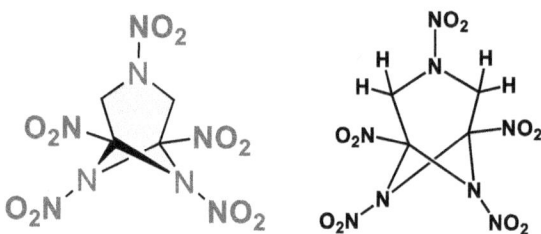

Figure 14.2: Molecular structure of UIX, $C_4H_4N_8O_{10}$ (left: structural drawing, right: cdx-type of file).

UIX has not yet been synthesized and has the chemical formula $C_4H_4N_8O_{10}$ and therefore an oxygen balance of $\Omega(CO_2) = 0\%$. The corresponding smiles code is:

[H]C(N([N+]([O−])=O)C([H])([H])C1([N+]([O−])=O)N2[N+]([O−])=O)([H])C2(N1[N+]([O−])=O)[N+]([O−])=O

A CBS-4M calculation using the G16W software resulted in a true minimum (NIMAG = 0) and a
    CBS-4 Enthalpy = −1342.937563H.
The enthalpies of the gaseous species M can now be calculated using the method of the atomization energies [727–729] (eq. (14.1)):

$$\Delta_f H^\circ(g,m) = H^\circ_{(molecule)} - \sum H^\circ_{(atoms)} + \sum \Delta_f H^\circ_{(atoms)},$$ (14.1)

Table 14.2: Estimated values for the $\Delta_f H^\circ_m$ (s) for 2:1 salts using values calculated at the CBS-QB3 level of theory where necessary, as well as the experimentally determined $\Delta_f H^\circ_m$ (s) values previously reported in the literature.

Compound	$\Delta_f H^\circ_m$/kJ/mol CBS-QB3: Cation *Anion* ΔH_{latt}/kJ/mol (Jenkins)	$\Delta_f H^\circ_m$/kJ/mol CBS-B3/ Jenkins	$\Delta_f H^\circ_m$/kJ/mol $\Delta H_{salt\ reaction}$ Salt	$\Delta_f H^\circ_m$/kJ/mol Neutral adduct	Average of salt + adduct	80:20	Exptl. value	Δ (exptl. and 80:20 values)
TKX-50	2 × +669.2 1 × +541.05 −1,488.95	+390.5	+168.3	+240.49	+204.4	+182.74	+193 (average)	−10.26
TKX-50	2 × +669.2 1 × +541.05 −1,488.95	+390.5	+168.3	+252.39	+210.3	+185.12	+193 (average)	−7.88
GZT	2 × 570.31 1 × +790.5 −1,287.98	+643.14	+374.2	+532.37	+453.3	+405.83	+410, +452, +387	−4.17, +18.83, −46.17
AG₂AzT	2 × 668.9 1 × +790.5 −1,246.01	+882.29	−509.8	+737.9	+664.4	+620.22	+462, +434, +782	+158.22, +186.22, −161.78
DAG₂AzT	2 × 813.1 1 × +790.5 −1,215.11	+1,201.59	+833.6	+885.56	+859.6	+843.99	+709	+134.99
TAG₂AzT	2 × 920.8 1 × +790.5 −1,177.40	+1,454.7	+1,047.8	+1,167.04	+1,107.42	+1,071.65	+1,075, +1,065	−3.35, +6.65

(continued)

Table 14.2 (continued)

Compound	$\Delta_f H^\circ_m$/kJ/mol CBS-QB3: Cation Anion ΔH_{latt}/kJ/mol (Jenkins)	$\Delta_f H^\circ_m$/kJ/mol CBS-B3/ Jenkins	$\Delta_f H^\circ_m$/kJ/mol $\Delta H_{salt\ reaction}$ Salt	$\Delta_f H^\circ_m$/kJ/mol Neutral adduct	Average of salt + adduct	80:20	Exptl. value	Δ (exptl. and 80:20 values)
(NH$_4$)$_2$AzT	2 × **632.12** 1 × +790.5 −1,467.43	+587.31	+417.08	+623.03	+520.1	+458.27	+443.9, +452, +551	+14.37, +6.27, −92.73
(N$_2$H$_5$)$_2$AzT	2 × **765.52** 1 × +790.5 —	Density Unknown	+645.06	+784.88	+714.97	+673.02	+659 +858	+14.02, −184.98
G$_2$CO$_3$	2 × **+570.31** 1 × −232.2 −1,415.0	−505.68	−1,044.6	−902.36	−973.48	−1,016.15	−971.1	−45.05
(NH$_4$)$_2$SO$_4$	2 × **+632.12** 1 × −607.3 −1,806.74	−1,149.8	−1,196.9	−989.22	−1,093.1	−1,155.36	−1,180.9	+25.54
(NH$_3$OH)$_2$SO$_4$	2 × **+669.2** 1 × −607.3 −1,712.2	−981.1	−1,200.0	−1,143.6	−1,171.8	−1,188.72	−181.98	−6.74

The calculated CBS-4M value for UIX as well as the relevant atoms H, C, N, and O, are summarized in Table 14.3.

Therefore, in Table 14.3 we already have the $H°_{(molecules)}$ and $H°_{(atom)}$ values (given in a.u. = atomic units; 1 a.u. = 1 H = 627.089 kcal/mol). The values for $\Delta_f H°_{(atoms)}$ are easily obtained from the literature and are summarized in Table 14.4.

Table 14.3: CBS-4M values for NG, the ions NH_4^+ and $N(NO_2)_2^-$, and the relevant atoms H, C, N, and O.

	$-H^{298}$ / a.u.	$-G^{298}$ / a.u.
UIX	1342.937563	1343.006108
H	0.500991	0.514005
C	37.786156	37.803062
N	54.522462	54.539858
O	74.991202	75.008515

Table 14.4: Literature values for $\Delta_f H°_{(atoms)}$ (in kcal/mol).

	Reference	NIST [199]
H	52.6	52.1
C	170.2	171.3
N	113.5	113.0
O	60.0	59.6

With eq. (14.1), we now obtain a gas phase enthalpy of formation for UIX of
$$\Delta_f H°(UIX, g) = + 306.8 \text{ kJ/mol.}$$
Using the RoseBoom ML code for calculation, the sublimation enthalpy for UIX, we obtain
$$\Delta_{sub} H(UIX) = 146.6 \text{ kJ/mol.}$$

With this value we obtain an enthalpy of formation in the solid state of
$$\Delta_f H°(UIX, s) = + 160.2 \text{ kJ/mol.}$$
The density of UIX can be predicted using different methods (see above). The obtained results are as follows:
Keshavarz: 1.87 g/cm^3
Holden: 2.05 g/cm^3
RoseBoom ML: 1.95 g/cm^3

If we assume an "average" density of $\rho = 1.95$ g/cm^3, we now have all the input data we need for a calculation using a thermodynamic code, for example, EXPLO5:

Formula: $C_4H_4N_8O_{10}$
Density: $\rho = 1.95$ g/cm^3
Enthalpy of formation: $\Delta H°_f(UIX, s) = + 160$ kJ/mol[1].

With these input data, an EXPLO5 (V7.01.01) calculation was performed. The obtained results are summarized in the following section:

REACTANT INFORMATION:
1. UIX, 100 %

C(4.000) H(4.000) N(8.000) O(10.000)

Molecular weight   = 324.13
Theoretical maximum density   = 1.95 g/cm3
Density   = 1.95 g/cm^3 (100.00 % TMD)
Initial pressure   = 0.1 MPa
Oxygen balance   = 0 %
Enthalpy of formation   = 493.64 kJ/kg (160.00 kJ/mol)
Energy of formation   = 577.76 kJ/kg (187.27 kJ/mol)

DETONATION PARAMETERS (at the C-J point):
Heat of detonation   = −6,522.276 kJ/kg
Detonation temperature   = 4,483.971 K
Detonation pressure   = 40.34728 Gpa
Detonation velocity   = 9,452.55 m/s
Particle velocity   = 2,188.918 m/s
Sound velocity   = 7,263.631 m/s
Exponent 'Gamma'   = 3.318357
Density of all products   = 2.537639 g/cm^3
Density of gaseous products   = 2.53761 g/cm^3
Specific volume of all products   = 0.3940671cm^3/g
Specific volume of gaseous prod. = 0.3940716cm^3/g
Volume of gas at STP   = 685.972 dm^3/kg
Moles of gaseous products   = 30.605 mol/kg
Moles of condensed products   = 0.0 mol/kg
Mean molecular mass of gas. prod.   = 32.67445g/mol
Mean molecular mass of cond. prod. = 0g/mol
Mean molecular mass of all prod.   = 32.67445g/mol
Entropy of products   = 6.5815 kJ/kg K
Internal energy of products   = 8,917.971 kJ/kg, i.e., 17.39004 kJ/cm^3

Compression energy	= 2,395.694 kJ/kg, i.e., 4.671604 kJ/cm^3
TNT equivalent from Qd	= 149.51 %
TNT equivalent from pCJ	= 217.40 %

14.2 Summary

Developing new energetic materials (explosives and propellants) in the lab is slow, costly, and risky. To streamline research, scientists rely on predictive models to identify the most promising compounds before synthesis. Over the past 60 years, both empirical and thermodynamic codes have been developed for this purpose. Empirical methods like EMDB 2.1 and RoseBoom are faster but less accurate, while thermodynamic codes like ICT, CHEETAH, and EXPLO5 require precise input data – such as density and enthalpy of formation – for reliable predictions. Key properties for evaluating explosives include detonation velocity, pressure, heat release, and gas volume, while propellants are assessed by combustion heat, temperature, and specific impulse. Estimating density and enthalpy of formation for unknown compounds is challenging. Density can be approximated using empirical methods (e.g., Keshavarz, Holden, or RoseBoom's machine learning approach), but calculating enthalpies of formation is more complex. For neutral compounds, high-level quantum chemistry methods like CBS-QB3 are recommended, combined with machine learning for sublimation enthalpies.

For salts, the process is trickier due to unreliable lattice energy estimates. Traditional methods like the Jenkins equation, often fail, especially for 1:2 or 2:1 salts (e.g., TKX-50). An alternative approach involves calculating reaction enthalpies and weighting results from different methods (80% experimentally derived, 20% computational). Tables summarizing these values for various salts highlight discrepancies between predicted and experimental data.

A case study on the hypothetical explosive UIX ($C_4H_4N_8O_{10}$) demonstrates this predictive process. Using CBS-4M calculations and machine learning for sublimation enthalpy, researchers estimated its solid-state enthalpy of formation (+160.2 kJ/mol) and density (~1.95 g/cm^3). These inputs fed into EXPLO5 predicted a detonation velocity of 9,452 m/s and a pressure of 40.3 GPa – showing potential as a high-performance explosive.

Overall, computational methods help prioritize synthesis targets, but limitations remain, especially for complex salts. Continued refinement of predictive models is essential for advancing energetic materials research.

Problems

(Hint: the necessary information for some problems are given in Appendix A)

Chapter 1

Use the following equations to calculate the crystal density of the specified energetic compounds:

(1) Equation (1.1) and Table 1.1:

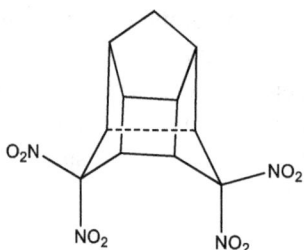

(2) Equations (1.1) and (1.5), as well as Tables 1.1 and 1.2:

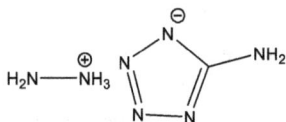

(3) Equations (1.11) and (1.11a) to (1.11g):

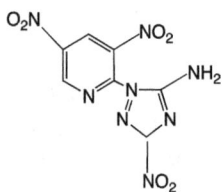

(4) Equations (1.12) and (1.12a) to (1.12e), or (1.13), or (1.14):

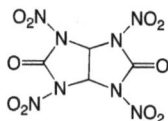

(5) Equation (1.15): $C(NO_2)_3CH_2CO_2(CH_2)_3CO_2CH_2C(NO_2)_3$

https://doi.org/10.1515/9783112206768-015

(6) Equation (1.16):

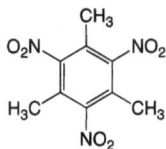

(7) Equation (1.17):

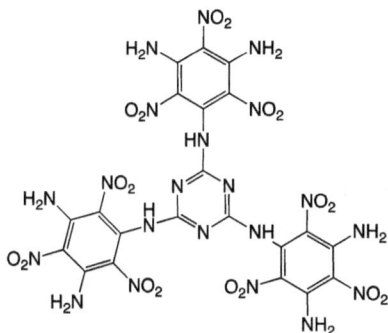

(8) Equation (1.18):

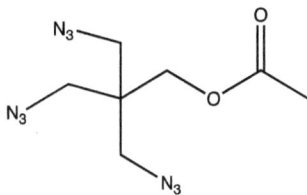

(9) Equation (1.19):

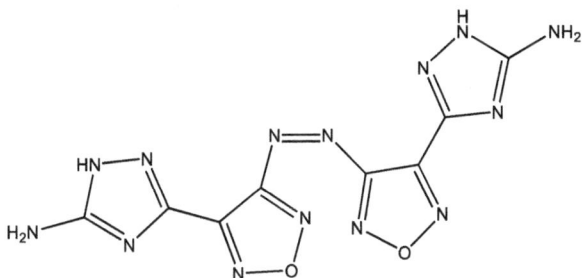

Chemical Formula: $C_8H_6N_{14}O_2$
Molecular Weight: 330.23

(10) Equation (1.19):

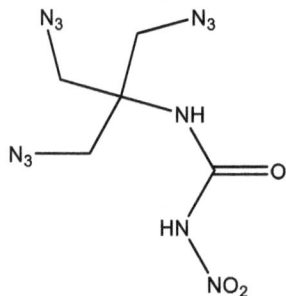

Chemical Formula: $C_5H_8N_{12}O_3$
Molecular Weight: 284.20

(11) Equations (1.21) and (1.22): Aminoguanidinium nitroformate
(12) Equation (1.23): 5-Azido-4-methyltetrazolium nitrate
(13) Equation (1.24):

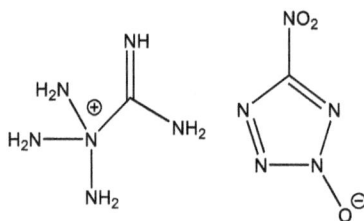

(14) Equation (1.24), the calculated ρ_{PM6} = 1.903 g/cm^3:

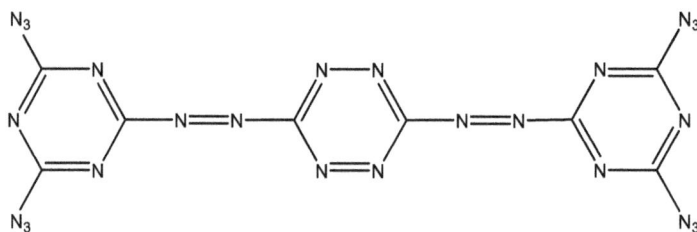

Chemical Formula: C_8N_{26}
Molecular Weight: 460.27

(15) Equation (1.24), the calculated $\rho_{PM6} = 1.375$ g/cm^3:

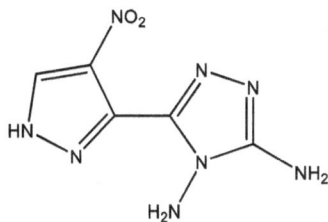

Chemical Formula: $C_5H_6N_8O_2$
Molecular Weight: 210.16

(16) Equation (1.25):

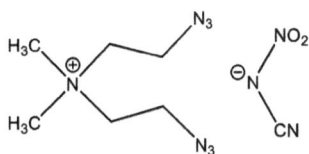

(17) Equation (1.25):

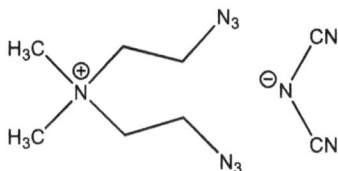

(18) Equation (1.25):

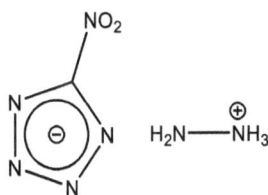

(19) Equation (1.25):

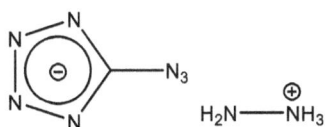

Chapter 2

Use the following equations to calculate the condensed-phase heat of formation for the specified energetic compounds:

(1) Equation (2.6):

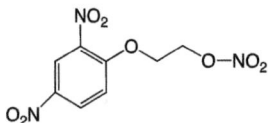

(2) Equation (2.7):

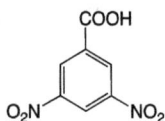

(3) Equation (2.9): $C(NO_2)_3CH_2CH_2C(=O)OCH_2C(NO_2)_3$

(4) Equation (2.10):

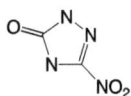

(5) Equation (2.11):

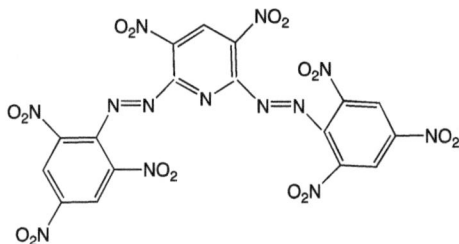

(6) Equation (2.12): 4-Nitrocinnamic acid

(7) Equation (2.13):

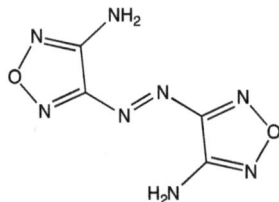

(8) Equations (2.14) and (2.15) with

$$\left(\Delta_f H^{\theta}(g)\right)_{\text{B3LYP}} = 1,077.4 \text{ kJ/mol} \text{ and } \left(\Delta_f H^{\theta}(g)\right)_{\text{PM6}} = 1,039.5 \text{ kJ/mol,}$$

respectively:

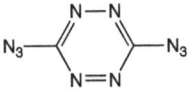

(9) Equation (2.17):

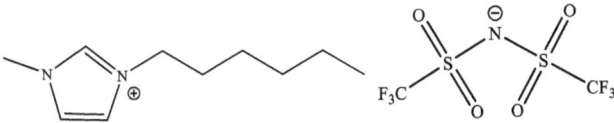

(10) Equation (2.18):

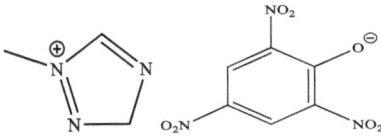

Chapter 3

Use the following equations to calculate the melting point of the specified energetic compounds:

(1) Equation (3.4):

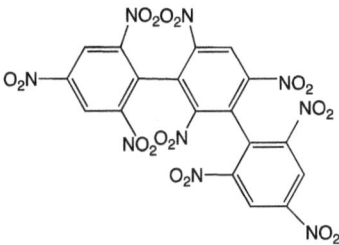

(2) Equation (3.5):

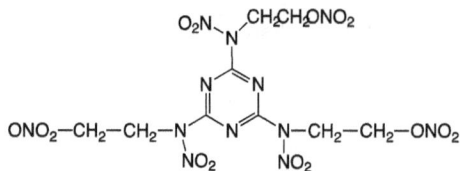

(3) Equation (3.6): $[CH_3(CH_2)_6]_4NNO_3$

(4) Equation (3.7):

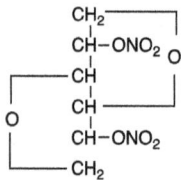

(5) Equations (3.8)–(3.10):

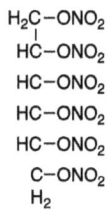

(6) Equations (3.11)–(3.13):

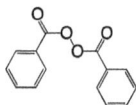

(7) Equation (3.14): 4-Chlorobenzoyl azide

(8) Equations (3.15)–(3.17):

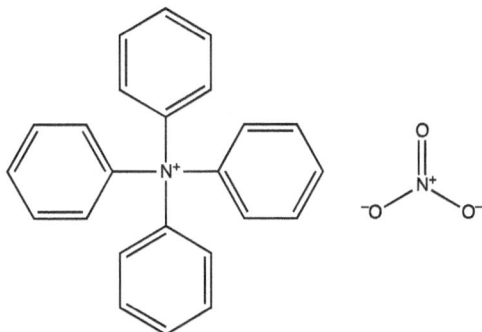

(9) Equation (3.18): (2-Tridecylpropane-1,3-diyl)dicyclohexane
(10) Equation (3.19): 1-Butyl-3-methylimidazolium trifluoromethanesulfonate
(11) Equation (3.20): 1,2-methyl-3-propylimidazolium bis((perfluoroethane)sulfonyl) imide
(12) Equation (3.21): 1-Nonyl-3-methylimidazolium hexafluorophosphate

Chapter 4

Use the following equations to calculate the enthalpy and entropy of fusion for the specified energetic compounds:
(1) Equation (4.1): *N*-Methyl-*N*,2,4,6-tetranitroaniline (tetryl)
(2) Equation (4.2):

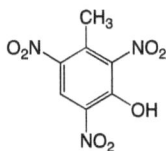

(3) Equation (4.3):

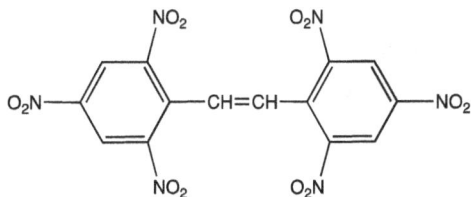

(4) Equation (4.4):

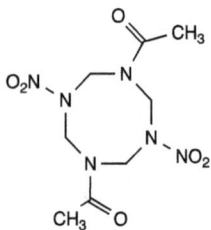

(5) Equation (4.5):

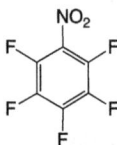

(6) Equations (4.6) and (4.7): 1,7-Diazido-2,4,6-trinitro-2,4,6-triazaheptane
(7) Equation (4.8):

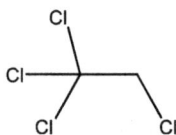

(8) Equations (4.15) and (4.16):

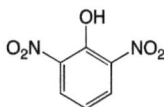

(9) Equations (4.17), (4.18), and (4.19) for 1-(2-azidoethyl)-3-azido-1,2,4-triazolium nitrate

Chapter 5

Use the following equations to calculate the heat of sublimation for the specified energetic compounds:
(1) Equation (5.2): 2,2,2-Trinitro-1-phenylethane

(2) Equation (5.3):

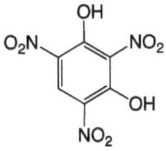

(3) Equation (5.4):

$$CH_3{-}CH_2{-}\underset{\underset{NO_2}{|}}{N}{-}CH_2{-}CH_3$$

(4) Equation (5.5):

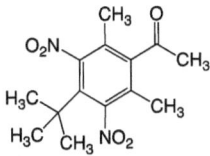

(5) Equation (5.6): *N*-(4-Nitrophenyl)-*N*-phenylamine

Chapter 6

Use the following equations to calculate the impact sensitivity for the specified energetic compounds:
(1) Equations (6.5) or (6.6): *N*-(2-propyl)-trinitroacetamide
(2) Equations (6.7): Bis-(2,2-dinitropropyl)-carbonate
(3) Equation (6.8): 1-Picrylimidazole
(4) Equation (6.9): 5-Nitro-1-picryl-4-picrylaminopyrazole
(5) Equation (6.10): *N*-Nitro-*N*-(3,3,3-trinitropropyl)-2,2,2-trinitroethyl carbamate
(6) Equations (6.11) and (6.12): Trinitroethyl-bis-(trinitroethoxy)-acetate
(7) Equation (6.13): Ammonium 1-hydroxy-1H,2'H-[5,5'-bitetrazol]-2'-olate
(8) Equation (6.14): CL-20 (Hexanitrohexaazaisowurtzitane or HNIW)

Chapter 7

Use the following equations to calculate the electric spark sensitivity of the specified energetic compounds:

(1) Equation (7.2):

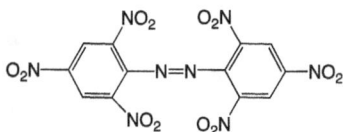

(2) Equation (7.3):

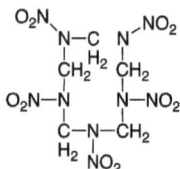

(3) Equation (7.4):

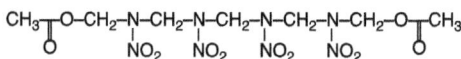

(4) Equation (7.5): 1-(2,4,6-Trinitrophenyl)-5,7-dinitrobenzotriazole (BTX)
(5) Equation (7.6) with $E_{ES, PNA}(RDAD) = 10.61$ J: 2,4,6-Tris(2,4,6-trinitrophenyl)-1,3,5-triazine (TPT)
(6) Equation (7.7): 1,3,3-Trinitroazetidine (TNAZ)
(7) Equation (7.8) with $E_{ES, PNA}(RDAD) = 4.70$ J: 2,4,6,8,10,12-Hexanitro-2,4,6,8,10, 12-hexaazaisowurtzitane (HNIW)
(8) Equation (7.9): Hydroxylammonium 5,7-dinitrobenzo[d][1,2,3]triazol-1-ide

(9) ESD from eq. (7.10):

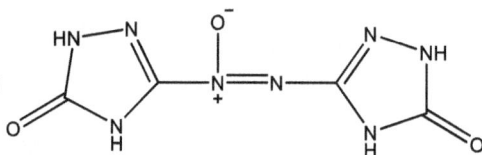

Chapter 8

Use the following equations to calculate the electric spark sensitivity of the specified energetic compounds:
(1) Equations (8.1)–(8.3): Octol-75/25
(2) Equation (8.4), if the values of ρ_0 and ρ_{TM} are 1.682 and 1.72 g/cm^3, respectively: PBX-9205

(3) Equation (8.5): Picric acid
(4) Equation (8.5): Pentolite 50/50 (PETN/TNT 50/50)

Chapter 9

(1) Use eq. (9.1) to calculate the friction sensitivity of 1-(2-nitratoethylnitramino)-2,4,6-trinitrobenzene.
(2) Use eq. (9.2) to calculate the friction sensitivity of hydroxylammonium 5,5'-dinitro-2H,2'H-[3,3'-bi(1,2,4-triazole)]-2,2'-bis(olate).

Chapter 10

Use the following equations to calculate the activation energies of low-temperature non-autocatalyzed thermolysis, heat of decomposition, onset temperature, and deflagration temperature for the specified energetic compounds:
(1) Equation (10.1): 1,1,1-Trinitrobutane
(2) Equation (10.2):

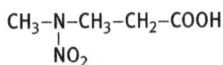

$$CH_3-N-CH_3-CH_2-COOH$$
$$NO_2$$

(3) Equation (10.3):

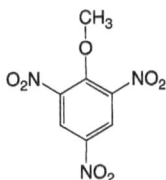

(4) Equation (10.4):

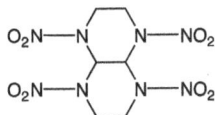

(5) Equation (10.6): 3,5-Dinitrobenzoic acid
(6) Equation (10.7): Tert-amyl peroxy-2-ethylhexyl carbonate

(7) Equation (10.8):

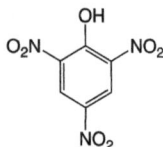

(8) Equation (10.9): Ethyl-3,3-di-(tert-amyl peroxy) butyrate
(9) Equation (10.10): 4-Azido-1-methyl-2-oxo-1,2-dihydroquinoline-3-carbaldehyde
(10) Equation (10.11):

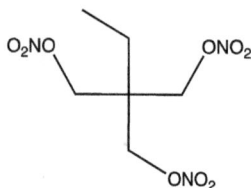

(11) Equation (10.12): 2,3-Dimethyl-1H-imidazol-3-ium bis((trifluoromethyl)sulfonyl) amide
(12) Equation (10.13): 1-Methyl-3-(pent-2-ynyl)-1H-imidazol-3-ium azide
(13) Equation (10.14): 1-(2-(5-Nitro-2H-tetrazol-2-yl)ethyl)-1H-tetrazol-5-amine
(14) Equation (10.15): Bis(nitrofuranzano)furoxan

Chapter 11

Use the following equations to calculate the relationships between the sensitivities of the specified energetic compounds:
(1) Equation (11.1) if E_a is 186.2 kJ/mol: 1,4-Dinitro-1.4-diazabutane (EDNA)
(2) Equation (11.2) if E_{IS} = 8.58 J:

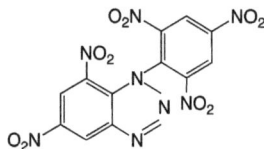

(3) Equation (11.3) if E_{IS} = 10.20 J: 1,5-Endomethylene-3,7-dinitro-1,3,5-tetraazacyclooctane
(4) Equation (11.4) if E_a is 178.8 kJ/mol: N,N-dimethyl-N,N-dinitroethanediamide
(5) Equation (11.5): 1,3-Dichloro-2,4,6-trinitrobenzene (DCTB)
(6) Equation (11.6) if E_a is 140 kJ/mol: 1,3,5-Trinitro-2-oxo-1,3,5-triazacyclohexane (keto-RDX)
(7) Equation (11.7) if $P_{90\% TMD}$ is 70.38 kbar: 1,3,5-Triamino-2,4,6-trinitrobenzene (TATB)

(8) Equations (11.8), (11.9), and (11.10) if E_{ES} is 120.17 J: 1,3,5-Triamino-2,4,6-trinitroben-zene (TATB)

Chapter 12

Use the following equations to calculate the desired property for the specified energetic compounds:

(1) Density of $[Na_2DNGTz(H_2O)_6]_n$ using eq. (12.1)
(2) $\Delta_f H^\theta(c)$ of $[Cu_3(MA)(N_3)_3]$ using eq. (12.1)
(3) Equation (12.3) for

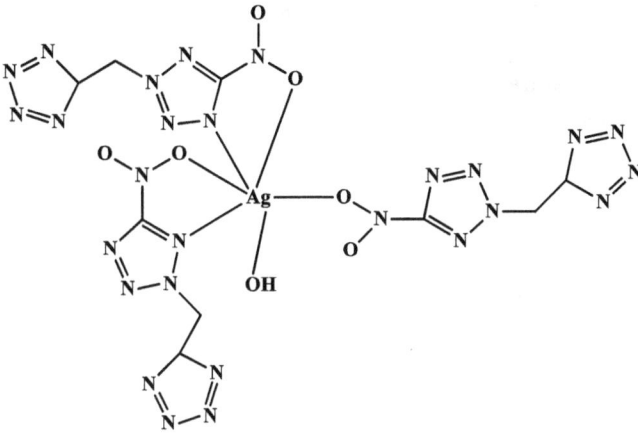

(4) Equation (12.3) for

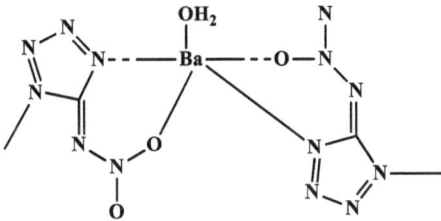

Chapter 13

Calculate the desired property for the specified energetic polymers using the following equations:

(1) Solubility of poly-BAMO with the following molecular structure using eq. (13.1)

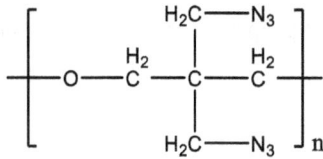

(2) Intrinsic viscosity of poly-BAMO with solvent dimethyl sulfone ($\delta_D = 19.0$, $\delta_P = 19.4$
and $\delta_H = 12.3$ MPa$^{1/2}$) using eq. (13.2)

(3) Glass transition temperature of poly-NIMMO with the following molecular structure using eqs. (13.3)–(13.6):

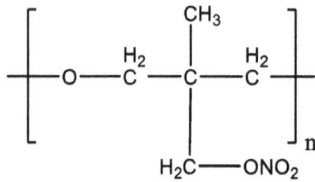

(4) Refractive index for poly-NIMMO using eq. (13.7).

Answers to Problems

Chapter 1

(1)   1.704 g/cm^3
(2)   1.555 g/cm^3
(3)   1.736 g/cm^3
(4)   2.028 g/cm^3
(5)   1.669 g/cm^3
(6)   1.467 g/cm^3
(7)   1.996 g/cm^3
(8)   1.29 g/cm^3
(9)   1.814 g/cm^3
(10) 1.535 g/cm^3
(11) 1.755 and 1.754 g/cm^3
(12) 1.631 g/cm^3
(13) 1.722 g/cm^3
(14) 1.717 g/cm^3
(15) 1.863 g/cm^3
(16) 1.290 g/cm^3
(17) 1.185 g/cm^3
(18) 1.755 g/cm^3
(19) 1.627 g/cm^3

Chapter 2

(1)   −77.6 kJ/mol
(2)   −440.2 kJ/mol
(3)   −463.2 kJ/mol
(4)   −135.1 kJ/mol
(5)   664.7 kJ/mol
(6)   −398.5 kJ/mol
(7)   527.0 kJ/mol
(8)   1,094.4 and 1,095.2 kJ/mol
(9)   −2,019.8 kJ/mol

https://doi.org/10.1515/9783112206768-016

Chapter 3

(1) 688 K
(2) 358.3 K
(3) 355.2 K
(4) 441.7 K
(5) 387.1 K
(6) 367 K
(7) 337 K
(8) 423.9 K
(9) 276.8 K
(10) 276.1 K
(11) 284.4 K
(12) 284.3 K

Chapter 4

(1) 24.88 kJ/mol
(2) 26.87 kJ/mol
(3) 39.06 kJ/mol
(4) 28.19 kJ/mol
(5) 17.00 kJ/mol
(6) 47.86 kJ/mol
(7) 10.90 kJ/mol
(8) 57.9 J/(K · mol)
(9) $\Delta_{fus}H = 15.744$ kJ/mol, $\Delta_{fus}S = 0.063143$ kJ/mol K; $T_m = 249.4$ K

Chapter 5

(1) 88.60 kJ/mol
(2) 114.49 kJ/mol
(3) 61.77 kJ/mol
(4) 115.5 kJ/mol
(5) 125.6 kJ/mol
(6) 47.86 kJ/mol
(7) 57.9 kJ/mol

Chapter 6

(1) 95 cm
(2) 72 cm
(3) 284 cm
(4) 9 cm
(5) 9 cm
(6) 9 cm
(7) 33 J
(8) 13 cm

Chapter 7

(1) 7.54 J
(2) 2.65 J
(3) 14.51 J
(4) 141.5 mJ
(5) 323.3 mJ
(6) 78.3 mJ
(7) 486.2 mJ
(8) 1,380 mJ
(9) 1.10 J

Chapter 8

(1) $P_{90\%\,TMD} = 11.34$ kbar, $P_{95\%\,TMD} = 16.54$ kbar, and $P_{98\%\,TMD} = 22.13$ kbar
(2) 50.80 mm
(3) 2.50 mm
(4) 6.66 mm

Chapter 9

(1) 326.1 N
(2) 336 N

Chapter 10

(1) 185.41 kJ/mol
(2) 175.48 kJ/mol
(3) 113.12 kJ/mol
(4) 215.44 kJ/mol
(5) 671 kJ/mol
(6) −975.6 J/g
(7) 487.0 K
(8) 411.1 K
(9) 401.51 K
(10) 499 K
(11) 122.7 kJ/mol
(12) 398 K
(13) 493 K
(14) 514 K

Chapter 11

(1) 7.43 J
(2) 6.85 J
(3) 16.50 J
(4) 6.48 J
(5) 2.51 J
(6) 97.8 N
(7) 3.92 mm
(8) 68.24, 121.92, and 164.61 kbar

Chapter 12

(1) 1.62 g/cm^3
(2) 1,926 kJ/mol
(3) 215 K
(4) 346 K

Chapter 13

(1) 31.86 MPa$^{1/2}$
(2) 194.8 cm^3/g
(3) 249 K
(4) 1.492

List of symbols

a	Number of carbon atoms
a'	Number of carbon atoms divided by molecular weight of explosive
a_{ani}	Number of carbon atoms in the anion
a_{cat}	Number of carbon atoms in the cation
a_i	Number of the ith atom group A_i in eq. (5.2)
A_i	Contribution of the first-, second-, or third-order group of type i
AMMO	3-Azido-methyl-3-methyl oxetane
A_s^+	Portion of the cation surface that has a positive electrostatic potential
A_s^-	Portion of the anion surface that has a negative electrostatic potential
AM1	Semiempirical method
AMW	Average molecular weight
ANN	Artificial neural network
AP	Ammonium perchlorate
b	Number of hydrogen atoms
B01[O–Si]	Binary indicators for O–Si bonds at a topological distance of 1
BAMO	3,3-Bis(azidomethyl)oxetane
b_{ani}	Number of hydrogen atoms in the anion
b_{cat}	Number of hydrogen atoms in the cation
b'	Number of hydrogen atoms divided by molecular weight of explosive
B_i	Number of ith atoms in one mole
BPNN	Backpropagation neural network
BRSP3	Total number of nonring, nonterminal, and branched sp^3 atoms
c	Number of nitrogen atoms
c_{ani}	Number of nitrogen atoms in the anion
c'	Number of nitrogen atoms divided by molecular weight of explosive
C	Capacitance
C_j	A correction for interactions in cycles
$C_{CH_2NNO_2 \geq 3, C(=O)(O \, or \, NH)}$	Presence of methylenenitramine greater than, or equal to three in cyclic nitramines or the presence of COO or CONH functional groups in eq. (7.3)
CHEETAH	Thermochemical computer code
C_{De}	Contribution of specific polar groups attached to aromatic rings in eq. (5.5)
C_{In}	Presence of some molecular parameters in eq. (5.5)
C_{NG}	Negative contribution of structural parameter of crystal density in eq. (1.16)
c_{mater}	Speed of light in the material
C_{NH_2}	Contribution of amino groups in nitroaromatics
$C_{-NO_2(-ONO_2)}$	Correcting function for the enthalpy of fusion in eq. (4.4)
CMOS	Complementary metal-oxide-semiconductor
C_{PG}	Positive contribution of s tructural fragments of crystal density in eq. (1.16)
$C_{R,OR}$	Presence of alkyl (–R) or alkoxy groups (–OR) in eq. (7.2)
CSD	Cambridge Structural Database
C_{poly}	Concentration of polymer
C_{SFG}	Contribution of specific functional groups in eq. (3.5)
C_{SG}	Contribution of certain polar groups in eq. (5.4)
C_{Shock}	Correcting function in eq. (8.5)
C_{SPG}	Contribution of specific polar groups attached to an aromatic ring in eq. (4.2)
C_{SSP}	Contribution of specific groups in eq. (4.3)
c_{vacum}	Speed of light in a vacuum

https://doi.org/10.1515/9783112206768-017

CVsim(GK)	Coefficient of variance of similarity values
d	Number of oxygen atoms
d_{ani}	Number of oxygen atoms in the anion
d_{cat}	Number of oxygen atoms in the cation
d_c	Critical diameter
d'	Number of oxygen atoms divided by molecular weight of explosive
d''	Number of oxygen or sulfur
D_k	A correction for atomic functional groups of kth type
DCF	Variable for decreasing the heat content of an energetic compound in eq. (2.15)
DF	Variable for decreasing the heat content of an energetic compound in eq. (2.14)
DFT	Density functional theory
DMA	Dynamic mechanical analysis
DMP	Diminishing intermolecular interaction for decreasing crystal density in eq. (1.15)
DMSO	Dimethyl sulfoxide
DRI	Refractive index reduction function in eq. (13.7)
DSC	Differential scanning calorimetry
DT	Deflagration temperature
DT_g	Correcting function in eq. (13.3) to reduce overestimated outputs
DTA	Differential thermal analysis
DSSP	Decreasing effects of melting point of some specific structural features in eq. (3.4)
DSSPH	Decreasing sensitivity structural parameter in eq. (6.10)
e	Number of fluorine atoms
e_{ani}	Number of fluorine atoms in the anion
E	Variable for decreasing heat content of an energetic compound in eq. (2.7)
E_a	Activation energy of low-temperature non-autocatalyzed thermolysis
$E_{a,IL}$	Activation energy of thermolysis of the desired ionic liquid or salt
$E_{a,IL}^{+}$	Correcting function for increasing activation energy of thermolysis of the desired ionic liquid or salt in eq. (10.11)
$E_{a,IL}^{-}$	Correcting function for decreasing activation energy of thermolysis of the desired ionic liquid or salt in eq. (10.11)
E_{Ar}	Presence of aromatic carbon-carbon double bonds in repeating unit structures
E_{cor}^{+}	Correcting function for increasing E_{ES} in eq. (11.2)
E_{cor}^{-}	Correcting function for decreasing E_{ES} in eq. (11.2)
E_D	Decreasing structural parameter of crystal density in eq. (1.13)
EDPHT	Computer code based on empirical correlations
E_{ES}	Electrostatic or electric spark sensitivity
$E_{ES,NTA}$ (ESZ KTTV)	Electric spark sensitivity of nitramines based on the ESZ KTTV system
$E_{ES,PNA}$ (ESZ KTTV)	Electric spark sensitivity of polynitro arenes based on the ESZ KTTV system
$E_{ES,NTA}$ (RDAD)	Electric spark sensitivity based on the RDAD system
$E_{ES,PNA}$ (RDAD)	Electrostatic sensitivity based on the RDAD system for polynitro arenes
$E_{ES,NTA}^{+}$	Correcting function in eq. (7.6)
$E_{ES,NTA}^{-}$	Correcting function that can decrease the predicted results on the basis of the RDAD system in eq. (7.8) for nitramines
$E_{ES,PNA}^{+}$	Correcting function in eq. (7.5)

$E_{ES,PNA}^{+'}$	Correcting function that can increase the predicted results on the basis of the RDAD system in eq. (7.6)
$E_{ES,PNA}^{-}$	Correcting function that can decrease the predicted results on the basis of the RDAD system in eq. (7.6) for polynitro arenes
E_{IS}^{++}	Correcting function for increasing E_{IS} in eq. (11.1)
E_{IS}^{--}	Correcting function for decreasing E_{IS} in eq. (11.1)
$E_{ES,Ar}^{+}$	Correcting function for increasing E_{ES} in eq. (11.5)
$E_{ES,Ar}^{-}$	Correcting function for decreasing E_{ES} in eq. (11.5)
$E_{ES,corr}^{+}$	Increasing electric spark sensitivity factor in eq. (11.4)
$E_{ES,NAr,NiA}^{+}$	Correcting function for increasing E_{ES} in eq. (11.3)
$E_{ES,NAr,NiA}^{-}$	Correcting function for decreasing E_{ES} in eq. (11.3)
E_{Ether}	Presence of ether groups in repeating unit structures
E_I	Increasing structural parameter of crystal density in eq. (1.15)
E_{IS}	Impact sensitivity in J
E_{IS}^{+}	Correcting function for increasing E_{ES} in eq. (7.4)
$E_{N\,excess}$	Existence of excess nitrogen atoms for cyclic nitramines containing excess nitrogen atoms beside $-NNO_2$ and $-NO_2$
E_{NH}	Presence of -NH- groups in repeating unit structures
$E_{NH_2,NH}$	Presence of $-NH_2/-NH-$ groups
E_{NNO_2,ONO_2,NO_2}	Presence of $-NNO_2/-ONO_2/-NO_2$ groups
$E_{N^+O^-}$	Presence of $N=N^+-O^-$ fragments
E_{NonAr}	Presence of non-aromatic carbon-carbon double bonds in repeating unit structures
E_{nonadd}^{+}	Contribution of nonadditive structural parameters for increasing E_a in eq. (10.4)
E_{nonadd}^{-}	Contribution of nonadditive structural parameters for decreasing E_a in eq. (10.4)
$E_{a\,CH/NNO_2}^{0}$	A parameter that indicates the existence of α-C–H in nitroaromatic compounds or the $N-NO_2$ functional group
E_{OH}	Presence of hydroxyl group in repeating unit structures
ESD	Electrostatic discharge
E_{Si-O}	Presence of $-SiO-$ groups in repeating unit structures
$ES_{IL}\,(mJ)$	Sensitivity toward electrical discharge of a desired energetic ionic compound in mJ
ES_{IL}^{+}	Correcting function for increasing electric spark of quaternary ammonium-based energetic ionic liquids or salts based on ESZ KTTV
ES_{IL}^{-}	Correcting function for decreasing electric spark of quaternary ammonium-based energetic ionic liquids or salts based on ESZ KTTV
EILoS	Energetic ionic liquids or salts
EMOFs	Energetic metal-organic frameworks
E–P–S	Evans–Polanyi–Semenov
ESZ KTTV	An instrument for measuring electric spark sensitivity
EXPLO5	Thermochemical computer code
f	Number of chlorine atoms
f_{ani}	Number of chlorine atoms in the anion
F	Variable for increasing the heat content of an energetic compound in eq. (2.7)
F^{+}	Existence of specific molecular moieties for increasing value of $(\log H_{50})_{core}$
F^{-}	Existence of specific molecular moieties for decreasing value of $(\log H_{50})_{core}$
$F_{attract}$	Attractive intermolecular forces in eq. (5.6)

F_{nonadd}^{+}	Contribution of non-additive structural parameters for increasing the deflagration temperature of eq. (10.10)
F_{nonadd}^{-}	Contribution of non-additive structural parameters for decreasing the deflagration temperature of eq. (10.10)
F_{repul}	Repulsive intermolecular forces in eq. (5.6)
FS	Friction sensitivity
FS^{+}	Correcting function for increasing friction sensitivity in eq. (11.6)
FS^{-}	Correcting function for decreasing friction sensitivity in eq. (11.6)
$FS_{IL}(N)$	Friction sensitivity of quaternary ammonium-based energetic ionic liquids in newton
FS_{IL}^{+}	Correcting function for increasing friction sensitivity of ionic liquids in eq. (9.2)
FS_{IL}^{-}	Correcting function for decreasing friction sensitivity of ionic liquids in eq. (9.2)
g	Number of aluminum atoms
g_{ani}	Number of aluminum atoms in the anion
GATS1p	Geary autocorrelation weighted by polarizability
g_j	Number of the special group G_j in eq. (5.2)
G_{50}	Barrier thickness required to inhibit detonation in the test explosive half the time for gap test
GA	Genetic algorithm
GAV	Group additivity values
GIPF	Generalized interaction property function
GNN	Graph Neural Networks
H_{50}	Impact drop height
HEDMs	High energy-density materials
I_{ani}	Number of boron atoms in anion
$H_{fus,i}$	Contribution of molecular fragment or group i to the enthalpy of fusion in eq. (4.1)
h_{ani}	Number of sulfur atoms in anion
HNCSILs	High-nitrogen-containing salts and ionic liquids
HOMO	Highest occupied molecular orbital
HRIPs	High-refractive-index polymers
HTPB	Hydroxy-terminated polybutadiene
ICF	Variable for increasing heat content of an energetic compound in eq. (2.15)
IF	Variable for increasing heat content of an energetic compound in eq. (2.14)
IML	Interactive machine learning
IMP	Increment of intermolecular interaction for increasing crystal density in eq. (1.17)
IPF	Inefficient packing factor in eq. (3.13)
IRI	Refractive index enhancement function in eq. (13.7)
IS_{IL}^{+}	Correction function of friction sensitivity that depends on stabilizing structural parameters in cations or anions
IS_{IL}^{-}	Correction function of friction sensitivity that depends on destabilizing structural parameters in cations or anions
ISPBKW	Computer code for calculation of the specific impulse using BKW-EOS
ISSP	Increasing effects of melting point of some specific structural features in eq. (3.4)
ISSPH	Increasing sensitivity structural parameter in eq. (6.10)

IT_g	Correcting function in eq. (13.3) to increase underestimated outputs
k	Number of halogen atoms
k_h	Huggins constant
KNN	k-nearest neighbors
L_j	Number of cycles of jth order in one mole
l_{ani}	Number of boron atoms in anion
LANL	Los Alamos National Laboratory
LED	Light-emitting diode
LINSP3	Number of nonring, nonterminal, and nonbranched sp^3 atoms
$(\log T_g)_{Elem}$	Elemental composition contributions to $logT_g$ in eq. (13.3)
$(logT_g)_{NonAr,Ar}$	Unsaturated double bond contributions to $logT_g$ in eq. (13.3)
$(logT_g)_{PG}$	polar group contributions to $logT_g$ in eq. (13.3)
M	Molecular mass of the molecule in g/molecule
MAE	Mean absolute error
$m_{cyc}^{>6}$	Cyclic nitramines with rings which are larger than six membered rings and which contain only carbon and nitrogen atoms
MCD	Molecular cyclized degree
MD	Molecular Dynamics
MESP	Molecular surface electrostatic potential
Mi	Mean first ionization potential
ML	Machine learning
$MLFER\_E$	Molecular linear free energy relations
MLP	Multilayer Perceptron
ML-QSPR	Conventional QSPR approaches and machine learning-based QSPR
MLR	Multilinear regression
MM (MM2, MM3, . . .)	Molecular mechanics methods
Mp	Mean atomic polarizability
MPNN	Message Passing Neural Networks
MSE	Mean squared error
MSST	Multiscale shock technique
Mw	Molecular weight of the desired energetic compound
Mw'	Molecular weight of the desired energetic compound under certain conditions in eqs. (5.5) and (5.6)
Mw_{cat}	Molecular weights of cation
Mw_{ani}	Molecular weights of anion
$n_{c,i}$ or $n'_{c,j}$	Occurrence of the groups i in the cation of the desired ionic liquid
$n_{a,j}$ or $n'_{a,i}$	Occurrence of the groups j in the anion of the desired ionic liquid
n_{CN}	Number of the –CN group
NC	Nitrocellulose
n_{EDNA}	Number of N,N'-(ethane-1,2-diyl)dinitramide (EDNA) moieties $(O_2NNCH_xCH_xNNO_2)$ in cyclic nitramines
n_{H_2O}	Number of the H_2O molecules in the crystal of the salt
N_j	Number of atoms in a cycle of jth order
N_k	Number of atoms in the group of k type
n_{N_3}	Number of the –N_3 group
n_{NH}	Number of the >NH group
n_{NH_2}	Number of the –NH_2 group
n_{NH_3}	Number of the NH_3 molecules in the crystal of the salt
n_{NH_4}	Number of ammonium cations

$n_{NH, OH}$	Number of the >NH (and –OH) group
n'_{Ar}	The number of aromatic rings under certain conditions for eq. (2.7)
$n_{NNO_2}^{>3, linear}$	Number of –NNO$_2$ groups for those acyclic linear nitramines containing more than three nitramine groups in eq. (4.4)
NASA-CEC-71	Computer program for calculation of complex chemical equilibrium compositions
NLR	Non-linear regression
NRHECs	Nitrogen-rich heterocyclic energetic compounds
NSWC	Naval Surface Warfare Center
OB_{100}	Oxygen balance
OECD	Read-Across Structure–Property Relationship
p	Number of cations
P_{FS}^+	Correcting parameter for increasing friction sensitivity in eq. (9.1)
P_{FS}^-	Correcting parameter for decreasing friction sensitivity in eq. (9.1)
P_x	Presence of x
$P_{90\% TMD}$	Pressure in kbar required to initiate material pressed to 90 % of theoretical maximum density
$P_{95\% TMD}$	Pressure in kbar required to initiate material pressed to 95 % of theoretical maximum density
$P_{98\% TMD}$	Pressure in kbar required to initiate material pressed to 98 % of theoretical maximum density
$P_{>5}$	Existence of cyclic nitramines that contain more than five member ring as well as cage nitramines
PHA	Polyhydroxyalkanoate
PLS	Partial least squares
PLSR	Partial least squares regression
PM3, PM6, PM7, . . .	Semi-empirical methods
PSO	Particle Swarm Optimization
q	Number of anions
QBMSST	Modified quantum-bath-coupled version of MSST
Q_{corr}	The corrected heats of detonation on the basis of Kamlet's method
QM	Quantum mechanical method
QSPR	Quantitative Structure–Property Relationships
q-RASPR	Read-Across Structure–Property Relationship
R^2	Coefficient of determination
RA function (GK) or RA function (LK)	A composite q-RASPR descriptor
RBFNN	Radial-based function neural network
RF	Random Forest
RI	Refractive index
R_k	Number of groups of k-type in one mole
RDAD	An instrument for measuring electric spark sensitivity
RED	Relative Energy Difference
RING	Number of single, fused, or conjugated ring systems
RMSE	Root Mean Square Error
ROT	Extra entropy produced by freely rotating sp^3 atoms
V_m	Volume inside the 0.001 a.u. isosurface of electron density surrounding the molecule
SA	Molecular surface area

SADT	Self-accelerating decomposition temperature
SMILES	Simplified Molecular Input Line Entry System
SMM	Soviet manometric method
SP2	Number of nonring, and nonterminal sp^2 atoms
SPG	Contribution of specific polar groups of melting point in eq. (3.13)
SPARC	SPARC Performs Automated Reasoning in Chemistry
$SpMAD\_A$	Spectral mean absolute deviation from the adjacency matrix
SSP_i	Contribution of structural parameters in eq. (3.13)
SVR	Support Vector Regression
SVM	Support vector machine
T^+	Increasing parameter of melting point in eq. (3.6)
T^-	Decreasing parameter of melting point in eq. (3.6)
T_{add}	Additive function of melting point of eq. (3.7)
$T_{add, elem}$	Contribution of the elemental composition as an additive part in eq. (3.14)
T_{core}	Core contribution of elemental composition for prediction of melting point in eq. (3.10)
$T_{corr, strut}$	Non-additive part of the melting point in eq. (3.14)
$T_{correcting}$	Correcting function of melting point in eq. (3.10)
T_{dmax}	Temperature of maximum mass loss
T_{dmax}^+	Correcting function for increasing T_{dmax} in eq. (10.9)
T_{dmax}^-	Correcting function for decreasing T_{dmax} in eq. (10.9)
T_m	Melting point
$T_{c,i}$	Contribution of cation groups i for prediction of melting point
$T_{a,j}$	Contribution of anion groups j for prediction of melting point
T_g	Glass transition temperature
TMA	Thermomechanical analysis
$T_{m, peroxide}$	Melting point of peroxide compound
$T_{m, cyc hyd}^+$	Correction term for increment of melting point of cyclic saturated and unsaturated hydrocarbons in eq. (3.17)
$T_{m, cyc hyd}^-$	Correction term for decreasing melting point of cyclic saturated and unsaturated hydrocarbons in eq. (3.17)
$T_{m, IL}^+$	Positive adjusting function in eq. (3.19)
$T_{m, IL}^-$	Negative adjusting function in eq. (3.19)
$T_{non-add}$	Non-additive function of melting point of eq. (3.7)
$T_{o,p}$	A parameter of eq. (3.3) that can be applied in disubstituted benzene ring
T_{onset}	Onset temperature
$T_{onset, azole}$	Onset decomposition temperature of the desired azole based energetic compound
$T_{onset, azole}^+$	Correcting parameter for increasing onset decomposition temperature in eq. (10.13)
$T_{onset, azole}^-$	Correcting parameter for decreasing onset decomposition temperature in eq. (10.13)
$T_{onset, IL}$	Onset decomposition temperature of imidazolium based energetic ionic liquids or salts
$T_{onset, IL}^+$	Correcting parameter for increasing onset decomposition temperature in eq. (10.12)
$T_{onset, IL}^-$	Correcting parameter for decreasing onset decomposition temperature in eq. (10.12)
T_{SFG}	Contribution of a specific functional group in eq. (3.3)

T_{struc}^{+}	Increasing structural parameters of melting point in eq. (3.16)
T_{struc}^{-}	Decreasing structural parameters of melting point in eq. (3.16)
T_{PC}	Increasing non-additive parameter of melting point in eq. (3.9)
T_{NC}	Decreasing non-additive parameter of melting point in eq. (3.9)
TET	Correcting parmeter in eq. (1.20) for tetrazole salts containing 1-N-oxide and 2-N-oxide fragment
TGA	Thermogravimetry
TMD	Theoretical maximum density
U	Voltage
UPPER	Unified Physical Property Estimation Relationships
V	Total volume
V_{PM6}	Quantum-mechanically calculated molecular volume of an isolated molecule, obtained using the PM6 method
V_s	Electrostatic potential on the surface
V^{+}	Effective volume of cation
V^{-}	Effective volume of anion
$\bar{V}_s^{+}$	Average of the positive values of V_s
$\bar{V}_s^{-}$	Average of the negative values of V_s
Void$_{theo}$	Theoretical calculated percent void
XGBoost	Extreme Gradient Boosting
$(x\,Gap)_{90\%\,TMD}$	Thickness of aluminum in mm at 90 % of TMD
$(\log H_{50})_{core}$	Impact drop height on the basis of elemental composition in eq. (6.12)
α	Molecular polarizability
δ	Solubility parameter
δ_D	Dispersion component of δ
δ_H	Hydrogen bonding component of δ
δ_P	Polar component of δ
δ^{+}	Correcting function in eq. (13.1) to adjust underestimating results
δ^{-}	Correcting function in eq. (13.1) to adjust overestimating results
ΔH_{decom}	Heat of decomposition in kJ/mol
$\Delta H'_{decom}$	Heat of decomposition in J/g
ΔH_{decom}^{+}	Contribution of non-additive structural parameters for increasing ΔH_{decom} in eq. (10.5)
ΔH_{decom}^{-}	Contribution of non-additive structural parameters for decreasing ΔH_{decom} in eq. (10.5)
$\Delta H_{Inc,fus}$	Contribution of structural parameter for increasing enthalpy of fusion in eq. (4.5)
$\Delta H_{Dec,fus}$	Contribution of structural parameter for decreasing enthalpy of fusion in eq. (4.5)
$(\Delta_{fus}H)_{add}$	Additive contribution of elemental composition in eq. (4.6)
$(\Delta_{fus}H)_{Non\text{-}add}^{Inc}$	Non-additive contribution for increasing effects of specific groups in eq. (4.6)
$(\Delta_{fus}H)_{Non\text{-}add}^{Dec}$	Non-additive contribution for decreasing effects of specific groups in eq. (4.6)
$\Delta_{fus}H_{corr}^{in}$	A correction term for increasing effects of specific groups in eq. (4.8)
$(\Delta_{fus}H_{corr}^{in}$	A correction term for decreasing effects of specific groups in eq. (4.8)
$\Delta_f H^{\theta}(c)[TAILoS]$	Condensed phase heat of formation of of triazolium-based energetic ionic liquids or salts
$\Delta_f H(g)$	Gas phase heat of formation
$\Delta_f H^{\theta}(g)$	Standard gas phase heat of formation
$\Delta_f H^{\theta}(l)$	Standard liquid phase heat of formation

$\Delta_f H(s)$	Solid phase heat of formation
$\Delta_f H^\theta(s)$	Standard solid phase heat of formation
$\Delta_f H^\theta_{IEC}$	Correcting function for increasing heat content of an energetic compound in eq. (2.13)
$\Delta_f H^\theta_{DEC}$	Correcting function for decreasing heat content of an energetic compound in eq. (2.13)
$\Delta_f H^\theta_{add,DHC}$	Correcting additive function for decreasing heat content of an energetic compound in eq. (2.14)
$\Delta_f H^\theta_{nonadd,DHC}$	Correcting non-additive function for decreasing heat content of an energetic compound in eq. (2.14)
$\Delta_f H^\theta_{add,IHC}$	Correcting additive function for increasing heat content of an energetic compound in eq. (2.14)
$\Delta_f H^\theta_{nonadd,IHC}$	Correcting non-additive function for increasing heat content of an energetic compound in eq. (2.14)
$\Delta_f H^\theta_{Inc}[IMILoS]$	Increasing and decreasing heat contents in imidazolium-based ionic liquids or salts in eq. (2.19)
$\Delta_f H^\theta_{Dec}[IMILoS]$	Decreasing heat contents in imidazolium-based ionic liquids or salts in eq. (2.19)
$\Delta_f H^\theta_{Inc}[TAILoS]$	Increasing function in the triazolium-based energetic ionic liquids or salts in eq. (2.20)
$\Delta_f H^\theta_{Dec}[TAILoS]$	Decreasing function in the triazolium-based energetic ionic liquids or salts in eq. (2.20)
$\Delta_f H^\theta(g)$	Standard heat of formation of a specific compound in the gas phase
$[\Delta_f H^\theta(g)]_{B3LYP/6-31G*}$	Gas phase heat of formation in kJ/mol using the B3LYP/6-31G* method
$[\Delta_f H^\theta(g)]_{PM3}$	Gas phase heat of formation in kJ/mol using the PM3 method
$[\Delta_f H^\theta(g)]_{PM6}$	Gas phase heat of formation in kJ/mol using the PM6 method
$\Delta_f H^\theta(c)$	Standard heat of formation of a specific compound in the condensed phase (solid or liquid)
$\Delta_f H^\theta(c)[EILoS]$	Condensed phase heat of formation of EILoS
$\Delta_f H^\theta(c)[IMILoS]$	Condensed phase heat of formation of imidazolium-based ionic liquids or salts
$\Delta_f H^\theta(c)[TAILoS]$	Condensed phase heat of formation of triazolium-based energetic ionic liquids or salts
$\sum_i \Delta_f H^\theta_i(g)[cation]$	Sum of gas-phase heats of formation of cations
$\sum_j \Delta_f H^\theta_j(g)[anion]$	Sum of gas-phase heats of formation of anions
$\Delta_{fus}H$	Enthalpy or heat of fusion
$\Delta_{fus}H_{a,j}$	Enthalpy or heat of fusion of group "j" appears in the anion
$\Delta_{fus}H_{c,i}$	Enthalpy or heat of fusion of group "i" appears in the cation
$\Delta_{vap}H^\theta$	Standard heat of vaporization.
$\Delta_{fus}S$	Entropy of fusion
ΔH_{PT}	Heat of phase transition
$\Delta_{fus}S_{add}$	Additive contribution of entropy of fusion in eq. (4.14)
$\Delta_{fus}S_{nonadd}$	Non-additive contribution of entropy of fusion in eq. (4.14)
ρ	Crystal density in g/cm^3
ρ'	Uncorrected crystal density in g/cm^3
ρ^+	Increasing crystal density parameter in eq. (1.22)
ρ^-	Decreasing crystal density parameter in eq. (1.22)
ρ_0	Loading density
ρ^+_{azide}	Increasing crystal density parameter of azide compounds in eq. (1.18)

ρ_{azide}^-	Decreasing crystal density parameter of azide compounds in eq. (1.18)
ρ^{Inc}	Increasing crystal density parameter of high-nitrogen organic materials in eq. (1.20)
ρ^{Dec}	Increasing crystal density parameter of high-nitrogen organic materials in eq. (1.20)
ρ_{nonadd}^+	Increasing crystal density parameter of high-nitrogen organic materials in eq. (1.19)
ρ_{nonadd}^-	Decreasing crystal density parameter of high-nitrogen organic materials in eq. (1.19)
ρ_{PM6}	Calculated density using the PM6 method
$\rho_{tetrazolium\ nitrate}^-$	Decreasing structural parameter in tetrazolium nitrate salts
ρ_{TM}	Theoretical maximum density
σ	Molecular rotational symmetry
σ'	Pseudosymmetry
σ_{tot}^2	Total variance of the electrostatic potential on the 0.001 a.u. molecular surface
ν	Degree of balance between the positive and negative potentials on the molecular surface
ε	Molecular eccentricity
ε_0	Dielectric permittivity of free space
ε_{ar}	Aromatic eccentricity
ε_{al}	Aliphatic eccentricity
ε_r	Relative permittivity (dielectric constant)
ϕ	Flexibility number
ζ	Normalizing phonon-vibration couplings
Π	Average deviation of electrostatic potential
$(C-N(NO_2)-C)_{pure}$	Contribution of $C-N(NO_2)-C$ linkage in pure nitramines
P_{FS}^+	Correcting parameter of increasing friction sensitivity in eq. (9.1)
P_{FS}^-	Correcting parameter of decreasing friction sensitivity in eq. (9.1)
$\lvert n_{TNB}-2 \rvert$	Absolute value of the number of 1,3,5-trinitrobenzene minus two in eq. (10.7)
λ'	Positive and negative contributions of various structural parameters to obtain more reliable T_{onset} in eq. (10.8)
λ_{sym}	Peroxides containing the same fragments attached to the $-O-O-$ bond in eq. (10.8)
$[\eta]$	Intrinsic viscosity
η_{sp}	Specific viscosity
η_{rel}	Relative change in viscosity

A Glossary of compound names and heats of formation for pure as well as composite explosives

Abbreviation	Full name or composition	Chemical formula	$\Delta_f H^\theta$ (c) (kJ mol⁻¹)	
ABH	Azobis (2,2',4,4',6,6'-hexanitrobiphenyl)	$C_{24}H_6N_{14}O_{24}$	485.34	[730]
Alex 20	44/32/20/4 RDX/TNT/Al/Wax	$C_{1.783}H_{2.469}N_{1.613}O_{2.039}Al_{0.7335}$	−7.61	
Alex 32	37/28/31/4 RDX/TNT/Al/Wax	$C_{1.647}H_{2.093}N_{1.365}O_{1.744}Al_{1.142}$	−9.33	
AMATEX-20	42/20/38 AN/RDX/TNT	$C_{1.44}H_{1.38}N_{1.04}O_{1.54}(AN)_{0.53}$	−95.77	
AMATEX-40	21/41/38 AN/RDX/TNT	$C_{1.73}H_{1.95}N_{1.61}O_{2.11}(AN)_{0.26}$	−197.49	
AMATOL80/20	80/20 AN/TNT	$C_{0.62}H_{0.44}N_{0.26}O_{0.53}(AN)_1$	−371.25	
AN	Ammonium nitrate	NH_4NO_3 or $H_4N_2O_3$	−365.14	[512]
AN/Al (90/10)	—	$Al_{0.37}(AN)_{1.125}$ or $H_{4.5}N_{2.25}O_{3.37}Al_{0.37}$	−412.42	
AN/Al (80/20)	—	$Al_{0.74}(AN)_1$ or $H_4N_2O_3Al_{0.74}$	−368.32	
AN/Al (70/30)	—	$Al_{1.11}(AN)_{0.875}$ or $H_{3.5}N_{1.75}O_{2.62}Al_{1.11}$	−324.55	
BTF	Benzotris(1,2,5-oxadiazole-1-oxide)	$C_6N_6O_6$	602.50	[730]
COMP A-3	91/9 RDX/WAX	$C_{1.87}H_{3.74}N_{2.46}O_{2.46}$	11.88	[730]
COMP B	63/36/1 RDX/TNT/wax	$C_{2.03}H_{2.64}N_{2.18}O_{2.67}$	5.36	[512]
COMP C-3	77/4/10/5/1/3 RDX/TNT/DNT/MNT/NC/Tetryl	$C_{1.90}H_{2.83}N_{2.34}O_{2.60}$	−26.99	[730]
COMP C-4	91/5.3/2.1/1.6 RDX/TNT/MNT/NC	$C_{1.82}H_{3.54}N_{2.46}O_{2.51}$	13.93	[730]
Cyclotol-50/50	50/50 RDX/TNT	$C_{2.22}H_{2.45}N_{2.01}O_{2.67}$	0.04	
Cyclotol-60/40 (or COMP B-3)	60/40 RDX/TNT	$C_{2.04}H_{2.50}N_{2.15}O_{2.68}$	4.81	[512]
Cyclotol-65/35	65/35 RDX/TNT	$C_{1.96}H_{2.53}N_{2.22}O_{2.68}$	8.33	
Cyclotol-70/30	70/30 RDX/TNT	$C_{1.87}H_{2.56}N_{2.29}O_{2.68}$	11.13	
Cyclotol-75/25	75/25 RDX/TNT	$C_{1.78}H_{2.58}N_{2.36}O_{2.69}$	13.4	[512]
Cyclotol-77/23	77/23 RDX/TNT	$C_{1.75}H_{2.59}N_{2.38}O_{2.69}$	14.98	
Cyclotol-78/22	78/22 RDX/TNT	$C_{1.73}H_{2.59}N_{2.40}O_{2.69}$	15.52	
DATB	1,3-Diamino-2,4,6-trinitrobenzene	$C_6H_5N_5O_6$	−98.74	[730]
Destex	74.766/18.691/4.672/1.869 TNT/Al/Wax/Graphite	$C_{2.791}H_{2.3121}N_{0.987}O_{1.975}Al_{0.6930}$	−34.39	

(continued)

https://doi.org/10.1515/9783112206768-018

(continued)

Abbreviation	Full name or composition	Chemical formula	$\Delta_f H^0$ (c) (kJ mol⁻¹)	
DIPAM (Dipicramide)	2,2′,4,4′,6,6′-Hexanitro-[1,1-biphenyl]-3,3′-diamine	$C_{12}H_6N_8O_{12}$	-14.90	[731]
DIPAM (Dipicramide)	2,2′,4,4′,6,6′-Hexanitro-[1,1-biphenyl]-3,3′-diamine	$C_{12}H_6N_8O_{12}$	-28.45	[730]
EXP D	Ammonium picrate or Explosive D	$C_6H_6N_4O_7$	-393.30	[730]
EDC-11	64/4/30/1/1 HMX/RDX/TNT/Wax/Trylene	$C_{1.986}H_{2.78}N_{2.23}O_{2.63}$	4.52	
EDC-24	95/5 HMX/Wax	$C_{1.64}H_{3.29}N_{2.57}O_{2.57}$	18.28	
HBX-3	31/29/35/5/0.5 RDX/TNT/AL/WAX/CaCl₂	$C_{1.66}H_{2.18}N_{1.21}O_{1.60}Al_{1.29}Ca_{0.005}Cl_{0.009}$	-8.71	[732]
HMX	Cyclotetramethylenetetranitramine	$C_4H_8N_8O_8$	74.98	[730]
HMX/Al (80/20)	—	$C_{1.08}H_{2.16}N_{2.16}O_{2.16}Al_{0.715}$	20.21	
HMX/Al (70/30)	—	$C_{0.944}H_{1.888}N_{1.888}O_{1.888}Al_{1.11}$	17.66	
HMX/Al (60/40)	—	$C_{0.812}H_{1.624}N_{1.624}O_{1.624}Al_{1.483}$	15.19	
HMX/Exon (90.54/9.46)	—	$C_{1.43}H_{2.61}N_{2.47}O_{2.47}F_{0.15}Cl_{0.10}$	-1,026.80	
HNAB	2,2′,4,4′,6,6′-Hexanitroazobenzene	$C_{12}H_4N_8O_{12}$	284.09	[730]
Liquid TNT	—	$C_7H_5N_3O_6$	-53.26	
LX-04	85/15 HMX/Viton	$C_{5.485}H_{9.2229}N_8O_8F_{1.747}$	-89.96	[512]
LX-07	90/10 HMX/Viton	$C_{1.48}H_{2.62}N_{2.43}O_{2.43}F_{0.35}$	-51.46	[512]
LX-09	93/4.6/2.4 HMX/DNPA/FEFO	$C_{1.43}H_{2.74}N_{2.59}O_{2.72}F_{0.02}$	8.38	[512]
LX-10	95/5 HMX/Viton	$C_{1.42}H_{2.66}N_{2.57}O_{2.57}F_{0.17}$	-13.14	[512]
LX-11	80/20 HMX/Viton	$C_{1.61}H_{2.53}N_{2.16}O_{2.16}F_{0.70}$	-128.57	[512]
LX-14	95.5/4.5 HMX/Estane 5702-F1	$C_{1.52}H_{2.92}N_{2.59}O_{2.66}$	6.28	[730]
LX-15	95/5 HNS-I/Kel-F 800	$C_{3.05}H_{1.29}N_{1.27}O_{2.53}Cl_{0.04}F_{0.3}$	-18.16	[512]
LX-17	92.5/7.5 TATB/Kel-F 800	$C_{2.29}H_{2.18}N_{2.15}O_{2.15}Cl_{0.054}F_{0.2}$	-100.58	[512]
MEN-II	72.2/23.4/4.4 Nitromethane/Methanol/Ethylene diamine	$C_{2.06}H_{7.06}N_{1.33}O_{3.10}$	-310.87	[342]
MINOL-2	40/40/20 AN/TNT/Al	$C_{1.23}H_{0.88}N_{0.53}O_{1.06}Al_{0.74}(AN)_{0.5}$	-194.26	[342]
NM	Nitromethane	$C_1H_3N_1O_2$	-112.97	[730]
NONA	2,2′,2″,4,4′,4″,6,6′,6″-Nonanitro-m-terphenyl	$C_{18}H_5N_9O_{18}$	114.64	[730]
NQ	Nitroguanidine	$CH_4N_4O_2$	-92.47	[342]
NM/UP (60/40)	60/40 Nitromethane/UP; UP = 90/10 CO(NH₂)₂ HClO₄/H₂O	$C_{1.207}H_{4.5135}N_{1.432}O_{3.309}Cl_{0.2341}$	11.51	
Octol-76/23	76.3/23.7 HMX/TNT	$C_{1.76}H_{2.58}N_{2.37}O_{2.69}$	12.76	

Octol-75/25	75/25 HMX/TNT	$C_{1.78}H_{2.58}N_{2.36}O_{2.69}$	11.63	[512]
Octol-60/40	60/40 HMX/TNT	$C_{2.04}H_{2.50}N_{2.15}O_{2.68}$	4.14	
PBX-9007	90/9.1/0.5/0.4 RDX/Polystyrene/DOP/Resin	$C_{1.97}H_{3.22}N_{2.43}O_{2.44}$	29.83	[730]
PBX-9010	90/10 RDX/Kel-F	$C_{1.39}H_{2.43}N_{2.43}O_{2.43}Cl_{0.09}F_{0.26}$	-32.93	[512]
PBX-9011	90/10 HMX/Estane	$C_{1.73}H_{3.18}N_{2.45}O_{2.61}$	-16.95	[730]
PBX-9205	92/6/2 RDX/Polystyrene/DOP	$C_{1.83}H_{3.14}N_{2.49}O_{2.51}$	24.31	[730]
PBX-9407	94/6 RDX/Exon 461	$C_{1.41}H_{2.66}N_{2.54}O_{2.54}Cl_{0.07}F_{0.09}$	3.39	[512]
PBX-9501	95/2.5/2.5 HMX/Estane/BDNPA-F	$C_{1.47}H_{2.86}N_{2.60}O_{2.69}$	9.62	[730]
PBX-9502	95/5 TATB/Kel-F 800	$C_{2.3}H_{2.23}N_{2.21}O_{2.21}Cl_{0.04}F_{0.13}$	-87.15	[512]
PBX-9503	15/80/5 HMX/TATB/KEL-F 800	$C_{2.16}H_{2.28}N_{2.26}O_{2.26}Cl_{0.038}$	-74.01	[512]
PBXC-9	75/20/5 HMX/Al/Viton	$C_{1.15}H_{2.14}N_{2.03}O_{2.03}F_{0.17}Al_{0.74}$	113.01	
PBXC-116	86/14 RDX/Binder	$C_{1.968}H_{3.7463}N_{2.356}O_{2.4744}$	4.52	
PBXC-117	71/17/12 RDX/Al/Binder	$C_{1.65}H_{3.1378}N_{1.946}O_{2.048}Al_{0.6303}$	-65.56	
PBXC-119	82/18 HMX/Binder	$C_{1.817}H_{4.1073}N_{2.2149}O_{2.6880}$	18.28	
Pentolite-50/50	50/50 TNT/PETN	$C_{2.33}H_{2.37}N_{1.29}O_{3.22}$	-100.01	
PETN	Pentaerythritol tetranitrate	$C_5H_8N_4O_{12}$	-538.48	[512]
PF	1-Fluoro-2,4,6-trinitrobenzene	$C_6H_2N_3O_6F$	-224.72	
RDX	Cyclomethylenetrinitramine	$C_3H_6N_6O_6$	61.55	[512]
RDX/Al (90/10)	–	$C_{1.215}H_{2.43}N_{2.43}O_{2.43}Al_{0.371}$	24.89	
RDX/Al (80/20)	–	$C_{1.081}H_{2.161}N_{2.161}O_{2.161}Al_{0.715}$	22.13	
RDX/Al (70/30)	–	$C_{0.945}H_{1.89}N_{1.89}O_{1.89}Al_{1.11}$	19.37	
RDX/Al (60/40)	–	$C_{0.81}H_{1.62}N_{1.62}O_{1.62}Al_{1.483}$	16.61	
RDX/Al (50/50)	–	$C_{0.675}H_{1.35}N_{1.35}O_{1.35}Al_{1.853}$	13.85	
RDX/TFNA (65/35)	–	$C_{1.54}H_{2.64}N_{2.2}O_{2.49}F_{0.44}$	-823.83	
RDX/Exon (90.1/9.9)	–	$C_{1.44}H_{2.6}N_{2.44}O_{2.44}F_{0.17}Cl_{0.11}$	-195.48	
TATB	1,3,5-Triamino-2,4,6-trinitrobenzene	$C_6H_6N_6O_6$	-154.18	[512]
TATB/HMX/Kel-F (45/45/10)	–	$C_{1.88}H_{2.37}N_{2.26}O_{2.26}F_{0.28}Cl_{0.06}$	-478	
Tetryl	N-Methyl-N-nitro-2,4,6-trinitroaniline	$C_7H_5N_5O_8$	19.54	[512]
TFENA	2,2,2-Trifluoroethylnitramine	$C_2H_3N_2O_2F_3$	-694.54	
TFET	2,4,6-Trinitrophenyl-2,2,2-trifluoroethylnitramine	$C_8H_4N_5O_8F_3$	-576.8	
TNT	2,4,6-Trinitrotoluene	$C_7H_5N_3O_6$	-67.07	[54]

(continued)

(continued)

Abbreviation	Full name or composition	Chemical formula	$\Delta_f H^\theta$ (c) (kJ mol^{-1})
TNTAB	Trinitrotriazidobenzene	$C_6N_{12}O_6$	1,129.68 [730]
TNT/Al (89.4/10.6)	–	$C_{2.756}H_{1.969}N_{1.181}O_{2.362}Al_{0.393}$	−24.73
TNT/Al (78.3/21.7)	–	$C_{2.414}H_{1.724}N_{1.034}O_{2.069}Al_{0.804}$	−21.63
Toluene/Nitromethane (14.5/85.5)	–	$C_{2.503}H_{5.461}N_{1.4006}O_{2.8013}$	−160.71
Torpex	42/40/18 RDX/TNT/Al	$C_{1.8}H_{2.015}N_{1.663}O_{2.191}Al_{0.6674}$	−0.17
Tritonal	80/20 TNT/Al	$C_{2.465}H_{1.76}N_{1.06}O_{2.11}Al_{0.741}$	−23.64

B Calculation of the gas-phase standard enthalpies of formation

The gas-phase standard enthalpies of formation at 298 K, for $C_aH_bN_cO_d$ species, can be obtained by correcting the standard enthalpies of formation at 0 K (eq. (B.1)) [375, 733].

$$\Delta_f H^\theta(g)(C_aH_bN_cO_d, 298\text{ K}) = \Delta_f H^\theta(g)(C_aH_bN_cO_d, 0\text{ K})$$

$$+ \left[H^\theta(C_aH_bN_cO_d, 298\text{ K}) - H^\theta(C_aH_bN_cO_d, 0\text{ K})\right]$$

$$- a \cdot \left[H^\theta(C, 298\text{ K}) - H^\theta(C, 0\text{ K})\right]_{St}$$

$$- b \cdot \left[H^\theta(H, 298\text{ K}) - H^\theta(H, 0\text{ K})\right]_{St}$$

$$- c \cdot \left[H^\theta(N, 298\text{ K}) - H^\theta(N, 0\text{ K})\right]_{St}$$

$$- d \cdot \left[H^\theta(O, 298\text{ K}) - H^\theta(O, 0\text{ K})\right]_{St}, \tag{B.1}$$

whereby the terms in the square brackets are the heat capacity corrections and indicate the enthalpy changes due to raising the temperature from 0 K to 298 K. The $\left[H^\theta(298\text{ K}) - H^\theta(0\text{ K})\right]$ corrections for $C_aH_bN_cO_d$ molecules can be extracted from the output of the Gaussian program package [734] and were taken from the CRC Handbook [351] for the standard states of elements.

The standard enthalpies of formation at 0 K were calculated by subtracting the calculated values of the atomization energies of $C_aH_bN_cO_d$ compounds from the experimental enthalpies of formation of the isolated atoms (eq. (B.2)) [375, 733].

$$\Delta_f H^\theta(C_aH_bN_cO_d, 0\text{ K}) = a \cdot \Delta_f H^\theta(C, 0\text{ K})$$

$$+ b \cdot \Delta_f H^\theta(H, 0\text{ K})$$

$$+ c \cdot \Delta_f H^\theta(N, 0\text{ K})$$

$$+ d \cdot \Delta_f H^\theta(O, 0\text{ K})$$

$$- \Delta_a H^\theta(C_aH_bN_cO_d, 0\text{ K}). \tag{B.2}$$

The $\Delta_f H^\theta(\text{atom}, 0\text{ K})$ values, that is, the atomic enthalpies of formation of the elements in their standard states at 0 K, were obtained from standard thermodynamic tables [735, 736]. The $\Delta_a H^\theta(C_aH_bN_cO_d, 0\text{ K})$ values, that is, the atomization (dissociation) energies of $C_aH_bN_cO_d$ compounds, were obtained by subtracting the quantum mechanical energy of the molecule (electronic energy + zero-point energy) from the quantum mechanical energies (electronic energies) of the atoms (eq. (B.3)) [375, 733]:

https://doi.org/10.1515/9783112206768-019

$$\Delta_a H^\theta (C_a H_b N_c O_d, 0\,K) = a \cdot E_e(C) + b \cdot E_e(H) + c \cdot E_e(N) + d \cdot E_e(O)$$
$$- E_0(C_a H_b N_c O_d). \tag{B.3}$$

The $E_e(\text{atom})$ and $E_0(C_a H_b N_c O_d)$ values were extracted from the output of the Gaussian program package [734].

Table B.1: The B3LYP/6-31G* calculated total energies and formation enthalpies (at 0 K) for 100 energetic materials with high nitrogen contents.

Formula	Name	E_e (Hartrees)[a]	E_0 (Hartrees)[b]	$H^\theta_{298} - H^\theta_0$ (kJ/mol)[c]	$\Delta_f H^\theta(g)$ (0 K) (kJ/mol)[d]
CHN$_5$O$_2$	5-Nitro-1H-tetrazole	−462.73092	−462.68250	20.3	381.2
CHN$_7$	5-Azido-1H-tetrazole	−421.83034	−421.78146	20.9	694.1
CH$_2$N$_2$	Cyanamide	−148.78006	−148.74651	13.6	166.1
CH$_2$N$_4$	1H-Tetrazole	−258.25090	−258.20483	14.0	347.6
CH$_2$N$_4$O	1H-Tetrazol-5-ol	−333.47117	−333.42062	16.8	187.1
CH$_2$N$_6$O$_2$	Nitroguanylazide	−518.07074	−518.00826	27.4	435.0
CH$_2$N$_6$O$_2$	5-Nitroaminotetrazole	−518.08423	−518.01890	24.1	407.1
CH$_2$N$_6$O$_2$	5-Nitriminotetrazole	−518.08498	−518.01950	23.7	405.5
CH$_3$N$_5$	1H-Tetrazol-5-amine	−313.60883	−313.54613	18.5	360.7
CH$_3$N$_5$	1H-Tetrazol-1-amine	−313.57178	−313.50903	18.3	458.1
CH$_4$N$_4$O$_2$	1-Nitroguanidine	−409.83771	−409.76022	23.9	134.9
CH$_4$N$_6$	1H-Tetrazol-1,5-diamine	−368.93317	−368.85361	22.5	462.5
CH$_5$N$_3$	Guanidine	−205.36258	−205.28765	18.0	88.0
CH$_5$N$_5$O$_2$	N-Nitro-N′-aminoguanidine	−465.15588	−465.06137	27.8	253.4
C$_2$N$_6$O$_3$	5,6-(3,4-Furazano)-1,2,3,4-tetrazine-1,3-dioxide	−630.05517	−630.00396	24.0	726.6
C$_2$N$_{10}$	3,6-Diazido-1,2,4,5-tetrazine	−623.50507	−623.44884	29.7	1,110.1
C$_2$HN$_5$	1H-Tetrazole-5-carbonitrile	−350.48131	−350.43683	18.2	521.9
C$_2$H$_2$N$_6$	3-Azido-1H-1,2,4-triazole	−405.83770	−405.77581	21.6	541.1
C$_2$H$_2$N$_8$	1H,1′H-5,5′-Bitetrazole	−515.31771	−515.24441	24.2	695.6
C$_2$H$_3$N$_3$	1H-1,2,3-Triazole	−242.22232	−242.16432	14.9	286.1
C$_2$H$_3$N$_3$	1H-1,2,4-Triazole	−242.24927	−242.19042	14.9	217.6
C$_2$H$_3$N$_5$O$_2$	5-Amino-3-nitro-1H-1,2,4-triazole	−502.09665	−502.01906	26.0	238.8
C$_2$H$_3$N$_5$O$_2$	5-Nitro-1H-1,2,4-triazol-3-amine	−502.09856	−502.02088	25.7	234.0
C$_2$H$_3$N$_5$O$_2$	N-Nitro-1H-1,2,4-triazol-3-amine	−502.08080	−502.00273	25.2	281.7
C$_2$H$_3$N$_7$	3-Azido-4H-1,2,4-triazol-4-amine	−461.14442	−461.06652	26.3	687.0
C$_2$H$_3$N$_9$	Di(1H-Tetrazol-5-yl)amine	−570.66899	−570.57935	28.5	725.3
C$_2$H$_4$N$_4$	1H-1,2,4-Triazole-3-amine	−297.60778	−297.53239	19.2	228.9
C$_2$H$_4$N$_4$	4H-1,2,4-Triazol-4-amine	−297.55301	−297.47836	19.2	370.7
C$_2$H$_4$N$_4$	1-Methyl-1H-tetrazole	−297.56746	−297.49387	20.0	330.0
C$_2$H$_4$N$_4$	5-Methyl-1H-tetrazole	−297.57591	−297.50238	19.8	307.7
C$_2$H$_4$N$_4$	2-Methyl-2H-tetrazole	−297.57441	−297.50014	19.5	313.6
C$_2$H$_4$N$_4$	Dicyandiamide	−297.59010	−297.51710	22.3	269.0
C$_2$H$_4$N$_4$O	5-Methoxy-1H-tetrazole	−372.77902	−372.70045	22.4	193.7
C$_2$H$_4$N$_4$O	3,4-Diaminofurazan	−372.75583	−372.67773	23.4	253.3

Table B.1 (continued)

Formula	Name	E_e (Hartrees)[a]	E_0 (Hartrees)[b]	$H_{298}^{\theta} - H_0^{\theta}$ (kJ/mol)[c]	$\Delta_f H^{\theta}$(g) (0 K) (kJ/mol)[d]
$C_2H_4N_6$	1,2,4,5-tetrazine-3,6-diamine	−407.05804	−406.97417	24.4	453.8
$C_2H_4N_6O_2$	3,6-Diamino-1,2,4,5-tetrazine 1,4-dioxide	−557.40466	−557.31215	29.1	378.5
$C_2H_4N_6O_2$	1-Methyl-5-nitriminotetrazole	−557.40420	−557.31112	29.6	381.2
$C_2H_4N_6O_2$	2-Methyl-5-nitraminotetrazole	−557.40175	−557.30855	30.1	387.9
$C_2H_4N_8O_2$	1,3-Diazido-2-nitro-2-azapropane	−666.81480	−666.71334	37.1	709.9
$C_2H_4N_{10}$	5,5′-Hydrazotetrazole	−625.97430	−625.86810	32.7	876.4
$C_2H_5N_5$	1-Methyl-1H-tetrazol-5-amine	−352.92560	−352.83517	24.1	343.1
$C_2H_5N_5$	2-Methyl-1H-tetrazol-5-amine	−352.93288	−352.84193	23.8	325.3
$C_2H_5N_5$	2-Methyl-2H-tetrazol-5-amine	−352.93288	−352.84193	23.8	325.3
$C_2H_5N_5$	5-Methylamino-1H-tetrazole	−352.91787	−352.82735	23.9	363.6
$C_2H_5N_7$	5-Guanylaminotetrazole	−462.44641	−462.34413	27.7	391.6
$C_2H_6N_8$	3,6-Dihydrazino-1,2,4,5-tetrazine	−517.66931	−517.55236	33.7	754.1
C_3N_{12}	2,4,6-Triazido-1,3,5-triazine	−771.15932	−771.08616	37.1	1,129.5
$C_3HN_{11}O_2$	6-Nitroamino-2,4-diazido[1,3,5] triazine	−867.39800	−867.30913	41.4	880.4
$C_3H_2N_6$	Tetrazolo[1,5-b][1,2,4]triazine	−443.91956	−443.85135	20.2	650.5
$C_3H_2N_6O_2$	2-(5-Nitro-1H-tetrazol-1-yl) acetonitrile	−594.26953	−594.19472	29.7	561.1
$C_3H_3N_5$	2-Methyl-2H-tetrazole-5-carbonitrile	−389.80767	−389.73514	23.9	480.0
$C_3H_3N_9O_2$	5-((5-Nitro-2H-tetrazol-2-yl) methyl)-1H-tetrazole	−759.10801	−759.00420	35.8	729.7
$C_3H_4N_6$	5,8-Dihydrotetrazolo[1,5-b][1,2,4] triazine	−445.10641	−445.01526	24.2	653.7
$C_3H_4N_6$	3-Azido-5-methyl-1H-1,2,4-triazole	−445.16235	−445.07337	28.1	501.1
$C_3H_4N_6O_2$	4-Nitro-5,6-dihydro-4H-imidazo [1,2-d]tetrazole	−595.48066	−595.38046	29.3	506.9
$C_3H_4N_8O_4$	5,8-Dinitro-5,6,7,8-tetrahydrotetrazolo[1,5-b][1,2,4] triazine	−855.27837	−855.16010	39.2	656.8
$C_3H_4N_8O_4$	1-Nitroguanyl-3-nitro-5-amino -1,2,4-triazole	−855.37877	−855.26092	42.5	392.1
$C_3H_5N_5O$	5-Acetamidotetrazole	−466.26538	−466.16560	28.9	189.3
$C_3H_5N_5O$	4,6-Diamino-1,3,5-triazin-2(1H)-one	−466.35186	−466.24972	28.5	−31.5
$C_3H_5N_7$	3-Azido-5-methyl-4H-1,2,4-triazol -4-amine	−500.46950	−500.36435	32.5	646.3
$C_3H_6N_4$	1,5-Dimethyl-1H-tetrazole	−336.89210	−336.79098	25.4	291.2
$C_3H_6N_4$	2,5-Dimethyl-2H-tetrazole	−336.89818	−336.79677	25.6	276.1
$C_3H_6N_6$	1,3,5-Triazine-2,4,6-triamine	−446.49172	−446.37702	29.5	137.4
$C_3H_6N_6$	5,6,7,8-Tetrahydrotetrazolo[1,5-b] [1,2,4]triazine	−446.33547	−446.22037	26.5	548.7
$C_3H_6N_6O_2$	1,6-Dimethyl-5-nitraminotetrazole	−596.71121	−596.59053	35.2	388.9

Table B.1 (continued)

Formula	Name	E_e (Hartrees)[a]	E_0 (Hartrees)[b]	$H_{298}^{\theta} - H_0^{\theta}$ (kJ/mol)[c]	$\Delta_f H^{\theta}$(g) (0 K) (kJ/mol)[d]
$C_3H_6N_8$	1-(2-Azidoethyl)-1H-tetrazol-5-amine	−555.81867	−555.69632	35.5	683.9
$C_3H_6N_{10}O_4$	1,5-Diazido-2,4-dinitro-2,4-diazapentane	−965.95284	−965.80396	52.5	784.6
$C_3H_7N_5$	1-Methyl-5-methylaminotetrazole	−392.23449	−392.11628	29.6	346.3
$C_3H_7N_5$	5-(Dimethylamino)-tetrazole	−392.22789	−392.10972	29.1	363.6
$C_3H_7N_5O$	2-(5-Amino-1H-tetrazol-1-yl)ethan-1-ol	−467.44441	−467.32120	32.0	214.3
$C_4H_2N_{10}O_5$	5,5′-Dinitro-3,3′–azoxy-1,2,4-triazole	−1,076.90170	−1,076.78507	47.5	681.0
$C_4H_4N_8$	4,4′-Azobis-1,2,4-triazole	−592.68104	−592.57609	31.4	873.8
$C_4H_4N_8O_2$	3,3′-Diamino-4,4′-azofurazan	−743.06077	−742.94723	38.2	711.4
$C_4H_4N_8O_3$	3,3′-Diamino-4,4′-azoxyfurazan	−818.23777	−818.11919	40.4	665.9
$C_4H_4N_{12}$	6,6′-(Diazene-1,2-diyl)bis(1,2,4,5-tetrazin-3-amine)	−811.63562	−811.51265	43.5	1,184.5
$C_4H_4N_{14}$	3,6-Bis(1H-1,2,3,4-tetrazol-5-ylamino)-1,2,4,5-tetrazine	−921.17726	−921.03922	44.7	1,186.8
$C_4H_6N_4$	2-Methyl-5-vinyl-2H-tetrazole	−374.97864	−374.87194	27.2	386.5
$C_4H_6N_6$	3-Azido-5-ethyl-4H-1,2,4-triazole	−484.45817	−484.34170	32.8	537.9
$C_4H_6N_6$	3-(5-Amino-1H-tetrazol-1-yl)propanenitrile	−484.47524	−484.35765	32.9	496.1
$C_4H_6N_6$	3-(5-Amino-2H-tetrazol-2-yl)propanenitrile	−484.48380	−484.36570	32.6	474.9
$C_4H_6N_8$	1,2-Di(1H-tetrazol-5-yl)ethane	−593.94862	−593.81921	34.8	669.0
$C_4H_6N_8O_2$	4,4′-Hydrazobis-(1,2,5-oxadiazol-3-amine)	−744.28136	−744.14276	40.3	631.6
$C_4H_6N_{10}$	(cis)1,1′-Dimethyl-5,5′-azotetrazole	−703.38229	−703.24518	41.3	935.4
$C_4H_6N_{10}$	(trans)1,1′-Dimethyl-5,5′-azotetrazole	−703.38843	−703.25123	41.6	919.5
$C_4H_6N_{10}$	2,2′-Dimethyl-5,5′-azotetrazole	−703.39638	−703.25871	41.5	899.9
$C_4H_6N_{12}O_4$	3,6-Bis(nitroguanyl)-1,2,4,5-tetrazine	−1,113.66254	−1,113.49599	56.3	660.3
$C_4H_7N_5$	5-Amino-1-(2-propenyl)-1H-tetrazole	−430.31587	−430.19265	31.2	453.6
$C_4H_7N_5$	5-Amino-2-(2-propenyl)-2H-tetrazole	−430.32384	−430.20011	30.9	434.0
$C_4H_7N_9$	Bis(2-methyl-2H-tetrazol-5-yl)amine	−649.31166	−649.16587	40.0	668.0
$C_4H_8N_8O_2$	1,5-Diazido-3-nitro-3-azapentane	−745.44366	−745.28588	47.8	689.3
$C_4H_8N_{10}$	3,6-Diguanidino-1,2,4,5-tetrazine	−704.68475	−704.52319	44.5	639.0
$C_4H_8N_{10}O_4$	1,6-Diazido-2,5-dinitro-2,5-diazahexane	−1,005.26729	−1,005.09000	57.5	774.9
$C_4H_8N_{12}O_6$	1,7-Diazido-2,4,6-trinitro-2,4,6-triazaheptane	−1,265.08978	−1,264.89317	67.0	863.0

Table B.1 (continued)

Formula	Name	E_e (Hartrees)[a]	E_0 (Hartrees)[b]	$H^\theta_{298} - H^\theta_0$ (kJ/mol)[c]	$\Delta_f H^\theta$(g) (0 K) (kJ/mol)[d]
$C_5H_9N_9$	N,2-Dimethyl-N-(2-methyl-2H-tetrazol-5-yl)-2H-tetrazol-5-amine	−688.61739	−688.44403	45.8	679.0
C_6N_{20}	4,4′,6,6′-Tetra(azido)azo-1,3,5-triazine	−1,323.37055	−1,323.24159	63.9	1,933.0
$C_6H_2N_{20}$	4,4′,6,6′-Tetra(azido)hydrazo-1,3,5-triazine	−1,324.63571	−1,324.48119	65.9	1,737.4
$C_6H_{10}N_{10}$	2,2′-Diethyl-5,5′-azotetrazole	−782.03174	−781.83751	50.7	862.9
$C_6H_{14}N_{16}$	1,2-Bis(4,6-dihydrazinyl-1,3,5-triazin-2-yl)hydrazine	−1,112.97843	−1,112.70251	74.8	987.7
$C_7H_7N_9$	3-Azido-6-(3,5-dimethylpyrazol-1-yl)-1,2,4,5-tetrazine	−763.55913	−763.40099	46.9	974.1
$C_8H_4N_{16}O_6$	Bis[4-aminofurazanyl-3-azoxy]azofurazan	−1,634.00217	−1,633.80967	75.4	1,590.5

[a]Electronic energy at 0 K. The E_e energies for atoms (in Hartrees) were calculated as: C(−37.846280), H(−0.500273), N(−54.584489), O(−75.060623).
[b]Molecular energy with zero-point energy correction.
[c]Enthalpy difference from 298 K to 0 K. The $[H^\theta(298\ K) - H^\theta(0\ K)]_{St}$ for atoms (in kJ/mol) were obtained as: C(1.050), H(4.234), N(4.335), O(4.340) [734].
[d]Enthalpy of formation at 0 K. The $\Delta_f H^\theta$ for atoms in their standard states at 0 K (in kJ/mol) were obtained as: C(711.38), H(216.03), N(470.57), O(246.84) [735, 736].

Table B.2: Predicted gas-phase standard enthalpies of formation for 100 energetic materials with high nitrogen contents.

Formula	Name	Predicted $\Delta_f H^\theta$(g)(298 K) (kJ/mol)	
		B3LYP/6-31 G*	PM6
CHN_5O_2	5-Nitro-1H-tetrazole	365.8	386.2
CHN_7	5-Azido-1H-tetrazole	679.4	672.6
CH_2N_2	Cyanamide	161.5	147.7
CH_2N_4	1H-Tetrazole	334.7	336.3
CH_2N_4O	1H-Tetrazol-5-ol	172.7	177.2
$CH_2N_6O_2$	Nitroguanylazide	418.2	365.8
$CH_2N_6O_2$	5-Nitroaminotetrazole	387.0	367.6
$CH_2N_6O_2$	5-Nitriminotetrazole	385.0	366.7
CH_3N_5	1H-Tetrazol-5-amine	343.7	345.0
CH_3N_5	1H-Tetrazol-1-amine	440.9	424.8
$CH_4N_4O_2$	1-Nitroguanidine	114.8	77.0
CH_4N_6	1H-Tetrazol-1,5-diamine	441.1	430.7
CH_5N_3	Guanidine	70.8	68.8
$CH_5N_5O_2$	N-Nitro-N′-aminoguanidine	228.7	176.3

Table B.2 (continued)

Formula	Name	Predicted $\Delta_f H^{\theta}$(g)(298 K) (kJ/mol)	
		B3LYP/6-31 G*	PM6
$C_2N_6O_3$	5,6-(3,4-Furazano)-1,2,3,4-tetrazine-1,3-dioxide	709.5	682.7
C_2N_{10}	3,6-Diazido-1,2,4,5-tetrazine	1,094.4	1,095.2
C_2HN_5	1H-Tetrazole-5-carbonitrile	512.1	510.4
$C_2H_2N_6$	3-Azido-1H-1,2,4-triazole	526.1	569.3
$C_2H_2N_8$	1H,1'H-5,5'-Bitetrazole	674.6	688.8
$C_2H_3N_3$	1H-1,2,3-Triazole	273.2	251.2
$C_2H_3N_3$	1H-1,2,4-Triazole	204.6	220.2
$C_2H_3N_5O_2$	5-Amino-3-nitro-1H-1,2,4-triazole	219.7	264.3
$C_2H_3N_5O_2$	5-Nitro-1H-1,2,4-triazol-3-amine	214.5	266.5
$C_2H_3N_5O_2$	N-Nitro-1H-1,2,4-triazol-3-amine	261.7	257.2
$C_2H_3N_7$	3-Azido-4H-1,2,4-triazol-4-amine	668.1	644.4
$C_2H_3N_9$	Di(1H-Tetrazol-5-yl)amine	700.0	727.7
$C_2H_4N_4$	1H-1,2,4-Triazole-3-amine	211.7	229.9
$C_2H_4N_4$	4H-1,2,4-Triazol-4-amine	353.6	316.5
$C_2H_4N_4$	1-Methyl-1H-tetrazole	313.6	326.5
$C_2H_4N_4$	5-Methyl-1H-tetrazole	291.1	289.4
$C_2H_4N_4$	2-Methyl-2H-tetrazole	296.7	325.4
$C_2H_4N_4$	Dicyandiamide	254.9	244.7
$C_2H_4N_4O$	5-Methoxy-1H-tetrazole	175.3	185.8
$C_2H_4N_4O$	3,4-Diaminofurazan	236.0	238.4
$C_2H_4N_6$	1,2,4,5-tetrazine-3,6-diamine	433.2	398.9
$C_2H_4N_6O_2$	3,6-Diamino-1,2,4,5-tetrazine 1,4-dioxide	353.8	391.9
$C_2H_4N_6O_2$	1-Methyl-5-nitriminotetrazole	357.1	356.5
$C_2H_4N_6O_2$	2-Methyl-5-nitraminotetrazole	364.2	374.0
$C_2H_4N_8O_2$	1,3-Diazido-2-nitro-2-azapropane	684.6	620.3
$C_2H_4N_{10}$	5,5'-Hydrazotetrazole	846.7	814.6
$C_2H_5N_5$	1-Methyl-1H-tetrazol-5-amine	322.3	337.2
$C_2H_5N_5$	2-Methyl-1H-tetrazol-5-amine	304.2	338.1
$C_2H_5N_5$	2-Methyl-2H-tetrazol-5-amine	304.2	338.1
$C_2H_5N_5$	5-Methylamino-1H-tetrazole	342.6	342.0
$C_2H_5N_7$	5-Guanylaminotetrazole	365.7	387.6
$C_2H_6N_8$	3,6-Dihydrazino-1,2,4,5-tetrazine	725.6	564.8
C_3N_{12}	2,4,6-Triazido-1,3,5-triazine	1,111.4	1,236.2
$C_3HN_{11}O_2$	6-Nitroamino-2,4-diazido[1,3,5]triazine	858.1	933.3
$C_3H_2N_6$	Tetrazolo[1,5-b][1,2,4]triazine	633.1	661.0
$C_3H_2N_6O_2$	2-(5-Nitro-1H-tetrazol-1-yl)acetonitrile	544.5	542.2
$C_3H_3N_5$	2-Methyl-2H-tetrazole-5-carbonitrile	466.4	494.5
$C_3H_3N_9O_2$	5-((5-Nitro-2H-tetrazol-2-yl)methyl)-1H-tetrazole	702.0	738.1
$C_3H_4N_6$	5,8-Dihydrotetrazolo[1,5-b][1,2,4]triazine	631.9	561.6
$C_3H_4N_6$	3-Azido-5-methyl-1H-1,2,4-triazole	483.2	521.8
$C_3H_4N_6O_2$	4-Nitro-5,6-dihydro-4H-imidazo[1,2-d]tetrazole	481.4	457.5
$C_3H_4N_8O_4$	5,8-Dinitro-5,6,7,8-tetrahydrotetrazolo[1,5-b][1,2,4]triazine	623.8	568.4
$C_3H_4N_8O_4$	1-Nitroguanyl-3-nitro-5-amino-1,2,4-triazole	362.5	355.1

Table B.2 (continued)

Formula	Name	Predicted $\Delta_f H^\theta$(g)(298 K) (kJ/mol)	
		B3LYP/6-31 G*	PM6
$C_3H_5N_5O$	5-Acetamidotetrazole	167.9	140.3
$C_3H_5N_5O$	4,6-Diamino-1,3,5-triazin-2(1H)-one	−53.3	−35.7
$C_3H_5N_7$	3-Azido-5-methyl-4H-1,2,4-triazol-4-amine	624.1	601.4
$C_3H_6N_4$	1,5-Dimethyl-1H-tetrazole	270.8	282.4
$C_3H_6N_4$	2,5-Dimethyl-2H-tetrazole	255.8	284.4
$C_3H_6N_6$	1,3,5-Triazine-2,4,6-triamine	112.3	177.8
$C_3H_6N_6$	5,6,7,8-Tetrahydrotetrazolo[1,5-b][1,2,4]triazine	520.6	473.8
$C_3H_6N_6O_2$	1,6-Dimethyl-5-nitraminotetrazole	360.8	355.5
$C_3H_6N_8$	1-(2-Azidoethyl)-1H-tetrazol-5-amine	656.1	645.9
$C_3H_6N_{10}O_4$	1,5-Diazido-2,4-dinitro-2,4-diazapentane	747.9	642.3
$C_3H_7N_5$	1-Methyl-5-methylaminotetrazole	321.4	335.1
$C_3H_7N_5$	5-(Dimethylamino)-tetrazole	338.2	337.3
$C_3H_7N_5O$	2-(5-Amino-1H-tetrazol-1-yl)ethan-1-ol	187.5	132.2
$C_4H_2N_{10}O_5$	5,5′-Dinitro-3,3′–azoxy-1,2,4-triazole	650.8	760.0
$C_4H_4N_8$	4,4′-Azobis-1,2,4-triazole	849.4	800.7
$C_4H_4N_8O_2$	3,3′–Diamino-4,4′-azofurazan	685.1	730.3
$C_4H_4N_8O_3$	3,3′–Diamino-4,4′-azoxyfurazan	637.5	676.2
$C_4H_4N_{12}$	6,6′-(Diazene-1,2-diyl)bis(1,2,4,5-tetrazin-3-amine)	1,154.9	1,075.4
$C_4H_4N_{14}$	3,6-Bis(1H-1,2,3,4-tetrazol-5-ylamino)-1,2,4,5-tetrazine	1,149.7	1,164.4
$C_4H_6N_4$	2-Methyl-5-vinyl-2H-tetrazole	366.7	386.6
$C_4H_6N_6$	3-Azido-5-ethyl-4H-1,2,4-triazole	515.1	488.9
$C_4H_6N_6$	3-(5-Amino-1H-tetrazol-1-yl) propanenitrile	473.3	456.0
$C_4H_6N_6$	3-(5-Amino-2H-tetrazol-2-yl) propanenitrile	451.9	456.2
$C_4H_6N_8$	1,2-Di(1H-tetrazol-5-yl)ethane	639.5	613.9
$C_4H_6N_8O_2$	4,4′-Hydrazobis-(1,2,5-oxadiazol-3-amine)	598.9	593.8
$C_4H_6N_{10}$	(cis)1,1′-Dimethyl-5,5′-azotetrazole	903.8	927.9
$C_4H_6N_{10}$	(trans)1,1′-Dimethyl-5,5′-azotetrazole	888.2	929.1
$C_4H_6N_{10}$	2,2′-Dimethyl-5,5′-azotetrazole	868.4	948.9
$C_4H_6N_{12}O_4$	3,6-Bis(nitroguanyl)-1,2,4,5-tetrazine	617.6	572.9
$C_4H_7N_5$	5-Amino-1-(2-propenyl)-1H-tetrazole	429.3	403.5
$C_4H_7N_5$	5-Amino-2-(2-propenyl)-2H-tetrazole	409.4	409.3
$C_4H_7N_9$	Bis(2-methyl-2H-tetrazol-5-yl)amine	635.2	722.3
$C_4H_8N_8O_2$	1,5-Diazido-3-nitro-3-azapentane	655.7	584.2
$C_4H_8N_{10}$	3,6-Diguanidino-1,2,4,5-tetrazine	602.1	604.3
$C_4H_8N_{10}O_4$	1,6-Diazido-2,5-dinitro-2,5-diazahexane	733.6	626.7
$C_4H_8N_{12}O_6$	1,7-Diazido-2,4,6-trinitro-2,4,6-triazaheptane	813.9	675.7
$C_5H_9N_9$	N,2-Dimethyl-N-(2-methyl-2H-tetrazol-5-yl)-2H-tetrazol-5-amine	642.4	724.6
C_6N_{20}	4,4′,6,6′-Tetra(azido)azo-1,3,5-triazine	1,903.9	2,102.8
$C_6H_2N_{20}$	4,4′,6,6′-Tetra(azido)hydrazo-1,3,5-triazine	1,701.9	1,898.1
$C_6H_{10}N_{10}$	2,2′-Diethyl-5,5′-azotetrazole	821.6	884.9
$C_6H_{14}N_{16}$	1,2-Bis(4,6-dihydrazinyl-1,3,5-triazin-2-yl)hydrazine	927.6	816.2
$C_7H_7N_9$	3-Azido-6-(3,5-dimethylpyrazol-1-yl)-1,2,4,5-tetrazine	944.9	914.3
$C_8H_4N_{16}O_6$	Bis[4-aminofurazanyl-3-azoxy]azofurazan	1,545.1	1,625.4

C Glossary of compound names, as well as the measured and calculated values of the condensed-phase heats of formation for some energetic ionic liquids and salts

Table C.1: Predictions of $\Delta_f H^\theta$(c)[IMILoS] for some different energetic imidazolium-based ionic liquids or salts compared to experimental data and the predicted quantum calculated values based on the Byrd and Rice method [723].

Name		$\Delta_f H^\theta$(c)[IMILoS] (kJ/mol)		
Cation	Anion	Exp. or pred.[a]	Equation (2.18)	Dev
1-Methyl-3-octylimidazolium	Di-1H-tetrazol-5-ylazanide	539.6 [225]	533.3	−6.3
1-Hexyl-3-methylimidazolium	Di-1H-tetrazol-5-ylazanide	615.9 [225]	607.1	−8.8
1-Butyl-3-methylimidazolium	Di-1H-tetrazol-5-ylazanide	682.3 [225]	680.9	−1.4
1-Ethyl-3-methylimidazolium	Di-1H-tetrazol-5-ylazanide	725.6 [225]	754.8	29.2
1,2,3-Trimethylimidazolium	Di-1H-tetrazol-5-ylazanide	652.3 [225]	663.4	11.1
1,2,3-Trimethylimidazolium	Di-1H-tetrazol-5-ylazanide	774.7 [225]	783.2	8.5
1,3-Dimethylimidazolium	Di-1H-tetrazol-5-ylazanide	752.3 [225]	816.1	63.8
1-Ethyl-3-methylimidazolium	[(Cyanoimino)methylene]azanide	235.3 [737]	222.2	−13.1
1-Butyl-3-methylimidazolium	[(Cyanoimino)methylene]azanide	206.2 [738]	221.2	15.0
1-Ethyl-3-methylimidazolium	Ethyl sulfate	−579.1 [739]	−584.3	−5.2
1-Butyl-3-methylimidazolium	(E)-5,5'-(Diazene-1,2-diyl)bis(tetrazol-1-ide)	−1,006 [740]	−938.9	67.1
1,3-Dimethyl-5-nitro-1H-imidazol-3-ium	(E)-5,5'-(Diazene-1,2-diyl)bis(tetrazol-1-ide)	1,345 [740]	1,241.2	−103.8
1-Ethyl-3-methylimidazolium	Bis[(trifluoromethyl)sulfonyl]azanide	−1,917.2	−1843.2	74.0
1-Hexyl-3-methylimidazolium	Bis[(trifluoromethyl)sulfonyl]azanide	−2,006.4	−2,019.8	−13.4
1-Methyl-3-octylimidazolium	Bis[(trifluoromethyl)sulfonyl]azanide	−2,052.0	−2,035.7	16.3
1-Decyl-3-methyl-1H-imidazol-3-ium	Bis[(trifluoromethyl)sulfonyl]azanide	−2,095.7	−2,051.7	44.0
1-Tetradecyl-3-methylimidazolium	Bis[(trifluoromethyl)sulfonyl]azanide	−4,871.5	−4,892.4	−20.9
1-Butyl-3-methylimidazolium	Hexafluorophosphate	−2,105.3	−2,148.5	−43.2
1-Hexyl-3-methylimidazolium	Hexafluorophosphate	−2,144.3	−2,164.4	−20.1
1-Methyl-3-octylimidazolium	Hexafluorophosphate	−2,186.2	−2,180.4	5.8
1-Decyl-3-methyl-1H-imidazol-3-ium	Hexafluorophosphate	−2,226.1	−2,196.3	29.8
1-Dodecyl-3-methylimidazolium	Hexafluorophosphate	−4,693.4	−4,678.4	15.0
1-Butyl-3-methylimidazolium	Tetrafluoroborate	−1,662.7	−1,647.7	15.0
1-Methyl-3-octylimidazolium	Tetrafluoroborate	−1,740.3	−1,893.1	−152.8

https://doi.org/10.1515/9783112206768-020

Table C.1 (continued)

| Name | | $\Delta_f H^\theta$ (c)[IMILoS] (kJ/mol) | | |
Cation	Anion	Exp. or pred.[a]	Equation (2.18)	Dev
1-Dodecyl-3-methylimidazolium	Tetrafluoroborate	−4,245.3	−4,238.5	6.8
1-Butyl-2,3-dimethylimidazolium	Bromide	−230.2	−220.6	9.6
1-Pentyl-2,3-dimethylimidazolium	Bromide	−247.0	251.5	498.5
1-Octyl-2,3-dimethylimidazolium	Bromide	−305.3	−313.8	−8.5
1,3-Dibutyl-1H-imidazol-3-ium	Bromide	−248.2	−252.0	−3.8
1,3-Pentyl-1H-imidazol-3-ium	Bromide	−289.6	−313.8	−24.2
1,3-Octyl-1H-imidazol-3-ium	Bromide	−322.8	−303.8	19.0

[a]In the cases where the experimental values were not available, the predicted quantum-calculated values based on the Byrd and Rice method [723] were used.

Table C.2: Predictions of $\Delta_f H^\theta$(c)[TAILoS] for some different energetic triazolium-based ionic liquids or salts compared to experimental data.

Cation	Anion	Exp.		Equation (2.18)	Dev.
3-Azido-3H-1,2,4-triazol-4-ium	Nitrate	218.8	[741]	201.9	−16.9
3-Azido-1H-1,2,4-triazol-4-ium	5-Nitrotetrazol-1-ide	979.9	[741]	968.3	−11.6
3-azido-3H-1,2,4-triazol-4-ium	4,5-Dinitroimidazol-1-ide	401.7	[741]	411.5	9.8
1H-1,2,4-triazol-4-ium	5-Nitrotetrazol-1-ide	409.6	[741]	433.5	23.9
1H-1,2,4-triazol-4-ium	4,5-Dinitroimidazol-1-ide	503.3	[741]	456.5	−46.8
1H-1,2,4-triazol-4-ium	2,4,6-Trinitrophenolate	497.9	[741]	493.8	−4.1
1-Amino-4H-1,2,4-triazol-1-ium	Nitrate	34.7	[741]	−58.1	−92.8
1-Amino-4H-1,2,4-triazol-1-ium	Perchlorate	356.9	[741]	403.3	46.4
1-Amino-4H-1,2,4-triazol-1-ium	Nitrate	−109.6	[741]	−58.1	51.5
1-Amino-4H-1,2,4-triazol-1-ium	Perchlorate	298.3	[741]	291.5	−6.8
1-Amino-4H-1,2,4-triazol-1-ium	5-Nitrotetrazol-1-ide	702.1	[741]	708.3	6.2
1-amino-4H-1,2,4-triazol-1-ium	4-Nitro-1,2,3-triazol-1-ide	141.4	[741]	143.4	1.9
1-Amino-4H-1,2,4-triazol-1-ium	4,5-Dinitroimidazol-1-ide	466.5	[741]	412.2	−54.3
1-Amino-4H-1,2,4-triazol-1-ium	2,4,6-Trinitrophenolate	469.0	[741]	449.5	−19.5
3-Amino-1H-1,2,4-triazol-1-ium	Nitrate	−171.1	[742]	−169.9	1.2
1,5-Diamino-4H-1,2,4-triazol-1-ium	Nitrate	−89.5	[741]	−94.5	−5.0
1,5-Diamino-4H-1,2,4-triazol-1-ium	Perchlorate	484.5	[741]	478.7	−5.8
1,5-Diamino-4H-1,2,4-triazol-1-ium	4,5-Dinitroimidazol-1-ide	405.8	[741]	487.6	81.7
3,4,5-Triamino-4H-1,2,4-triazol-1-ium	5-Nitrotetrazol-1-ide	528.0	[743]	510.6	−17.4
3-Azido-1-methyl-4H-1,2,4-triazol-1-ium	Nitrate	93.3	[741]	147.3	54.0
3-Azido-1-methyl-4H-1,2,4-triazol-1-ium	Perchlorate	574.9	[741]	608.7	33.8
3-Azido-1-methyl-1H-1,2,4-triazol-4-ium	5-Nitrotetrazol-1-ide	1,085.3	[741]	1,113.1	27.8
3-Azido-1-methyl-4H-1,2,4-triazol-1-ium	4,5-Dinitroimidazol-1-ide	700.8	[741]	617.6	−83.2
3-Azido-5-methyl-4H-1,2,4-triazol-1-ium	Nitrate	156.5	[741]	147.3	−9.2

Table C.2 (continued)

Cation	Anion	Exp.		Equation (2.18)	Dev.
1-Methyl-1H-1,2,4-triazol-4-ium	5-Nitrotetrazol-1-ide	808.3	[741]	777.7	−30.6
1-Methyl-3H-1,2,4-triazol-1-ium	4,5-Dinitroimidazol-1-ide	80.3	[741]	95.9	15.6
1-Methyl-3H-1,2,4-triazol-1-ium	2,4,6-Trinitrophenolate	409.6	[741]	439.2	29.6
1-Amino-4-methyl-4H-1,2,4-triazol-1-ium	Nitrate	−160.2	[741]	−112.8	47.5
1-Amino-4-methyl-4H-1,2,4-triazol-1-ium	Perchlorate	544.8	[741]	548.1	3.3
4-Amino-1-methyl-4H-1,2,4-triazol-1-ium	Nitrate	−172.8	[741]	−112.8	60.0
4-Amino-1-methyl-4H-1,2,4-triazol-1-ium	Perchlorate	215.1	[741]	236.9	21.8
4-Amino-1-methyl-4H-1,2,4-triazol-1-ium	3,5-Dinitro-1,2,4-triazol-1-ide	378.2	[741]	350.5	−27.7
3,4,5-Triamino-1-methyl-4H-1,2,4-triazol-1-ium	Azide	530.6	[744]	506.3	−24.3
3,4,5-Triamino-1-methyl-4H-1,2,4-triazol-1-ium	Nitrate	64.2	[744]	38.0	−26.2
3,4,5-Triamino-1-methyl-4H-1,2,4-triazol-1-ium	Perchlorate	498.4	[744]	499.4	1.0
3,4,5-Triamino-1-methyl-4H-1,2,4-triazol-1-ium	Dinitroamide	276.3	[744]	242.1	−34.2
3,4,5-Triamino-1-methyl-4H-1,2,4-triazol-1-ium	5-Nitrotetrazol-1-ide	519.0	[744]	605.0	86.0
3-Azido-1-(2-azidoethyl)-1H-1,2,4-triazol-4-ium	Nitrate	437.2	[745]	428.0	−9.2
3-Azido-1-(2-azidoethyl)-1H-1,2,4-triazol-4-ium	5-Nitrotetrazol-1-ide	1,138.7	[745]	1,194.4	55.7
1-(2-Azidoethyl)-1H-1,2,4-triazol-4-ium	Nitrate	27.2	[745]	92.6	65.4
1-(2-Azidoethyl)-1H-1,2,4-triazol-4-ium	Perchlorate	426.2	[745]	442.3	16.1
3-Azido-1,4-dimethyl-1H-1,2,4-triazol-4-ium	3,5-Dinitro-1,2,4-triazol-1-ide	792.9	[741]	787.3	−5.5
3-Azido-1,4-dimethyl-1H-1,2,4-triazol-4-ium	2,4,6-Trinitrophenolate	611.3	[741]	600.3	−11.0
1-(2-Azidoethyl)-1H-1,2,4-triazol-4-ium	5-Nitrotetrazol-1-ide	676.0	[745]	659.6	−16.4
1-(2-Azidoethyl)-1H-1,2,4-triazol-4-ium	4,5-Dinitroimidazol-1-ide	380.2	[745]	376.7	−3.5
1,4-Dimethyl-4H-1,2,4-triazol-1-ium	3,5-Dinitro-1,2,4-triazol-1-ide	118.0	[741]	108.8	−9.2
1,4-Dimethyl-4H-1,2,4-triazol-1-ium	2,4,6-Trinitrophenolate	340.2	[741]	384.6	44.4
4-Amino-1-(2-azidoethyl)-1H-1,2,4-triazol-4-ium	Nitrate	321.1	[745]	327.6	6.5
4-Amino-1-(2-azidoethyl)-1H-1,2,4-triazol-4-ium	Perchlorate	828.4	[745]	828.9	0.5
4-Amino-1-(2-azidoethyl)-1H-1,2,4-triazol-4-ium	3,5-Dinitro-1,2,4-triazol-1-ide	331.7	[745]	333.3	1.6
1-(2-Azidoethyl)-4-methyl-1H-1,2,4-triazol-4-ium	Nitrate	66.3	[745]	38.0	−28.3
1-(2-Azidoethyl)-4-methyl-1H-1,2,4-triazol-4-ium	Perchlorate	568.0	[745]	499.4	−68.6
4-(2-Azidoethyl)-1-methyl-1H-1,2,4-triazol-4-ium	Nitrate	77.7	[745]	38.0	−39.7

Table C.2 (continued)

Cation	Anion	Exp.		Equation (2.18)	Dev.
4-(2-Azidoethyl)-1-methyl-1H-1,2,4-triazol-4-ium	Perchlorate	541.1	[745]	499.4	−41.7
1-(2-Azidoethyl)-4-methyl-1H-1,2,4-triazol-4-ium	3,5-Dinitro-1,2,4-triazol-1-ide	395.2	[746]	389.5	−5.7
4-(2-Azidoethyl)-1-methyl-1H-1,2,4-triazol-4-ium	3,5-Dinitro-1,2,4-triazol-1-ide	771.3	[746]	732.7	−38.6
4-Methyl-1-propyl-1H-1,2,4-triazol-4-ium	3,5-Dinitro-1,2,4-triazol-1-ide	198.7	[746]	223.0	24.3

D Common ligand abbreviations in energetic metal-organic frameworks (EMOFs)

Table D.1: Abbreviations of different ligands in EMOFs.

Abbreviation	Ligand name
AG	Aminoguanidinium
ANTA	5-Amino-3-nitro-1H-1,2,4-triazole
3-atrz	3-Amine-1H-1,2,4-triazole
ata	5-Amino-1H-tetrazole
atrz	4,4′-Azo-1,2,4-triazole
atz	4-Amino-1,2,4-triazole
5-ATZ or HATZ	5-Aminotetrazole
azide	N_3 (azide)
bta	N,N-bis(tetrazol-5-yl)amine
btm^{2-}	N,N-bis(tetrazol-5-yl)amine (dianionic)
BTF	4,4′-Oxybis[3,3′-(1-hydroxytetrazolyl)]furazan
BTO	1H,1′H-5,5′-bitetrazole-1,1′-diolate
cyclo-N_5^-	Cyclo-pentazolate
DNABT	1,1′-Dinitramino-5,5′-bistetrazole
3,5-DNBA	3,5-Dinitrobenzoic acid
DNBT	5,5′-Dinitro-2H,2H′-3,3′-bi-1,2,4-triazole
DNBTO	5,5′-Dinitro-3,3′-bis-1,2,4-triazole-1-diol
DNGTz	3,6-Bis-nitroguanyl-1,2,4,5-tetrazine
dntrza	1-(Dinitromethyl)-1H-1,2,4-triazole-3-carboxylic acid
en	Ethylenediamine
H2BTA	N,N-Bis(1(2)H-tetrazol-5-yl)-amine
H2BTE	1,2-Bis(tetrazol-5-yl)ethane
H2btm	Bis(tetrazole) methane
H2BTO	1H,1′H-[5,5′-bitetrazole]-1,1′-diol
H2BTT	4,5-Bis(tetrazol-5-yl)-2H-1,2,3-triazole
H2dns	3,5-Dinitrosalicylic acid
H2DTTZ	4,5-Di(1H-tetrazol-5-yl)-2H-1,2,3-triazole
H2NPA	(E)-1,2-bis(3,5-dinitro-1H-pyrazol-4-yl)diazene
H2to	1,2,4-Triazol-5-one
H2tza	1H-tetrazole-5-acetic acid
H2TZEG	N-[2-(1H-tetrazol-5-yl)ethyl]glycine
H2tztr	3-(1H-tetrazol-5-yl)-1H-triazole
H2ZTO	4,4′-Azo-1,2,4-triazol-5-one
H3BTT	4,5-Bis(tetrazol-5-yl)-2H-1,2,3-triazole
H3DTTZ	4,5-Di(1H-tetrazol-5-yl)-2H-1,2,3-triazole
HABTNA	N-(5′-amino-1H,1′H-[3,3′-bi(1,2,4-triazol)]-5-yl)nitramine
HAFT	4-Amino-3-(5-tetrazolate)-furazan
HANTP	3,4-Initro-1-(1H-tetrazol-5-yl)-1H-pyrazol-5-amine
HATr	3-Hydrazino-4-amino-1,2,4-triazole
Hatza	(5-Amino-1H-tetrazol-1-yl)acetic acid
HCONHNH2	Formyl hydrazide

https://doi.org/10.1515/9783112206768-021

Table D.1 (continued)

Abbreviation	Ligand name
Hntz	3-Nitro-1H-1,2,4-triazole
HTZ	Tetrazole
IO_3	Iodate
K2DNMAF	Potassium 4,4'-bis(dinitromethyl)-3,3'-azofurazanate
MA	Melamine
MHT	5-(1-Methylhydrazinyl) tetrazole
Mtz	5-Methyltetrazole
Mtta	5-Methyl tetrazole
N_3	Azide
OH-Tz	Hydroxytetrazole
PA	Picrate
pn	1,2-Diaminopropane
tntrza	1-(Trinitromethyl)-1H-1,2,4-triazole-3-carboxylic acid
Tz	Tetrazole

Bibliography

[1]   Agrawal JP. High Energy Materials: Propellants, Explosives and Pyrotechnics. John Wiley &
      Sons; 2010.
[2]   Klapötke TM. Energetic Materials Encyclopedia. Second ed. Walter de Gruyter GmbH & Co KG; 2021.
[3]   Klapötke TM. Chemistry of High-Energy Materials. Seventh ed. Walter de Gruyter GmbH & Co
      KG; 2025.
[4]   Keshavarz MH, Klapötke TM. Energetic Compounds: Methods for Prediction of Their Performance.
      Walter de Gruyter GmbH; 2017.
[5]   Keshavarz MH, Klapötke TM. The Properties of Energetic Materials: Sensitivity, Physical and
      Thermodynamic Properties. Walter de Gruyter GmbH & Co KG; 2017.
[6]   Keshavarz MH. Combustible Organic Materials: Determination and Prediction of Combustion
      Properties. Walter de Gruyter GmbH & Co KG; 2018.
[7]   Keshavarz MH. Liquid fuels as jet fuels and propellants. Nova Science; 2018.
[8]   Zhang Q, Shreeve JM. Energetic ionic liquids as explosives and propellant fuels: a new journey of
      ionic liquid chemistry. Chemical Reviews. 2014;114:10527–74.
[9]   Bastea S, Fried L, Glaesemann K, Howard W, Souers P, Vitello P. CHEETAH 5.0 User's Manual,
      Lawrence Livermore National Laboratory; 2011.
[10] Klapötke TM. Chemistry of High-Energy materials. 3th ed. Walter de Gruyter GmbH & Co KG; 2015.
[11] Keshavarz MH, Abadi YH, Esmaeilpour K, Oftadeh M. Assessment of N-oxide in a new high
      performance energetic tetrazine derivative on its physical, thermodynamic, sensitivity combustion
      and detonation performance. Chemistry of Heterocyclic Compounds. 2017;53:797–801.
[12] Keshavarz MH, Azarniamehraban J, Atabak HH, Ferdowsi M. Recent developments for prediction of
      power of aromatic and non-aromatic energetic materials along with a novel computer code for
      prediction of their power. Propellants, Explosives, Pyrotechnics. 2016;41:942–8.
[13] Roknabadi AG, Keshavarz MH, Esmailpour K, Zamani M. High performance nitroazacubane
      energetic compounds: Structural, thermochemical and detonation characteristics. ChemistrySelect.
      2016;1:6735–40.
[14] Fotouhi-Far F, Bashiri H, Hamadanian M, Keshavarz MH. A new method for assessment of
      performing mechanical works of energetic compounds by the cylinder test. Zeitschrift für
      anorganische und allgemeine Chemie. 2016;642:1086–90.
[15] Roknabadi AG, Keshavarz MH, Esmailpour K, Zamani M. Structural, thermochemical and detonation
      performance of derivatives of 1,2,4,5-tetrazine and 1,4 N-oxide 1,2,4,5-tetrazine as new high-
      performance and nitrogen-rich energetic materials. Journal of the Iranian Chemical Society.
      2017;14:57–63.
[16] Keshavarz MH, Kamalvand M, Jafari M, Zamani A. An improved simple method for the calculation of
      the detonation performance of CHNOFCl, aluminized and ammonium nitrate explosives. Central
      European Journal of Energetic Materials. 2016;13:381–96.
[17] Rezaei AH, Keshavarz MH, Tehrani MK, Darbani SMR, Farhadian AH, Mousavi SJ, Mousaviazar
      A. Approach for determination of detonation performance and aluminum percentage of
      aluminized-based explosives by laser-induced breakdown spectroscopy. Applied Optics.
      2016;55:3233–40.
[18] Kim CK, Cho SG, Kim CK, Park HY, Zhang H, Lee HW. Prediction of densities for solid energetic
      molecules with molecular surface electrostatic potentials. Journal of Computational Chemistry.
      2008;29:1818–24.
[19] Politzer P, Murray JS. Energetic Materials: Part 1. Decomposition, Crystal and Molecular Properties.
      Elsevier; 2003.
[20] Rice BM, Hare JJ, Byrd EF. Accurate predictions of crystal densities using quantum mechanical
      molecular volumes. The Journal of Physical Chemistry A. 2007;111:10874–9.

https://doi.org/10.1515/9783112206768-022

[21] Qiu L, Xiao H, Gong X, Ju X, Zhu W. Crystal density predictions for nitramines based on quantum chemistry. Journal of Hazardous Materials. 2007;141:280–8.

[22] Politzer P, Martinez J, Murray JS, Concha MC, Toro-Labbe A. An electrostatic interaction correction for improved crystal density prediction. Molecular Physics. 2009;107:2095–101.

[23] Tarver CM. Density estimations for explosives and related compounds using the group additivity approach. Journal of Chemical and Engineering Data. 1979;24:136–45.

[24] Ammon HL. Updated atom/functional group and atom_code volume additivity parameters for the calculation of crystal densities of single molecules, organic salts, and multi-fragment materials containing H, C, B, N, O, F, S, P, Cl, Br, and I. Propellants, Explosives, Pyrotechnics. 2008;33:92–102.

[25] Willer RL. Calculation of the density and detonation properties of C, H, N, O and F compounds: use in the design and synthesis of new energetic materials. Journal of the Mexican Chemical Society. 2009;53:108–19.

[26] Keshavarz MH. Prediction of densities of acyclic and cyclic nitramines, nitrate esters and nitroaliphatic compounds for evaluation of their detonation performance. Journal of Hazardous Materials. 2007;143:437–42.

[27] Keshavarz MH. New method for calculating densities of nitroaromatic explosive compounds. Journal of Hazardous Materials. 2007;145:263–9.

[28] Keshavarz MH, Pouretedal HR. A reliable simple method to estimate density of nitroaliphatics, nitrate esters and nitramines. Journal of Hazardous Materials. 2009;169:158–69.

[29] Keshavarz MH. Novel method for predicting densities of polynitro arene and polynitro heteroarene explosives in order to evaluate their detonation performance. Journal of Hazardous Materials. 2009;165:579–88.

[30] Cho SG, Goh EM, Kim JK. Holographic QSAR models for estimating densities of energetic materials. Bulletin of the Korean Chemical Society. 2001;22:775–8.

[31] Gorb L, Hill FC, Kholod Y, Muratov EN, Kuz'min VE, Leszczynski J. Progress in Predictions of Environmentally Important Physicochemical Properties of Energetic Materials: Applications of Quantum-Chemical Calculations. In: Practical Aspects of Computational Chemistry II. Springer, 2012. pp. 335–59.

[32] Ahmed A, Sandler SI. Physicochemical properties of hazardous energetic compounds from molecular simulation. Journal of Chemical Theory and Computation. 2013;9:2389–97.

[33] Katritzky AR, Kuanar M, Slavov S, Hall CD, Karelson M, Kahn I, Dobchev DA. Quantitative correlation of physical and chemical properties with chemical structure: utility for prediction. Chemical Reviews. 2010;110:5714–89.

[34] Dearden JC, Rotureau P, Fayet G. QSPR prediction of physico-chemical properties for REACH. SAR and QSAR in Environmental Research. 2013;24:279–318.

[35] Keshavarz MH, Abadi YH, Esmaeilpour K, Damiri S, Oftadeh M. Introducing novel tetrazole derivatives as high performance energetic compounds for confined explosion and as oxidizer in solid propellants. Propellants, Explosives, Pyrotechnics. 2017;42:492–8.

[36] Keshavarz MH, Abadi YH, Esmaeilpour K, Damiri S, Oftadeh M. A novel class of nitrogen-rich explosives containing high oxygen balance to use as high performance oxidizers in solid propellants. Propellants, Explosives, Pyrotechnics. 2017;42:1155–60.

[37] Keshavarz MH, Esmaeilpour K, Oftadeh M, Abadi YH. Assessment of two new nitrogen-rich tetrazine derivatives as high performance and safe energetic compounds. RSC Advances. 2015;5:87392–9.

[38] Keshavarz MH, Abadi YH. Novel organic compounds containing nitramine groups suitable as high-energy cyclic nitramine compounds. ChemistrySelect. 2018;3:8238–44.

[39] Keshavarz MH, Abadi YH, Esmaeilpour K, Damiri S, Oftadeh M. Assessment of the effect of N-oxide group in a new high-performance energetic tetrazine derivative on its physicochemical and thermodynamic properties, sensitivity, and combustion and detonation performance. Chemistry of Heterocyclic Compounds. 2017;53:797–801.

[40] Keshavarz MH, Abadi YH, Esmaeilpour K, Damiri S, Oftadeh M. Novel high-nitrogen content energetic compounds with high detonation and combustion performance for use in plastic bonded explosives (PBXs) and composite solid propellants. Central European Journal of Energetic Materials. 2018;15:364–75.

[41] Keshavarz MH, Esmailpour K, Zamani M, Roknabadi AG. Thermochemical, sensitivity and detonation characteristics of new thermally stable high performance explosives. Propellants, Explosives, Pyrotechnics. 2015;40:886–91.

[42] Wang Q, Shao Y, Lu M. Azo1,3,4-oxadiazole as a Novel Building Block to Design High-Performance Energetic Materials. Crystal Growth & Design. 2019;19:839–44.

[43] Fershtat LL, Ovchinnikov IV, Epishina MA, Romanova AA, Lempert DB, Muravyev NV, Makhova NN. Assembly of nitrofurazan and nitrofuroxan frameworks for high-performance energetic materials. ChemPlusChem. 2017;82:1315–9.

[44] Dalinger IL, Kormanov AV, Suponitsky KY, Muravyev NV, Sheremetev AB. Pyrazole–tetrazole hybrid with trinitromethyl, fluorodinitromethyl, or (difluoroamino) dinitromethyl groups: high-performance energetic materials. Chemistry–An Asian Journal. 2018;13:1165–72.

[45] Sun S, Lu M. Conjugation in multi-tetrazole derivatives: a new design direction for energetic materials. Journal of Molecular Modeling. 2018;24:173.

[46] Yang J, Gong X, Mei H, Li T, Zhang J, Gozin M. Design of zero oxygen balance energetic materials on the basis of Diels–Alder chemistry. The Journal of Organic Chemistry. 2018;83:14698–702.

[47] Ammon HL. New atom/functional group volume additivity data bases for the calculation of the crystal densities of C-, H-, N-, O-, F-, S-, P-, Cl-, and Br-containing compounds. Structural Chemistry. 2001;12:205–12.

[48] Ye C, Shreeve JM. New atom/group volume additivity method to compensate for the impact of strong hydrogen bonding on densities of energetic materials. Journal of Chemical & Engineering Data. 2008;53:520–4.

[49] Smirnov AS, Smirnov SP, Pivina TS, Lempert DB, Maslova LK. Comprehensive assessment of physicochemical properties of new energetic materials. Russian Chemical Bulletin. 2016;65:2315–32.

[50] Hofmann DWM. Fast estimation of crystal densities. Acta Crystallographica Section B: Structural Science. 2002;58:489–93.

[51] Ghule VD, Nirwan A, Devi A. Estimating the densities of benzene-derived explosives using atomic volumes. Journal of Molecular Modeling. 2018;24:50.

[52] Ye C, Shreeve JM. Rapid and accurate estimation of densities of room-temperature ionic liquids and salts. The Journal of Physical Chemistry A. 2007;111:1456–61.

[53] Beaucamp S, Marchet N, Mathieu D, Agafonov V. Calculation of the crystal densities of molecular salts and hydrates using additive volumes for charged groups. Acta Crystallographica Section B: Structural Science. 2003;59:498–504.

[54] Meyer R, Köhler J, Homburg A. Explosives. 7th ed. John Wiley & Sons; 2016.

[55] Hong D, Li Y, Zhu S, Zhang L, Pang C. Three insensitive energetic co-crystals of 1-nitronaphthalene, with 2,4,6-trinitrotoluene (TNT), 2,4,6-trinitrophenol (picric acid) and D-mannitol hexanitrate (MHN). Central European Journal of Energetic Materials. 2015;12:47–62.

[56] Rice BM, Sorescu DC. Assessing a generalized CHNO intermolecular potential through ab initio crystal structure prediction. The Journal of Physical Chemistry B. 2004;108:17730–9.

[57] Murray JS, Brinck T, Politzer P. Relationships of molecular surface electrostatic potentials to some macroscopic properties. Chemical Physics. 1996;204:289–99.

[58] Pan J-F, Lee Y-W. Crystal density prediction for cyclic and cage compounds. Physical Chemistry Chemical Physics. 2004;6:471–3.

[59] Kim CK, Lee KA, Hyun KH, Park HJ, Kwack IY, Kim CK, Lee HW, Lee BS. Prediction of physicochemical properties of organic molecules using van der Waals surface electrostatic potentials. Journal of Computational Chemistry. 2004;25:2073–9.

[60] Moxnes JF, Hansen FK, Jensen TL, Sele ML, Unneberg E. A computational study of density of some high energy molecules. Propellants, Explosives, Pyrotechnics. 2017;42:204–12.

[61] Kim CK, Cho SG, Kim CK, Kim M-R, Lee HW. Prediction of physicochemical properties of organic molecules using semi-empirical methods. Bulletin of the Korean Chemical Society. 2013;34:1043.

[62] Rice BM, Byrd EFC. Evaluation of electrostatic descriptors for predicting crystalline density. Journal of Computational Chemistry. 2013;34:2146–51.

[63] Kim CK, Cho SG, Li J, Park BH, Kim CK. Prediction of crystal density and explosive performance of high-energy-density molecules using the modified MSEP scheme. Bulletin of the Korean Chemical Society. 2016;37:1683–9.

[64] Wang G, Xu Y, Xue C, Ding Z, Liu Y, Liu H, Gong X. Prediction of the Crystalline Densities of Aliphatic Nitrates by Quantum Chemistry Methods. Central European Journal of Energetic Materials. 2019;16:412–32.

[65] Wang L, Zhang M, Chen J, Su L, Zhao S, Zhang C, Liu J, Chen C. Corrections of Molecular Morphology and Hydrogen Bond for Improved Crystal Density Prediction. Molecules. 2020;25:161.

[66] Keshavarz MH, Jafari M, Ebadpour R. Recent advances for assessment of the condensed phase heat of formation of high-energy content organic compounds and ionic liquids (or salts) to introduce a new computer code for design of desirable compounds. Fluid Phase Equilibria. 2020;533:112913.

[67] Nirwan A, Devi A, Ghule VD. Assessment of density prediction methods based on molecular surface electrostatic potential. Journal of Molecular Modeling. 2018;24:166.

[68] Ghule VD, Nirwan A. Role of forcefield in density prediction for CHNO explosives. Structural Chemistry. 2018;29:1375–82.

[69] Lal S, Gao H, Jean'ne MS. New approach for predicting crystal densities of energetic materials. New Journal of Chemistry. 2024;48(41):17947–52.

[70] Lu T, Chen F. Multiwfn: A multifunctional wavefunction analyzer. Journal of Computational Chemistry. 2012;33(5):580–92.

[71] Keshavarz MH, Klapötke TM, Sućeska M. Energetic materials designing bench (EMDB), version 1.0. Propellants, Explosives, Pyrotechnics. 2017;42:854–6.

[72] Fathollahi M, Sajady H. Prediction of density of energetic cocrystals based on QSPR modeling using artificial neural network. Structural Chemistry. 2018;29:1119–28.

[73] Rahimi R, Keshavarz MH, Akbarzadeh AR. Prediction of the density of energetic materials on the basis of their molecular structures. Central European Journal of Energetic Materials. 2016;13:73–101.

[74] Pagoria PF, Lee GS, Mitchell AR, Schmidt RD. A review of energetic materials synthesis. Thermochimica Acta. 2002;384:187–204.

[75] Keshavarz MH, Soury H, Motamedoshariati H, Dashtizadeh A. Improved method for prediction of density of energetic compounds using their molecular structure. Structural Chemistry. 2015;26:455–66.

[76] Zohari N, Sheibani N. Link between density and molecular structures of energetic azido compounds as green plasticizers. Zeitschrift für anorganische und allgemeine Chemie. 2016;642:1472–9.

[77] Yang J, Zhang X, Gao P, Gong X, Wang G. Exploring highly energetic aliphatic azido nitramines for plasticizers. RSC Advances. 2014;4:53172–9.

[78] Keshavarz MH, Ebadpour R, Jafari M. A simple approach for predicting the density of high nitrogen organic compounds as materials for providing clean products and enormous energy release. Central European Journal of Energetic Materials. 2020;17(2):296–317.

[79] Qu Y, Zeng Q, Wang J, Ma Q, Li H, Li H, Yang G. Furazans with azo linkages: stable CHNO energetic materials with high densities, highly energetic performance, and low impact and friction sensitivities. Chemistry–A European Journal. 2016;22(35):12527–32.

[80] Tang Y, Shreeve JM. Nitroxy/azido-functionalized triazoles as potential energetic plasticizers. Chemistry – A European Journal. 2015;21(19):7285–91.

[81]   Stewart JJ. Optimization of parameters for semiempirical methods V: modification of NDDO approximations and application to 70 elements. Journal of Molecular modeling. 2007;13(12):1173–213.

[82]   Keshavarz MH, Ebadpour R, Jafari M. Reliable prediction of crystal density of high nitrogen-containing organic compounds as powerful, less sensitive, eco-friendly energetic materials for dependable assessment of their performance. Fluid Phase Equilibria. 2023;565:113653.

[83]   Keshavarz MH, Rahimi R, Akbarzadeh AR. Two novel correlations for assessment of crystal density of hazardous ionic molecular energetic materials using their molecular structures. Fluid Phase Equilibria. 2015;402:1–8.

[84]   Zohari N, Abrishami F, Ebrahimikia M. Investigation of the effect of various substituents on the density of tetrazolium nitrate salts as green energetic materials. Zeitschrift für anorganische und allgemeine Chemie. 2016;642:749–60.

[85]   Zohari N, Bajestani IR. A novel correlation for predicting the density of tetrazole-N-oxide salts as green Energetic Materials through Their molecular structure. Central European Journal of Energetic Materials. 2018;15:629–51.

[86]   He P, Zhang JG, Yin X, Wu JT, Wu L, Zhou ZN, Zhang TL. Chemistry – A European Journal, Chem Eur J. 2016;22:7670–85.

[87]   Singh RP, Verma RD, Meshri DT, Shreeve JM. Energetic nitrogen-rich salts and ionic liquids. Angewandte Chemie International Edition. 2006;45(22):3584–601.

[88]   Badgujar D, Talawar M, Zarko V, Mahulikar P. Recent advances in safe synthesis of energetic materials: an overview. Combustion, Explosion, and Shock waves. 2019;55(3):245–57.

[89]   He P, Han J, Wu J, Mei H, Zhang J. Computational insight into a new family of functionalized tetrazole-N-oxides as high-energy density materials. New Journal of Chemistry. 2019;43(42):16454–60.

[90]   Li Y, Oommen C, Sarathy SM. Developing a theoretical approach for accurate determination of the density and thermochemical properties of energetic ionic liquids. Propellants, Explosives, Pyrotechnics. 2020;45(12):1949–58.

[91]   Jafari M, Hossein Keshavarz M, Reza Motamedi M, Hesamodin Hosseini S. An improved correlation for reliable assessment of the detonation performance of non-ideal explosives containing metals and the other solid particulates. Zeitschrift für anorganische und allgemeine Chemie. 2021;647(6):673–80.

[92]   Merone GM, Tartaglia A, Rosato E, D'Ovidio C, Kabir A, Ulusoy HI, Savini F, Locatelli M. Ionic liquids in analytical chemistry: applications and recent trends. Current Analytical Chemistry. 2021;17(9):1340–55.

[93]   Keshavarz MH, Pouretedal HR, Saberi E. A simple method for prediction of density of ionic liquids through their molecular structure. Journal of Molecular Liquids. 2016;216:732–7.

[94]   Xu J, Zheng S, Huang S, Tian Y, Liu Y, Zhang H, Sun J. Host–guest energetic materials constructed by incorporating oxidizing gas molecules into an organic lattice cavity toward achieving highly-energetic and low-sensitivity performance. Chemical Communications. 2019;55(7):909–12.

[95]   Snyder CJ, Imler GH, Chavez DE, Parrish DA. Synergetic Explosive Performance through Cocrystallization. Crystal Growth & Design. 2021;21(3):1401–5.

[96]   Keshavarz MH, Makvandi L. Assessment of recent researches for reliable prediction of density of organic compounds as well as ionic liquids and salts containing energetic groups at room temperature. Propellants, Explosives, Pyrotechnics. 2020;45(11):1680–90.

[97]   Keshavarz MH, Ebadpour R, Jafari M. An easy pathway for reliable assessment of crystal density of high nitrogen-containing salts and ionic liquids without using complex descriptors. Fluid Phase Equilibria. 2023;565:113667.

[98]   Joo Y-H, Gao H, Parrish DA, Cho SG, Goh EM, Shreeve JM. Energetic salts based on nitroiminotetrazole-containing acetic acid. Journal of Materials Chemistry. 2012;22(13):6123–30.

[99]   Huang Y, Zhang Y, Shreeve JM. Nitrogen-rich salts based on energetic nitroaminodiazido[1, 3, 5] triazine and guanazine. Chemistry – A European Journal. 2011;17(5):1538–46.

[100] Li G, Zhang C. Review of the molecular and crystal correlations on sensitivities of energetic materials. Journal of Hazardous Materials. 2020;398:122910.

[101] Li G, Zhang C. Review of the molecular and crystal correlations on sensitivities of energetic materials. Journal of Hazardous Materials. 2020;398:122910.

[102] Graser J, Kauwe SK, Sparks TD. Machine learning and energy minimization approaches for crystal structure predictions: a review and new horizons. Chemistry of Materials. 2018;30(11):3601–12.

[103] Tian X-l, Song S-w, Chen F, Qi X-j, Wang Y, Zhang Q-h. Machine learning-guided property prediction of energetic materials: Recent advances, challenges, and perspectives. Energetic Materials Frontiers. 2022;3(3):177–86.

[104] Deng Q, Hu J, Wang L, Liu Y, Guo Y, Xu T, Pu X. Probing impact of molecular structure on bulk modulus and impact sensitivity of energetic materials by machine learning methods. Chemometrics and Intelligent Laboratory Systems. 2021;215:104331.

[105] Agrawal A, Choudhary A. Perspective: Materials informatics and big data: Realization of the "fourth paradigm" of science in materials science. APL Materials. 2016;4(5):053208.

[106] Butler KT, Davies DW, Cartwright H, Isayev O, Walsh A. Machine learning for molecular and materials science. Nature. 2018;559(7715):547–55.

[107] Wang H, Ji Y, Li Y. Simulation and design of energy materials accelerated by machine learning. Wiley Interdisciplinary Reviews: Computational Molecular Science. 2020;10(1):e1421.

[108] Balakrishnan S, VanGessel FG, Boukouvalas Z, Barnes BC, Fuge MD, Chung PW. Locally optimizable joint embedding framework to design nitrogen-rich molecules that are similar but improved. Molecular Informatics. 2021;40(7):2100011.

[109] Casey AD, Son SF, Bilionis I, Barnes BC. Prediction of energetic material properties from electronic structure using 3D convolutional neural networks. Journal of Chemical Information and Modeling. 2020;60(10):4457–73.

[110] Fortunato ME, Coley CW, Barnes BC, Jensen KF, editors. Machine learned prediction of reaction template applicability for data-driven retrosynthetic predictions of energetic materials. In: AIP Conference Proceedings. AIP Publishing, 2020.

[111] Barnes BC, editor. Deep learning for energetic material detonation performance. In: AIP Conference Proceedings. AIP Publishing, 2020.

[112] Ma H, Bian Y, Rong Y, Huang W, Xu T, Xie W, Ye G, Huang J. Cross-dependent graph neural networks for molecular property prediction. Bioinformatics. 2022;38(7):2003–9.

[113] Chen C, Zuo Y, Ye W, Li X, Deng Z, Ong SP. A critical review of machine learning of energy materials. Advanced Energy Materials. 2020;10(8):1903242.

[114] Chen C, Zuo Y, Ye W, Li X, Deng Z, Ong SP. A critical review of machine learning of energy materials. Advanced Energy Materials. 2020;10(8):1903242.

[115] Fathollahi M, Sajady H. Prediction of density of energetic cocrystals based on QSPR modeling using artificial neural network. Structural Chemistry. 2018;29:1119–28.

[116] Yang C, Chen J, Wang R, Zhang M, Zhang C, Liu J. Density prediction models for energetic compounds merely using molecular topology. Journal of Chemical Information and Modeling. 2021;61(6):2582–93.

[117] Nguyen P, Loveland D, Kim JT, Karande P, Hiszpanski AM, Han TY-J. Predicting energetics materials' crystalline density from chemical structure by machine learning. Journal of Chemical Information and Modeling. 2021;61(5):2147–58.

[118] Pandey SK, Roy K. Predicting the performance and stability parameters of energetic materials (EMs) using a machine learning-based q-RASPR approach. Energy Advances. 2024;3(6):1293–306.

[119] Elton DC, Boukouvalas Z, Butrico MS, Fuge MD, Chung PW. Applying machine learning techniques to predict the properties of energetic materials. Scientific Reports. 2018;8(1):9059.

[120] Agrawal JP, Hodgson R. Organic Chemistry of Explosives. John Wiley & Sons; 2007.

[121] Sikder AK, Maddala G, Agrawal JP, Singh H. Important aspects of behaviour of organic energetic compounds: a review. Journal of Hazardous Materials. 2001;84:1–26.

[122] Sikder A, Sikder N. A review of advanced high performance, insensitive and thermally stable energetic materials emerging for military and space applications. Journal of Hazardous Materials. 2004;112:1–15.

[123] Keshavarz MH. Research progress on heats of formation and detonation of energetic compounds. In: Brar SK, editor. Hazardous Materials: Types, Risks and Control. New York, New York: Nova Science Publishers Inc., 2011, pp. 339–59.

[124] Rogers DW, Zavitsas AA, Matsunaga N. Determination of enthalpies ('Heats') of formation. Wiley Interdisciplinary Reviews: Computational Molecular Science. 2013;3:21–36.

[125] Goodwin A, Marsh K, Wakeham W. Measurement of the thermodynamic properties of single phases. Elsevier; 2003.

[126] Sućeska M. EXPLO5–Computer program for calculation of detonation parameters. In: Proc. of 32nd Int. Annual Conference of ICT. Karlsruhe, Germany, 2001.

[127] Keshavarz MH, Motamedoshariati H, Moghayadnia R, Nazari HR, Azarniamehraban J. A new computer code to evaluate detonation performance of high explosives and their thermochemical properties, part I. Journal of Hazardous Materials. 2009;172:1218–28.

[128] Keshavarz MH, Motamedoshariati H, Moghayadnia R, Ghanbarzadeh M, Azarniamehraban J. A new computer code for assessment of energetic materials with crystal density, condensed phase enthalpy of formation, and activation energy of thermolysis. Propellants, Explosives, Pyrotechnics. 2013;38:95–102.

[129] Keshavarz MH, Ghani K, Asgari A. A suitable computer code for prediction of sublimation energy and deflagration temperature of energetic materials. Journal of Thermal Analysis and Calorimetry. 2015;121:675–81.

[130] Gordon S, McBride BJ. Computer program for calculation of complex chemical equilibrium compositions and applications. Part 1: Analysis. Washington DC: NASA Lewis Research Center; 1994.

[131] Ameen R, Fasila P, Biju A. Theoretical studies of azete based high energy density materials with trinitromethane functional group. Computational and Theoretical Chemistry. 2021;1203:113346.

[132] Wang L, Zhai L, She W, Wang M, Zhang J, Wang B. Synthetic strategies toward nitrogen-rich energetic compounds via the reaction characteristics of cyanofurazan/furoxan. Frontiers in Chemistry. 2022;10:871684.

[133] Sana M, Leroy G, Peeters D, Wilante C. The theoretical study of the heats of formation of organic compounds containing the substituents CH3, CF3, NH2, NF2, NO2, OH and F. Journal of Molecular Structure: THEOCHEM. 1988;164:249–74.

[134] Sprague JT, Tai JC, Yuh Y, Allinger NL. The MMP2 calculational method. Journal of Computational Chemistry. 1987;8:581–603.

[135] Lii JH, Allinger NL. Molecular mechanics. The MM3 force field for hydrocarbons. 3. The van der Waals' potentials and crystal data for aliphatic and aromatic hydrocarbons. Journal of the American Chemical Society. 1989;111:8576–82.

[136] Nevins N, Lii JH, Allinger NL. Molecular mechanics (MM4) calculations on conjugated hydrocarbons. Journal of Computational Chemistry. 1996;17:695–729.

[137] Allinger NL. Molecular Structure: Understanding Steric and Electronic Effects from Molecular Mechanics. John Wiley & Sons; 2010.

[138] Akutsu Y, Tahara S-Y, Tamura M, Yoshida T. Calculations of heats of formation for nitro compounds by semi-empirical MO methods and molecular mechanics. Journal of Energetic Materials. 1991;9:161–71.

[139] Mole SJ, Zhou X, Liu R. Density functional theory (DFT) study of enthalpy of formation. 1. Consistency of DFT energies and atom equivalents for converting DFT energies into enthalpies of formation. The Journal of Physical Chemistry. 1996;100:14665–71.

[140] Dewar MJ, Zoebisch EG, Healy EF, Stewart JJ. Development and use of quantum mechanical molecular models. 76. AM1: a new general purpose quantum mechanical molecular model. Journal of the American Chemical Society. 1985;107:3902–9.

[141] Stewart JJP. Optimization of parameters for semiempirical methods II. Applications, Journal of Computational Chemistry. 1989;10:221–64.

[142] Stewart JJP. Optimization of parameters for semiempirical methods V: modification of NDDO approximations and application to 70 elements. Journal of Molecular Modeling. 2007;13:1173–213.

[143] Stewart JJP. Optimization of parameters for semiempirical methods VI: more modifications to the NDDO approximations and re-optimization of parameters. Journal of Molecular Modeling. 2013;19:1–32.

[144] Chen C, Wu J. Correlations between theoretical and experimental determination of heat of formation of certain aromatic nitro compounds. Computers & Chemistry. 2001;25:117–24.

[145] Oftadeh M, Keshavarz MH, Khodadadi R. Prediction of the condensed phase enthalpy of formation of nitroaromatic compounds using the estimated gas phase enthalpies of formation by the PM3 and B3LYP methods. Central European Journal of Energetic Materials. 2014;11:143–56.

[146] Zheng P, Yang W, Wu W, Isayev O, Dral PO. Toward chemical accuracy in predicting enthalpies of formation with general-purpose data-driven methods. The Journal of Physical Chemistry Letters. 2022;13(15):3479–91.

[147] Westermayr J, Gastegger M, Schütt KT, Maurer RJ. Perspective on integrating machine learning into computational chemistry and materials science. The Journal of Chemical Physics. 2021;154(23).

[148] Collins EM, Raghavachari K. Effective molecular descriptors for chemical accuracy at dft cost: Fragmentation, error-cancellation, and machine learning. Journal of Chemical Theory and Computation. 2020;16(8):4938–50.

[149] Dandu N, Ward L, Assary RS, Redfern PC, Narayanan B, Foster IT, Curtiss LA. Quantum-chemically informed machine learning: prediction of energies of organic molecules with 10 to 14 non-hydrogen atoms. The Journal of Physical Chemistry A. 2020;124(28):5804–11.

[150] Collins EM, Raghavachari K. A fragmentation-based graph embedding framework for QM/ML. The Journal of Physical Chemistry A. 2021;125(31):6872–80.

[151] Spannaus A, Law KJ, Luszczek P, Nasrin F, Micucci CP, Liaw PK, Santodonato LJ, Keffer DJ, Maroulas V. Materials fingerprinting classification. Computer Physics Communications. 2021;266:108019.

[152] Zang X, Zhou X, Bian H, Jin W, Pan X, Jiang J, Koroleva MY, Shen R. Prediction and construction of energetic materials based on machine learning methods. Molecules. 2022;28(1):322.

[153] Wang X, He Y, Cao W, Guo W, Zhang T, Zhang J, Shu Q, Guo X, Liu R, Yao Y. Fast explosive performance prediction via small-dose energetic materials based on time-resolved imaging combined with machine learning. Journal of Materials Chemistry A. 2022;10(24):13114–23.

[154] Nguyen P, Loveland D, Kim JT, Karande P, Hiszpanski AM, Han TY-J. Predicting energetics materials' crystalline density from chemical structure by machine learning. Journal of Chemical Information and Modeling. 2021;61(5):2147–58.

[155] Chun S, Roy S, Nguyen YT, Choi JB, Udaykumar HS, Baek SS. Deep learning for synthetic microstructure generation in a materials-by-design framework for heterogeneous energetic materials. Scientific Reports. 2020;10(1):13307.

[156] Mathieu D. Molecular energies derived from deep learning: application to the prediction of formation enthalpies up to high energy compounds. Molecular Informatics. 2022;41(5):2100064.

[157] Smith JS, Nebgen BT, Zubatyuk R, Lubbers N, Devereux C, Barros K, Tretiak S, Isayev O, Roitberg AE. Approaching coupled cluster accuracy with a general-purpose neural network potential through transfer learning. Nature Communications. 2019;10(1):2903.

[158] Sheibani N. Applications of Predictive Modeling for Energetic Materials. In: Materials Informatics III: Polymers, Solvents and Energetic Materials. Springer, 2025. pp. 339–64.

[159] Keshavarz MH. Modeling the performance of energetic materials. Materials Informatics III: Polymers, Solvents and Energetic Materials. 2025:311–37.

[160] Chen C, Liu D, Deng S, Zhong L, Chan SHY, Li S, Hng HH. Accurate machine learning models based on small dataset of energetic materials through spatial matrix featurization methods. Journal of Energy Chemistry. 2021;63:364–75.

[161] Keshavarz MH, Tehrani MK, Pouretedal HR, Semnani A. New pathway for quick estimation of gas phase heat of formation of non-aromatic energetic compounds. Indian Journal of Engineering and Materials Sciences. 2006;13:542–8.

[162] Keshavarz MH, Tehrani MK. A new method for determining gas phase heat of formation of aromatic energetic compounds. Propellants, Explosives, Pyrotechnics. 2007;32:155–9.

[163] Gharagheizi F, Sattari M, Tirandazi B. Prediction of crystal lattice energy using enthalpy of sublimation: a group contribution-based model. Industrial & Engineering Chemistry Research. 2011;50:2482–6.

[164] Mathieu D. Simple alternative to neural networks for predicting sublimation enthalpies from fragment contributions. Industrial & Engineering Chemistry Research. 2012;51:2814–9.

[165] Meftahi N, Walker ML, Enciso M, Smith BJ. Predicting the enthalpy and Gibbs energy of sublimation by QSPR modeling. Scientific Reports. 2018;8:1–9.

[166] Suntsova MA, Dorofeeva OV. Prediction of enthalpies of sublimation of high-nitrogen energetic compounds: Modified Politzer model. Journal of Molecular Graphics and Modelling. 2017;72:220–8.

[167] Keshavarz MH, Yousefi MH. Heats of sublimation of nitramines based on simple parameters. Journal of Hazardous Materials. 2008;152:929–33.

[168] Keshavarz MH. Prediction of heats of sublimation of nitroaromatic compounds via their molecular structure. Journal of Hazardous Materials. 2008;151:499–506.

[169] Keshavarz MH. Improved prediction of heats of sublimation of energetic compounds using their molecular structure. Journal of Hazardous Materials. 2010;177:648–59.

[170] Keshavarz MH, Bashavard B, Goshadro A, Dehghan Z, Jafari M. Prediction of heats of sublimation of energetic compounds using their molecular structures. Journal of Thermal Analysis and Calorimetry. 2015;120:1941–51.

[171] Chickos JS, Gavezzotti A. Sublimation enthalpies of organic compounds: a very large database with a match to crystal structure determinations and a comparison with lattice energies. Crystal Growth & Design. 2019;19:6566–76.

[172] Atkins PW, De Paula J, Keeler J. Atkins' Physical Chemistry. Eleventh ed. Oxford university press; 2018.

[173] Byrd EFC, Rice BM. Improved prediction of heats of formation of energetic materials using quantum mechanical calculations. The Journal of Physical Chemistry A. 2006;110:1005–13.

[174] Ohlinger WS, Klunzinger PE, Deppmeier BJ, Hehre WJ. Efficient calculation of heats of formation. The Journal of Physical Chemistry A. 2009;113:2165–75.

[175] Politzer P, Murray JS. Quantitative treatments of solute/solvent interactions. Elsevier Science; 1994.

[176] Rice BM, Pai SV, Hare J. Predicting heats of formation of energetic materials using quantum mechanical calculations. Combustion and Flame. 1999;118:445–58.

[177] Rice BM, Hare J. Predicting heats of detonation using quantum mechanical calculations. Thermochimica Acta. 2002;384:377–91.

[178] Politzer P, Murray JS, Edward Grice M, Desalvo M, Miller E. Calculation of heats of sublimation and solid phase heats of formation. Molecular Physics. 1997;91:923–8.

[179] Pan Y, Zhu W, Xiao H. Molecular design on a new family of azaoxaadamantane cage compounds as potential high-energy density compounds. Canadian Journal of Chemistry. 2019;97:86–93.

[180] Khan RU, Zhu W. Designing and looking for novel low-sensitivity and high-energy cage derivatives based on the skeleton of nonanitro nonaaza pentadecane framework. Structural Chemistry. 2020;31:1387–402.

[181] Du M, Han T, Liu F, Wu H. Theoretical investigation of the structure, detonation properties, and stability of bicyclo [3.2.1] octane derivatives. Journal of Molecular Modeling. 2019;25:253.

[182] Duan B, Liu N, Wang B, Lu X, Mo H. Comparative theoretical studies on a series of novel energetic salts composed of 4,8-dihydrodifurazano [3,4-b,e] pyrazine-based anions and ammonium-based cations. Molecules. 2019;24:3213.

[183] Li X-H, Zhang C, Ju X-H. Theoretical screening of bistriazole-derived energetic salts with high energetic properties and low sensitivity. RSC advances. 2019;9:26442–9.

[184] Zhao S-X, Xia Q-Y, Zhang C, Xing X-L, Ju X-H. Theoretical design of bistetrazole diolate derivatives as novel non-nitro energetic salts with low sensitivity. Structural Chemistry. 2019;30:1015–22.

[185] de Oliveira RSS, Borges I. Correlation between molecular charge densities and sensitivity of nitrogen-rich heterocyclic nitroazole derivative explosives. Journal of Molecular Modeling. 2019;25:314.

[186] Jaidann M, Roy S, Abou-Rachid H, Lussier L-S. A DFT theoretical study of heats of formation and detonation properties of nitrogen-rich explosives. Journal of Hazardous Materials. 2010;176:165–73.

[187] Poling BE, Prausnitz JM, Paul OCJ, Reid RC. The Properties of Gases and Liquids. 5th ed. New York: McGraw-Hill; 2001.

[188] Argoub K, Benkouider AM, Yahiaoui A, Kessas R, Guella S, Bagui F. Prediction of standard enthalpy of formation in the solid state by a third-order group contribution method. Fluid Phase Equilibria. 2014;380:121–7.

[189] Salmon A, Dalmazzone D. Prediction of enthalpy of formation in the solid state (at 298.15 K) using second-order group contributions – Part 2: Carbon-hydrogen, carbon-hydrogen-oxygen, and carbon-hydrogen-nitrogen-oxygen compounds. Journal of Physical and Chemical Reference Data. 2007;36:19–58.

[190] Bourasseau S. A systematic procedure for estimating the standard heats of formation in the condensed state of non-aromatic polynitro-compounds. Journal of Energetic Materials. 1990;8:266–91.

[191] Jafari M, Keshavarz MH, Noorbala MR, Kamalvand M. A Reliable Method for Prediction of the Condensed Phase Enthalpy of Formation of High Nitrogen Content Materials through their Gas Phase Information. ChemistrySelect. 2016;1:5286–96.

[192] Keshavarz MH, Oftadeh M. New method for estimating the heat of formation of CHNO explosives in crystalline state. High Temperatures-High Pressures. 2004;36:499–504.

[193] Keshavarz MH, Sadeghi H. A new approach to predict the condensed phase heat of formation in acyclic and cyclic nitramines, nitrate esters and nitroaliphatic energetic compounds. Journal of Hazardous Materials. 2009;171:140–6.

[194] Keshavarz MH. Predicting condensed phase heat of formation of nitroaromatic compounds. Journal of Hazardous Materials. 2009;169:890–900.

[195] Keshavarz MH. Prediction of the condensed phase heat of formation of energetic compounds. Journal of Hazardous Materials. 2011;190:330–44.

[196] Keshavarz MH, Oftadeh M. New method for estimating the heat of formation of CHNO explosives in crystalline state. High Temperatures-High Pressures. 2004;35:499–504.

[197] Kamlet MJ, Jacobs SJ. The chemistry of detonation. 1. A simple method for calculating detonation properties of CHNO explosives. Journal of Chemical Physics. 1967;48:23–35.

[198] Keshavarz MH. A simple procedure for calculating condensed phase heat of formation of nitroaromatic energetic materials. Journal of Hazardous Materials. 2006;136:425–31.

[199] Afeefy H, Liebman J, Stein S, Linstrom P, Mallard W. NIST Chemistry WebBook, NIST Standard Reference Database Number 69, Linstrom PJ, Mallard WG, editors; 2011.

[200] Keshavarz MH. Theoretical prediction of condensed phase heat of formation of nitramines, nitrate esters, nitroaliphatics and related energetic compounds. Journal of Hazardous Materials. 2006;136:145–50.

[201] Nazari B, Keshavarz MH, Hamadanian M, Mosavi S, Ghaedsharafi AR, Pouretedal HR. Reliable prediction of the condensed (solid or liquid) phase enthalpy of formation of organic energetic materials at 298 K through their molecular structures. Fluid Phase Equilibria. 2016;408:248–58.

[202] Rice BM, Byrd EFC, Mattson WD. Computational aspects of nitrogen-rich HEDMs. In: High Energy Density Materials. Springer, 2007. pp. 153–94.

[203] Wei H, Zhang J, Shreeve Jn M. Synthesis, characterization, and energetic properties of 6-amino-tetrazolo [1,5-b]-1,2,4,5-tetrazine-7-N-oxide: A nitrogen-rich material with high density. Chemistry–An Asian Journal. 2015;10:1130–2.

[204] Chavez DE, Hiskey MA. 1,2,4,5-Tetrazine based energetic materials. Journal of Energetic Materials. 1999;17:357–77.

[205] Klapötke TM. The Synthesis Chemistry of Energetic Materials. In: Armstrong RW, editor. Energetics Science and Technology in Central Europe. Maryland: CALCE EPSC Press, 2012. pp. 57.

[206] Talawar MB, Sivabalan R, Asthana SN, Singh H. Novel ultrahigh-energy materials. Combustion, Explosion and Shock waves. 2005;41:264–77.

[207] Damse R, Ghosh M, Naik N, Sikder A. Thermoanalytical screening of nitrogen-rich compounds for ballistic requirements of gun propellant. Journal of Propulsion and Power. 2009;25:249–56.

[208] Nair U, Asthana S, Rao AS, Gandhe B. Advances in High Energy Materials (Review Paper). Defence Science Journal. 2010;60:137–51.

[209] Gao H, Shreeve Jn M. Azole-based energetic salts. Chemical Reviews. 2011;111:7377–436.

[210] Jafari M, Keshavarz MH. Simple approach for predicting the heats of formation of high nitrogen content materials. Fluid Phase Equilibria. 2016;415:166–75.

[211] Huynh MHV, Hiskey MA, Chavez DE, Naud DL, Gilardi RD. Synthesis, characterization, and energetic properties of diazido heteroaromatic high-nitrogen C–N compound. Journal of the American Chemical Society. 2005;127:12537–43.

[212] Williams MM, McEwan WS, Henry RA. The heats of combustion of substituted triazoles, tetrazoles and related high nitrogen compounds. The Journal of Physical Chemistry. 1957;61:261–7.

[213] Zhang Q, Shreeve Jn M. Energetic ionic liquids as explosives and propellant fuels: a new journey of ionic liquid chemistry. Chemical Reviews. 2014;114:10527–74.

[214] Sebastiao E, Cook C, Hu A, Murugesu M. Recent developments in the field of energetic ionic liquids. Journal of Materials Chemistry A. 2014;2:8153–73.

[215] Thomas E, Vijayalakshmi KP, George BK. Imidazolium based energetic ionic liquids for monopropellant applications: a theoretical study. RSC Advances. 2015;5:71896–902.

[216] Singh HJ, Mukherjee U. A computational approach to design energetic ionic liquids. Journal of Molecular Modeling. 2013;19:2317–27.

[217] Bhosale VK, Kulkarni PS. Hypergolic behavior of pyridinium salts containing cyanoborohydride and dicyanamide anions with oxidizer RFNA. Propellants, Explosives, Pyrotechnics. 2016;41:1013–9.

[218] Nimesh S, Ang HG. 1-(2H-tetrazolyl)-1,2,4-triazole-5-amine (TzTA) – a thermally stable nitrogen rich energetic material: synthesis, characterization and thermo-chemical analysis. Propellants, Explosives, Pyrotechnics. 2015;40:426–32.

[219] Keshavarz MH, Klapötke TM. Energetic Compounds: Methods for Prediction of Their Performance. 2nd ed. Walter de Gruyter GmbH & Co KG; 2020.

[220] Nirwan A, Ghule VD. Estimation of heats of formation for nitrogen-rich cations using G3, G4, and G4 (MP2) theoretical methods. Theoretical Chemistry Accounts. 2018;137:115.

[221] Gao H, Ye C, Piekarski CM, Shreeve JM. Computational characterization of energetic salts. The Journal of Physical Chemistry C. 2007;111:10718–31.

[222] Jenkins HDB, Tudela D, Glasser L. Lattice potential energy estimation for complex ionic salts from density measurements. Inorganic Chemistry. 2002;41:2364–7.

[223] Zhang X, Zhu W, Wei T, Zhang C, Xiao H. Densities, heats of formation, energetic properties, and thermodynamics of formation of energetic nitrogen-rich salts containing substituted protonated

and methylated tetrazole cations: A computational study. The Journal of Physical Chemistry C. 2010;114:13142–52.

[224] Keshavarz MH, Nazari B, Jafari M, Bakhtiari R. A simple approach for prediction of the condensed phase heat of formation of imidazolium-based ionic liquids or salts. ChemistrySelect. 2018;3:3505–10.

[225] Dong L-L, He L, Liu H-Y, Tao G-H, Nie F-D, Huang M, Hu C-W. Nitrogen-rich energetic ionic liquids based on the N,N-bis(1H-tetrazol-5-yl)amine anion – syntheses, structures, and properties. European Journal of Inorganic Chemistry. 2013;2013:5009–19.

[226] Jafari M, Davtalab M, Keshavarz MH, Esmaeilpour K, Mosaviazar A, Ghasemi MA, Amini M. Accurate prediction of the condensed phase (solid or liquid) heat of formation of triazolium-based energetic ionic salts at 298.15 K. Central European Journal of Energetic Materials. 2018;15:501–15.

[227] Xue H, Twamley B, Jeanne MS. Energetic salts of substituted 1,2,4-triazolium and tetrazolium 3,5-dinitro-1,2,4-triazolates. Journal of Materials Chemistry. 2005;15:3459–65.

[228] Berthod A, Ruiz-Ángel M, Carda-Broch S. Recent advances on ionic liquid uses in separation techniques. Journal of Chromatography A. 2018;1559:2–16.

[229] Al-Fakih A, Algamal Z, Lee M, Aziz M. A penalized quantitative structure–property relationship study on melting point of energetic carbocyclic nitroaromatic compounds using adaptive bridge penalty. SAR and QSAR in Environmental Research. 2018;29:339–53.

[230] Yagofarov MI, Solomonov BN. Calculation of the fusion enthalpy temperature dependence of polyaromatic hydrocarbons from the molecular structure: old and new approaches. The Journal of Chemical Thermodynamics. 2021;152:106278.

[231] Liu Y, Lai W, Yu T, Ma Y, Guo W, Ge Z. Melting point prediction of energetic materials via continuous heating simulation on solid-to-liquid phase transition. ACS Omega. 2019;4:4320–4.

[232] Ma Q, Zhang Z, Yang W, Li W, Ju J, Fan G. Strategies for constructing melt-castable energetic materials: A critical review. Energetic Materials Frontiers. 2021;2(1):69–85.

[233] Song S, Chen F, Wang Y, Wang K, Yan M, Zhang Q. Accelerating the discovery of energetic melt-castable materials by a high-throughput virtual screening and experimental approach. Journal of Materials Chemistry A. 2021;9(38):21723–31.

[234] Chen F, Song S, Wang Y, Liu Y, Zhang Q. Effects of alkyl chains on the physicochemical properties of nitroguanidine derivatives. Energetic Materials Frontiers. 2020;1(3–4):157–64.

[235] Anniyappan M, Vijay Varma K, Amit R, Nair J. 1-methyl-2, 4, 5-trinitroimidazole (MTNI), a melt-cast explosive: synthesis and studies on thermal behavior in presence of explosive ingredients. Journal of Energetic Materials. 2020;38(1):111–25.

[236] Hervé G, Roussel C, Graindorge H. Selective preparation of 3, 4, 5-trinitro-1H-pyrazole: A stable all-carbon-nitrated arene. Angewandte Chemie International Edition. 2010;49(18):3177–81.

[237] Sikder N, Sikder A, Bulakh N, Gandhe B. 1, 3, 3-Trinitroazetidine (TNAZ), a melt-cast explosive: synthesis, characterization and thermal behaviour. Journal of Hazardous Materials. 2004;113(1–3):35–43.

[238] Johnson EC, Sabatini JJ, Chavez DE, Sausa RC, Byrd EF, Wingard LA, Guzmàn PE. Bis (1, 2, 4-oxadiazole) bis (methylene) dinitrate: a high-energy melt-castable explosive and energetic propellant plasticizing ingredient. Organic Process Research & Development. 2018;22(6):736–40.

[239] Zhang J, Mitchell LA, Parrish DA, Shreeve Jn M. Enforced layer-by-layer stacking of energetic salts towards high-performance insensitive energetic materials. Journal of the American Chemical Society. 2015;137(33):10532–5.

[240] Manner VW, Cawkwell MJ, Kober EM, Myers TW, Brown GW, Tian H, Snyder CJ, Perriot R, Preston DN. Examining the chemical and structural properties that influence the sensitivity of energetic nitrate esters. Chemical Science. 2018;9(15):3649–63.

[241] McDonagh J, van Mourik T, Mitchell JB. Predicting melting points of organic molecules: Applications to aqueous solubility prediction using the general solubility equation. Molecular Informatics. 2015;34(11-12):715–24.

[242] Joback KG, Reid RC. Estimation of pure-component properties from group-contributions. Chemical Engineering Communications. 1987;57:233–43.

[243] Keshavarz MH, Pouretedal HR. New approach for predicting melting point of carbocyclic nitroaromatic compounds. Journal of Hazardous Materials. 2007;148:592–8.

[244] Lydersen A, Greenkorn RA, Hougen OA. Estimation of Critical Properties of Organic Compounds by the Method of Group Contributions. University of Wisconsin; 1955.

[245] Ambrose D. Correlation and Estimation of Vapour-Liquid Critical Properties. National Physical Library; 1978.

[246] Klincewicz KM, Reid RC. Estimation of critical properties with group contribution methods. AIChE Journal. 1984;30:137–42.

[247] Lyman WJ, Reehl WF, Rosenblatt DH, Rosenblatt DH. Handbook of Chemical Property Estimation Methods: Environmental Behavior of Organic Compounds. New York: McGraw-Hill; 1982.

[248] Constantinou L, Prickett SE, Mavrovouniotis ML. Estimation of thermodynamic and physical properties of acyclic hydrocarbons using the ABC approach and conjugation operators. Industrial & Engineering Chemistry Research. 1993;32:1734–46.

[249] Prickett SE, Constantinou L, Mavrovouniotis ML. Computational identification of conjugate paths for estimation of properties of organic compounds. Molecular Simulation. 1993;11:205–28.

[250] Constantinou L, Gani R. New group contribution method for estimating properties of pure compounds. AIChE Journal. 1994;40:1697–710.

[251] Constantinou L, Prickett SE, Mavrovouniotis ML. Estimation of properties of acyclic organic compounds using conjugation operators. Industrial & Engineering Chemistry Research. 1994;33:395–402.

[252] Marrero-Morejón J, Pardillo-Fontdevila E. Estimation of pure compound properties using group-interaction contributions. AIChE journal. 1999;45:615–21.

[253] Marrero J, Gani R. Group-contribution based estimation of pure component properties. Fluid Phase Equilibria. 2001;183:183–208.

[254] Simamora P, Yalkowsky SH. Group contribution methods for predicting the melting points and boiling points of aromatic compounds. Industrial & Engineering Chemistry Research. 1994;33:1405–9.

[255] Krzyzaniak JF, Myrdal PB, Simamora P, Yalkowsky SH. Boiling point and melting point prediction for aliphatic, non-hydrogen-bonding compounds. Industrial & Engineering Chemistry Research. 1995;34:2530–5.

[256] Alamdari RF, Keshavarz MH. A simple method to predict melting points of non-aromatic energetic compounds. Fluid Phase Equilibria. 2010;292:1–6.

[257] Agrawal PM, Rice BM, Thompson DL. Molecular dynamics study of the melting of nitromethane. The Journal of Chemical Physics. 2003;119:9617–27.

[258] Alavi S, Thompson DL. Simulations of the solid, liquid, and melting of 1-n-butyl-4-amino-1,2,4-triazolium bromide. The Journal of Physical Chemistry B. 2005;109:18127–34.

[259] Agrawal PM, Rice BM, Zheng L, Velardez GF, Thompson DL. Molecular dynamics simulations of the melting of 1,3,3-trinitroazetidine. Journal of Physical Chemistry B. 2006;110:5721.

[260] Siavosh-Haghighi A, Thompson DL. Melting point determination from solid–liquid coexistence initiated by surface melting. The Journal of Physical Chemistry C. 2007;111:7980–5.

[261] Jain A, Yalkowsky SH. Estimation of melting points of organic compounds-II. Journal of Pharmaceutical Sciences. 2006;95:2562–618.

[262] Jain A, Yalkowsky SH. Comparison of two methods for estimation of melting points of organic compounds. Industrial & Engineering Chemistry Research. 2007;46:2589–92.

[263] Evans DC, Yalkowsky SH. A simplified prediction of entropy of melting for energetic compounds. Fluid Phase Equilibria. 2011;303:10–4.

[264] Hilal SH, Carreira LA, Karickhoff SW. Estimation of chemical reactivity parameters and physical properties of organic molecules using SPARC. In: Murry PPaJS, editor. Theoretical and Computational Chemistry. Amsterdam: Elsevier, 1994. pp. 291–353.

[265] Whiteside T, Hilal S, Brenner A, Carreira L. Estimating the melting point, entropy of fusion, and enthalpy of fusion of organic compounds via SPARC. SAR and QSAR in Environmental Research. 2016;27:677–701.

[266] Admire B, Lian B, Yalkowsky SH. Estimating the physicochemical properties of polyhalogenated aromatic and aliphatic compounds using UPPER: Part 1. Boiling point and melting point. Chemosphere. 2015;119:1436–40.

[267] Lian B, Yalkowsky SH. Unified physicochemical property estimation relationships (UPPER). Journal of Pharmaceutical Sciences. 2014;103:2710–23.

[268] Admire B, Lian B, Yalkowsky SH. Estimating the physicochemical properties of polyhalogenated aromatic and aliphatic compounds using UPPER: Part 2. Aqueous solubility, octanol solubility and octanol–water partition coefficient. Chemosphere. 2015;119:1441–6.

[269] Alantary D, Yalkowsky SH. Estimating the physicochemical properties of polysubstituted aromatic compounds using UPPER. Journal of Pharmaceutical Sciences. 2018;107:297–306.

[270] Coley CW, Barzilay R, Green WH, Jaakkola TS, Jensen KF. Convolutional embedding of attributed molecular graphs for physical property prediction. Journal of Chemical Information and Modeling. 2017;57(8):1757–72.

[271] Mi W, Chen H, Zhu DA, Zhang T, Qian F. Melting point prediction of organic molecules by deciphering the chemical structure into a natural language. Chemical Communications. 2021;57(21):2633–6.

[272] Tetko IV, M Lowe D, Williams AJ. The development of models to predict melting and pyrolysis point data associated with several hundred thousand compounds mined from PATENTS. Journal of Cheminformatics. 2016;8:1–18.

[273] Young T, Hazarika D, Poria S, Cambria E. Recent trends in deep learning based natural language processing. IEEE Computational Intelligence Magazine. 2018;13(3):55–75.

[274] Zheng S, Yan X, Yang Y, Xu J. Identifying structure–property relationships through SMILES syntax analysis with self-attention mechanism. Journal of Chemical Information and Modeling. 2019;59(2):914–23.

[275] Sivaraman G, Jackson NE, Sanchez-Lengeling B, Vázquez-Mayagoitia Á, Aspuru-Guzik A, Vishwanath V, De Pablo JJ. A machine learning workflow for molecular analysis: application to melting points. Machine Learning: Science and Technology. 2020;1(2):025015.

[276] Kearnes S, McCloskey K, Berndl M, Pande V, Riley P. Molecular graph convolutions: moving beyond fingerprints. Journal of Computer-Aided Molecular Design. 2016;30:595–608.

[277] Galeazzo T, Shiraiwa M. Predicting glass transition temperature and melting point of organic compounds via machine learning and molecular embeddings. Environmental Science: Atmospheres. 2022;2(3):362–74.

[278] Song S, Wang Y, Chen F, Yan M, Zhang Q. Machine learning-assisted high-throughput virtual screening for on-demand customization of advanced energetic materials. Engineering. 2022;10:99–109.

[279] Fung V, Zhang J, Juarez E, Sumpter BG. Benchmarking graph neural networks for materials chemistry. npj Computational Materials. 2021;7(1):84.

[280] Jiang Y, Yang Z, Guo J, Li H, Liu Y, Guo Y, Li M, Pu X. Coupling complementary strategy to flexible graph neural network for quick discovery of coformer in diverse co-crystal materials. Nature Communications. 2021;12(1):5950.

[281] Song S, Wang Y, Tian X, He W, Chen F, Wu J, Zhang Q. Predicting the Melting Point of Energetic Molecules Using a Learnable Graph Neural Fingerprint Model. The Journal of Physical Chemistry A. 2023;127(19):4328–37.

[282] Brown RJC, Brown RFC. Melting point and molecular symmetry. Journal of Chemical Education. 2000;77:724.

[283] Carnelley T. XIII. Chemical symmetry, or the influence of atomic arrangement on the physical properties of compounds. The London, Edinburgh, and Dublin Philosophical Magazine and Journal of Science. 1882;13:112–30.

[284] Keshavarz MH. New method for predicting melting points of polynitro arene and polynitro heteroarene compounds. Journal of Hazardous Materials. 2009;171:786–96.

[285] Agrawal JP. Recent trends in high-energy materials. Progress in Energy and Combustion Science. 1998;24:1–30.

[286] Keshavarz MH. Approximate prediction of melting point of nitramines, nitrate esters, nitrate salts and nitroaliphatics energetic compounds. Journal of Hazardous Materials. 2006;138:448–51.

[287] Keshavarz MH, Gharagheizi F, Pouretedal HR. Improved reliable approach to predict melting points of energetic compounds. Fluid Phase Equilibria. 2011;308:114–28.

[288] Khozani MH, Keshavarz MH, Nazari B, Mohebbi M. Simple approach for prediction of melting points of organic molecules containing hazardous peroxide bonds. Journal of the Iranian Chemical Society. 2015;12:587–98.

[289] Nazari B, Hamadanian M, Keshavarz MH, Rezaei J. New method for assessment of melting points of organic azides using their molecular structures. Fluid Phase Equilibria. 2016;427:27–34.

[290] Hodgson HH, Norris W. Replacement of the diazonium by the azido group in acid solution. Journal of the Chemical Society (Resumed). 1949;162:762–3.

[291] Hamadanian M, Keshavarz MH, Nazari B, Mohebbi M. Reliable method for safety assessment of melting points of energetic compounds. Process Safety and Environmental Protection. 2016;103:10–22.

[292] Keshavarz MH, Maghsoodi NK, Shokrollahi A. A reliable model for assessment of melting points of cyclic hydrocarbons containing complex molecular structures, isomers and stereoisomers. Fluid Phase Equilibria. 2020;521:112692.

[293] Singh SK, Savoy AW. Ionic liquids synthesis and applications: An overview. Journal of Molecular Liquids. 2020;297:112038.

[294] Bera A, Agarwal J, Shah M, Shah S, Vij RK. Recent advances in ionic liquids as alternative to surfactants/chemicals for application in upstream oil industry. Journal of Industrial and Engineering Chemistry. 2020;82:17–30.

[295] Nasirpour N, Mohammadpourfard M, Heris SZ. Ionic liquids: Promising compounds for sustainable chemical processes and applications. Chemical Engineering Research and Design. 2020;160:264–300.

[296] Qin M, Zhong F, Sun Y, Tan X, Hu K, Zhang H, Kong M, Wang G, Zhuang L. Effect of cation substituent of dodecanesulfate-based anionic surface active ionic liquids on micellization: Experimental and theoretical studies. Journal of Molecular Liquids. 2020;303:112695.

[297] Low K, Kobayashi R, Izgorodina EI. The effect of descriptor choice in machine learning models for ionic liquid melting point prediction. The Journal of Chemical Physics. 2020;153:104101.

[298] Venkatraman V, Evjen S, Knuutila HK, Fiksdahl A, Alsberg BK. Predicting ionic liquid melting points using machine learning. Journal of Molecular Liquids. 2018;264:318–26.

[299] Chen L, Bryantsev VS. A density functional theory based approach for predicting melting points of ionic liquids. Physical Chemistry Chemical Physics. 2017;19:4114–24.

[300] Mehrkesh A, Karunanithi AT. New quantum chemistry-based descriptors for better prediction of melting point and viscosity of ionic liquids. Fluid Phase Equilibria. 2016;427:498–503.

[301] Sarker IH. Deep learning: a comprehensive overview on techniques, taxonomy, applications and research directions. SN Computer Science. 2021;2(6):1–20.

[302] Acar Z, Nguyen P, Lau KC. Machine-learning model prediction of ionic liquids melting points. Applied Sciences. 2022;12(5):2408.

[303] Valderrama JO, Cardona LF. Predicting the melting temperature and the heat of melting of ionic liquids. Journal of Ionic Liquids. 2021;1(1):100002.

[304] Lazzús JA. A group contribution method for predicting the freezing point of ionic liquids. Periodica Polytechnica Chemical Engineering. 2016;60:273–81.

[305] Zhang S, Lu X, Zhou Q, Li X, Zhang X, Li S. Ionic Liquids: Physicochemical Properties. Elsevier; 2009.

[306] Mital DK, Nancarrow P, Zeinab S, Jabbar NA, Ibrahim TH, Khamis MI, Taha A. Group contribution estimation of ionic liquid melting points: critical evaluation and refinement of existing models. Molecules. 2021;26(9):2454.

[307] Dzyuba SV, Bartsch RA. Influence of structural variations in 1-alkyl (aralkyl)-3-methylimidazolium hexafluorophosphates and bis (trifluoromethylsulfonyl) imides on physical properties of the ionic liquids. ChemPhysChem. 2002;3(2):161–6.

[308] Todeschini R, Consonni V. Handbook of Molecular Descriptors. John Wiley & Sons; 2008.

[309] Wang D, Yuan Y, Duan S, Liu R, Gu S, Zhao S, Liu L, Xu J. QSPR study on melting point of carbocyclic nitroaromatic compounds by multiple linear regression and artificial neural network. Chemometrics and Intelligent Laboratory Systems. 2015;143:7–15.

[310] Liu Y, Holder AJ. A quantum mechanical quantitative structure–property relationship study of the melting point of a variety of organosilicons. Journal of Molecular Graphics and Modelling. 2011;31:57–64.

[311] Liang G, Xu J, Liu L. QSPR analysis for melting point of fatty acids using genetic algorithm based multiple linear regression (GA-MLR). Fluid Phase Equilibria. 2013;353:15–21.

[312] Yan F, Xia S, Wang Q, Yang Z, Ma P. Predicting the melting points of ionic liquids by the Quantitative Structure Property Relationship method using a topological index. The Journal of Chemical Thermodynamics. 2013;62:196–200.

[313] Watkins M, Sizochenko N, Rasulev B, Leszczynski J. Estimation of melting points of large set of persistent organic pollutants utilizing QSPR approach. Journal of Molecular Modeling. 2016;22:55.

[314] Morrill JA, Byrd EF. Development of quantitative structure property relationships for predicting the melting point of energetic materials. Journal of Molecular Graphics and Modelling. 2015;62:190–201.

[315] Farahani N, Gharagheizi F, Mirkhani SA, Tumba K. Ionic liquids: Prediction of melting point by molecular-based model. Thermochimica Acta. 2012;549:17–34.

[316] Paduszyński K, Kłębowski K, Królikowska M. Predicting melting point of ionic liquids using QSPR approach: Literature review and new models. Journal of Molecular Liquids. 2021;344:117631.

[317] Yan F, Xia S, Wang Q, Yang Z, Ma P. Predicting the melting points of ionic liquids by the Quantitative Structure Property Relationship method using a topological index. The Journal of Chemical Thermodynamics. 2013;62:196–200.

[318] Niu H, Zhang Y, Jia Q, Wang Q, Yan F. Property estimation of organic compounds based on QSPR models with norm indices. Chemical Engineering Science. 2024;288:119835.

[319] Lotfi S, Ahmadi S, Kumar P. The Monte Carlo approach to model and predict the melting point of imidazolium ionic liquids using hybrid optimal descriptors. RSC advances. 2021;11(54):33849–57.

[320] Krossing I, Slattery JM, Daguenet C, Dyson PJ, Oleinikova A, Weingärtner H. Why are ionic liquids liquid? A simple explanation based on lattice and solvation energies. Journal of the American Chemical Society. 2006;128(41):13427–34.

[321] Liu X, Yin J, Zhang X, Qiu W, Jiang W, Zhang M, Zhu L, Li H, Li H. Rapid and accurate prediction of the melting point for imidazolium-based ionic liquids by artificial neural network. Chemistry. 2024;6(6):1552–71.

[322] Dai Z, Wang L, Lu X, Ji X. Melting points of ionic liquids: Review and evaluation. Green Energy & Environment. 2024;9(12):1802–11.

[323] Keshavarz MH, Pouretedal HR, Saberi E. A novel method for predicting melting point of ionic liquids. Process Safety and Environmental Protection. 2018;116:333–9.

[324] Aguirre CL, Cisternas LA, Valderrama JO. Melting-point estimation of ionic liquids by a group contribution method. International Journal of Thermophysics. 2012;33:34–46.

[325] Tarver CM, Chidester SK. On the Violence of High Explosive Reactions. Journal of Pressure Vessel Technology. 2005;127:39–48.

[326] Zeman S. Some predictions in the field of the physical thermal stability of nitramines. Thermochimica Acta. 1997;302:11–6.

[327] Zeman S. Sensitivities of High Energy Compounds. In: Klapötke TM, editor. High Energy Density Materials. Berlin Heidelberg: Springer, 2007. pp. 195–271.

[328] Zeman S, Krupka M. New aspects of impact reactivity of polynitro compounds, Part II. Impact sensitivity as "the First Reaction" of polynitro arenes. Propellants, Explosives, Pyrotechnics. 2003;28:249–55.

[329] Zeman S, Krupka M. New aspects of impact reactivity of polynitro compounds, Part III. Impact sensitivity as a function of the imtermolecular interactions. Propellants, Explosives, Pyrotechnics. 2003;28:301–7.

[330] Zeman S. Study of the Initiation Reactivity of Energetic Materials. In: Armstrong R, editor. Energetics Science and Technology in Central Europe. Maryland: CALCE EPSC Press, 2012. pp. 131–67.

[331] Goodarzi M, Chen T, Freitas MP. QSPR predictions of heat of fusion of organic compounds using Bayesian regularized artificial neural networks. Chemometrics and Intelligent Laboratory Systems. 2010;104:260–4.

[332] Atalar T, Zeman S. A new view of relationships of the N–N bond dissociation energies of cyclic nitramines. Part I. Relationships with heats of fusion. Journal of Energetic Materials. 2009;27:186–99.

[333] Chickos JS, Acree WE Jr, Liebman JF. Estimating solid–liquid phase change enthalpies and entropies. Journal of Physical and Chemical Reference Data. 1999;28:1535–673.

[334] Yu J, Sumathi R, Green WH. Accurate and efficient method for predicting thermochemistry of polycyclic aromatic hydrocarbons-bond-centered group additivity. Journal of the American Chemical Society. 2004;126:12685–700.

[335] Gharagheizi F, Salehi GR. Prediction of enthalpy of fusion of pure compounds using an artificial neural network-group contribution method. Thermochimica Acta. 2011;521:37–40.

[336] Mosaei Oskoei Y, Keshavarz MH. Improved method for reliable predicting enthalpy of fusion of energetic compounds. Fluid Phase Equilibria. 2012;326:1–14.

[337] Keshavarz MH. Prediction of enthalpy of fusion of non-aromatic energetic compounds containing nitramine, nitrate and nitro functional groups. Propellants, Explosives, Pyrotechnics. 2011;36:42–7.

[338] Keshavarz MH. A simple correlation for predicting heats of fusion of nitroaromatic carbocyclic energetic compounds. Journal of Hazardous Materials. 2008;150:387–93.

[339] Keshavarz MH. Predicting heats of fusion of nitramines. Indian Journal of Engineering and Materials Sciences. 2007;14:386–90.

[340] Semnani A, Keshavarz MH. Using molecular structure for reliable predicting enthalpy of melting of nitroaromatic energetic compounds. Journal of Hazardous Materials. 2010;178:264–72.

[341] Keshavarz MH. A new computer code for prediction of enthalpy of fusion and melting point of energetic materials. Propellants, Explosives, Pyrotechnics. 2015;40:150–5.

[342] Keshavarz MH, Akbarzadeh AR, Rahimi R, Jafari M, Pasandideh M, Sadeghi R. A reliable method for prediction of enthalpy of fusion in energetic materials using their molecular structures. Fluid Phase Equilibria. 2016;427:46–55.

[343] Ognichenko LN, Kuz'min VE, Gorb L, Muratov EN, Artemenko AG, Kovdienko NA, Polishchuk PG, Hill FC, Leszczynski J. New Advances in QSPR/QSAR Analysis of Nitrocompounds: Solubility, Lipophilicity, and Toxicity. In: Leszczynski J, Shukla MK, editors. Practical Aspects of Computational Chemistry II. Dordrecht, The Netherlands: Springer, 2012. pp. 279–334.

[344] Jain A, Yang G, Yalkowsky SH. Estimation of melting points of organic compounds. Industrial & Engineering Chemistry Research. 2004;43:7618–21.

[345] Yalkowsky SH. Carnelley's rule and the prediction of melting point. Journal of Pharmaceutical Sciences. 2014;103:2629–34.

[346] Naef R, Acree WE. Calculation of five thermodynamic molecular descriptors by means of a general computer algorithm based on the group-additivity method: standard enthalpies of vaporization, sublimation and solvation, and entropy of fusion of ordinary organic molecules and total phase-change entropy of liquid crystals. Molecules. 2017;22:1059.

[347] Alnemrat S, Hooper JP. Predicting Temperature-Dependent Solid Vapor Pressures of Explosives and Related Compounds Using a Quantum Mechanical Continuum Solvation Model. The Journal of Physical Chemistry A. 2013;117:2035–43.

[348] Politzer P, Ma Y, Lane P, Concha MC. Computational prediction of standard gas, liquid, and solid-phase heats of formation and heats of vaporization and sublimation. International Journal of Quantum Chemistry. 2005;105:341–7.

[349] Yan F, Zhang Y, Niu H, Feng X, Xiong J, Jia Q, Xia S, Wang Q. QSPR models for enthalpy and entropy of organic compounds based on a set of norm indices. Fluid Phase Equilibria. 2023;573:113869.

[350] Keshavarz MH, Maghsoodi NK, Shokrollahi A. An improved simple correlation for reliable prediction of the enthalpy of fusion of cyclic and acyclic hydrocarbons including different types of saturated and unsaturated aliphatic hydrocarbons. Fluid Phase Equilibria. 2020;525:112813.

[351] Haynes WM. CRC Handbook of Chemistry and Physics. CRC press; 2016.

[352] Keshavarz MH, Pouretedal HR. A new simple approach to predict entropy of fusion of nitroaromatic compounds. Fluid Phase Equilibria. 2010;298:24–32.

[353] Keshavarz MH, Zakinejad S, Esmailpour K. An improved simple method for prediction of entropy of fusion of energetic compounds. Fluid Phase Equilibria. 2013;340:52–62.

[354] Jain A, Yang G, Yalkowsky SH. Estimation of total entropy of melting of organic compounds. Industrial & Engineering Chemistry Research. 2004;43:4376–9.

[355] Dannenfelser R, Surendran N, Yalkowsky SH. Molecular symmetry and related properties. SAR and QSAR in Environmental Research. 1993;1:273–92.

[356] Dannenfelser R-M, Yalkowsky SH. Estimation of entropy of melting from molecular structure: A non-group contribution method. Industrial & Engineering Chemistry Research. 1996;35:1483–6.

[357] Yalkowsky SH, Alantary D. Estimation of melting points of organics. Journal of Pharmaceutical Sciences. 2018;107:1211–27.

[358] Coutinho JA, Carvalho PJ, Oliveira NM. Predictive methods for the estimation of thermophysical properties of ionic liquids. RSC Advances. 2012;2(19):7322–46.

[359] Valderrama JO, Cardona LF. Predicting the melting temperature and the heat of melting of ionic liquids. Journal of Ionic Liquids. 2021;1(1):100002.

[360] Cervinka C, Fulem M. State-of-the-art calculations of sublimation enthalpies for selected molecular crystals and their computational uncertainty. Journal of Chemical Theory and Computation. 2017;13(6):2840–50.

[361] Campbell CT, Sellers JR. Enthalpies and entropies of adsorption on well-defined oxide surfaces: Experimental measurements. Chemical Reviews. 2013;113(6):4106–35.

[362] Yurata T, Lei H, Tang L, Lu M, Patel J, Lim S, Piumsomboon P, Chalermsinsuwan B, Li Ce. Feasibility and sustainability analyses of carbon dioxide–hydrogen separation via de-sublimation process in comparison with other processes. International Journal of Hydrogen Energy. 2019;44(41):23120–34.

[363] Gharagheizi F, Sattari M, Tirandazi B. Prediction of crystal lattice energy using enthalpy of sublimation: a group contribution-based model. Industrial & Engineering Chemistry Research. 2011;50(4):2482–6.

[364] Foltz M. Aging of pentaerythritol tetranitrate (PETN), LLNL-TR-415057, Lawrence Livermore National Laboratory, Livermore; 2009.

[365] Bhattacharia SK, Maiti A, Gee RH, Weeks BL. Sublimation properties of pentaerythritol tetranitrate single crystals doped with its homologs. Propellants, Explosives, Pyrotechnics. 2012;37:563–8.

[366] Hikal WM, Weeks BL. Sublimation kinetics and diffusion coefficients of TNT, PETN, and RDX in air by thermogravimetry. Talanta. 2014;125:24`8.

[367] Bhattacharia S. Kinetics of Energetic Materials: Investigation of Sublimation and Decomposition. Chemistry Department, Texas Tech University; 2013. p. 120.

[368] Atkins PW, De Paula J. Physical chemistry. 9th ed. Oxford: Oxford University Press; 2010.

[369] Fried LE, Howard WM, Souers PC. CHEETAH 2.0 User's Manual (LLNL UCRL-MA-117541 Rev. 5), Lawrence Livermore National Laboratory, Livermore, CA; 1998.

[370] Mader CL. Numerical Modeling of Explosives and Propellants. 3th ed. Boca Raton: Taylor & Francis Group; 2008.

[371] Muthurajan H, Sivabalan R, Talawar MB, Asthana SN. Computer simulation for prediction of performance and thermodynamic parameters of high energy materials. Journal of Hazardous Materials. 2004;112:17–33.

[372] Chickos JS, Gavezzotti A. Sublimation enthalpies of organic compounds: a very large database with a match to crystal structure determinations and a comparison with lattice energies. Crystal Growth & Design. 2019;19(11):6566–76.

[373] Růžička K, Fulem M, Červinka C. Recommended sublimation pressure and enthalpy of benzene. The Journal of Chemical Thermodynamics. 2014;68:40–7.

[374] Růžička Kt, Koutek B, Fulem M, Hoskovec M. Indirect determination of vapor pressures by capillary gas–liquid chromatography: analysis of the reference vapor-pressure data and their treatment. Journal of Chemical & Engineering Data. 2012;57(5):1349–68.

[375] Curtiss LA, Redfern PC, Frurip DJ. Theoretical methods for computing enthalpies of formation of gaseous compounds. In: Lipkowitz KB, Boyd DB, editors. Reviews in Computational Chemistry. New York: Wiley, 2000. pp. 147–211.

[376] Motalov VB, Korobov MA, Dunaev AM, Dunaeva VV, Tyunina EY, Kudin LS. Refined data on the sublimation enthalpy and thermodynamic functions of L-and DL-methionine. Journal of Chemical & Engineering Data. 2022;67(6):1326–34.

[377] Sahu H, Shen K-H, Montoya JH, Tran H, Ramprasad R. Polymer structure predictor (PSP): a python toolkit for predicting atomic-level structural models for a range of polymer geometries. Journal of Chemical Theory and Computation. 2022;18(4):2737–48.

[378] Hu A, Larade B, Dudiy S, Abou-Rachid H, Lussier L-S, Guo H. Theoretical prediction of heats of sublimation of energetic materials using pseudo-atomic orbital density functional theory calculations. Propellants, Explosives, Pyrotechnics. 2007;32:331–7.

[379] Ghosh MK, Cho SG, Choi CH. A priori prediction of heats of vaporization and sublimation by EFP2-MD. The Journal of Physical Chemistry B. 2014;118:4876–82.

[380] Salahinejad M, Le TC, Winkler DA. Capturing the crystal: prediction of enthalpy of sublimation, crystal lattice energy, and melting points of organic compounds. Journal of Chemical Information and Modeling. 2013;53:223–9.

[381] Zeman S, Krupka M. Some predictions of the heats of fusion, heats of sublimation and lattice energies of energetic materials. Chinese Journal of Energetic Materials (HanNeng CaiLiao). 2002;10:27–33.

[382] Liu R, Tang Y, Tian J, Huang J, Zhang C, Wang L, Liu J. QSPR models for sublimation enthalpy of energetic compounds. Chemical Engineering Journal. 2023;474:145725.

[383] Wilding WV, Rowley RL, Oscarson JL. DIPPR® Project 801 evaluated process design data. Fluid Phase Equilibria. 1998;150:413–20.

[384] Smola AJ, Schölkopf B. A tutorial on support vector regression. Statistics and Computing. 2004;14:199–222.

[385] Breiman L. Random forests. Machine Learning. 2001;45:5–32.

[386] Friedman JH. Greedy function approximation: a gradient boosting machine. Annals of Statistics. 2001:1189–232.

[387] Taherkhani M, Safabakhsh R. A novel stability-based adaptive inertia weight for particle swarm optimization. Applied Soft Computing. 2016;38:281–95.

[388] Wahler S, Chung P, Klapötke TM. Training machine learning models based on the structural formula for the enthalpy of vaporization and sublimation and a thorough analysis of Trouton's rules. Journal of Energetic Materials. 2025;43(2):199–211.

[389] Liu Y, Tran H, Huang C, del Rio BG, Joseph VR, Losego M, Ramprasad R. Accelerated predictions of the sublimation enthalpy of organic materials with machine learning. Materials Genome Engineering Advances. 2025:e84.

[390] Cundall RB, Palmer TF, Wood CE. Vapour pressure measurements on some organic high explosives. Journal of the Chemical Society, Faraday Transactions 1: Physical Chemistry in Condensed Phases. 1978;74:1339–45.

[391] Zeman S. New aspects of initiation reactivities of energetic materials demonstrated on nitramines. Journal of Hazardous Materials. 2006;132:155–64.

[392] Keshavarz MH, Shokrolahi A, Esmailpoor K, Zali A, Hafizi HR, Azamiamehraban J. Recent developments in predicting impact and shock sensitivities of energetic materials. Chinese Journal of Energetic Materials (HanNeng CaiLiao). 2008;16:113–20.

[393] Türker L. Recent Developments in the Theory of Explosive Materials. In: Janssen TJ, editor. Explosive Materials: Classification, Composition and Properties. New York: Nova Publisher, 2011. pp. 1–52.

[394] Keshavarz MH. Important aspects of sensitivity of energetic compounds: a simple novel approach to predict electric spark sensitivity, Explosive materials: classification, composition and properties. New York: Nova Science Publishers; 2011. p. 103–23.

[395] Yan QL, Zeman S. Theoretical evaluation of sensitivity and thermal stability for high explosives based on quantum chemistry methods: a brief review. International Journal of Quantum Chemistry. 2013;113:1049–61.

[396] Zeman S, Jungová M. Sensitivity and performance of energetic materials. Propellants, Explosives, Pyrotechnics. 2016;41:426–51.

[397] Li G, Zhang C. Review of the molecular and crystal correlations on sensitivities of energetic materials. Journal of Hazardous Materials. 2020;398:122910.

[398] Politzer P, Murray JS. Impact sensitivity and the maximum heat of detonation. Journal of molecular modeling. 2015;21:1–11.

[399] Politzer P, Murray JS. Impact sensitivity and crystal lattice compressibility/free space. Journal of molecular modeling. 2014;20:1–8.

[400] Pospíšil M, Vávra P, Concha MC, Murray JS, Politzer P. Sensitivity and the available free space per molecule in the unit cell. Journal of Molecular Modeling. 2011;17:2569–74.

[401] Jungová M, Zeman S, Yan Q-L. Recent Advances in the Study of the Initiation of Nitramines by Impact Using Their 15N NMR Chemical Shifts. Central European Journal of Energetic Materials. 2014;11.

[402] Zhou Y, Du J-L, Long X-P, Shu Y-J. Impact sensitivity and nucleus-independent chemical shift for aromatic explosives. Molecular Simulation. 2013;39:716–20.

[403] Zohari N, Keshavarz MH, Seyedsadjadi SA. A link between impact sensitivity of energetic compounds and their activation energies of thermal decomposition. Journal of Thermal Analysis and Calorimetry. 2014;1–10.

[404] Keshavarz MH. A new general correlation for predicting impact sensitivity of energetic compounds. Propellants, Explosives, Pyrotechnics. 2013;38:754–60.

[405] Fang-Qiang Y, Cong-Zhong C, Shuai Z. Prediction of impact sensitivity of nitro energetic compounds by using structural parameters. Explosion and Shock Waves. 2013;1:012.

[406] Keshavarz MH, Motamedoshariati H, Moghayadnia R, Ghanbarzadeh M, Azarniamehraban J. Prediction of sensitivity of energetic compounds with a new computer code. Propellants, Explosives, Pyrotechnics. 2014;39:95–101.

[407] Kim H, Hwang S-N, Lee SK. QSPR analysis for the prediction of impact sensitivity of High Energy Density Materials (HEDM). Chinese Journal of Energetic Materials (HanNeng CaiLiao). 2012;275.

[408] Fayet G, Rotureau P. Development of simple QSPR models for the impact sensitivity of nitramines. Journal of Loss Prevention in the Process Industries. 2014;30:1–8.

[409] Prana V, Fayet G, Rotureau P, Adamo C. Development of validated QSPR models for impact sensitivity of nitroaliphatic compounds. Journal of Hazardous Materials. 2012;235:169–77.

[410] Fayet G, Rotureau P, Prana V, Adamo C. Global and local QSPR models to predict the impact sensitivity of nitro compounds. In: AIChE Spring Meeting 2012 & 8. Global Congress on Process Safety (GCPS). New York: AIChE, 2012.

[411] Fayet G, Rotureau P, Prana V, Adamo C. Global and local quantitative structure–property relationship models to predict the impact sensitivity of nitro compounds. Process Safety Progress. 2012;31:291–303.

[412] Xu J, Zhu L, Fang D, Wang L, Xiao S, Liu L, Xu W. QSPR studies of impact sensitivity of nitro energetic compounds using three-dimensional descriptors. Journal of Molecular Graphics and Modelling. 2012;36:10–9.

[413] Wang R, Wang YG, Liu H. Predicting Impact Sensitivity of Heterocyclic Nitroarenes from Molecular Structures Selected by Genetic Algorithm. In: Advanced Materials Research. Trans Tech Publ, 2012. pp. 2550–3.

[414] Wang R, Jiang J, Pan Y. Prediction of impact sensitivity of nonheterocyclic nitroenergetic compounds using genetic algorithm and artificial neural network. Journal of Energetic Materials. 2012;30:135–55.

[415] Owens FJ, Jayasuriya K, Abrahmsen L, Politzer P. Computational analysis of some properties associated with the nitro groups in polynitroaromatic molecules. Chemical Physics Letters. 1985;116:434–8.

[416] Murray JS, Lane P, Politzer P, Bolduc PR. A relationship between impact sensitivity and the electrostatic potentials at the midpoints of C–NO2 bonds in nitroaromatics. Chemical Physics Letters. 1990;168:135–9.

[417] Vaullerin M, Espagnacq A, Morin-Allory L. Prediction of explosives impact sensitivity. Propellants, Explosives, Pyrotechnics. 1998;23:237–9.

[418] Fukuyama I, Ogawa T, Miyake A. Sensitivity and evaluation of explosive substances. Propellants, explosives, pyrotechnics. 1986;11:140–3.

[419] Brinck T, Murray JS, Politzer P. Quantitative determination of the total local polarity (charge separation) in molecules. Molecular Physics. 1992;76:609–17.

[420] Xiao H-M, Fan J-F, Gu Z-M, Dong H-S. Theoretical study on pyrolysis and sensitivity of energetic compounds:(3) Nitro derivatives of aminobenzenes. Chemical Physics. 1998;226:15–24.

[421] Politzer P, Murray JS, Markinas PL. Organic Energetic Compounds. New York: Nova Science; 1996.

[422] Rice BM, Hare JJ. A quantum mechanical investigation of the relation between impact sensitivity and the charge distribution in energetic molecules. The Journal of Physical Chemistry A. 2002;106:1770–83.

[423] Jane P, Murray PL. Effects of strongly electron-attracting components on molecular surface electrostatic potentials: application to predicting impact sensitivities of energetic molecules. Molecular Physics. 1998;93:187–94.

[424] Brill T, Oyumi Y. Thermal decomposition of energetic materials. 9. A relationship of molecular structure and vibrations to decomposition: polynitro-3, 3,7,7-tetrakis (trifluoromethyl)-2,4,6,8-tetraazabicyclo (3.3.0) octanes. The Journal of Physical Chemistry. 1986;90:2679–82.

[425] Edwards J, Eybl C, Johnson B. Correlation between sensitivity and approximated heats of detonation of several nitroamines using quantum mechanical methods. International Journal of Quantum Chemistry. 2004;100:713–9.

[426] Ren F, Cao D, Shi W, Gao H. A theoretical prediction of the relationships between the impact sensitivity and electrostatic potential in strained cyclic explosive and application to H-bonded complex of nitrocyclohydrocarbon. Journal of Molecular Modeling. 2016;22:97.

[427] Oliveira MA, Borges I Jr. On the molecular origin of the sensitivity to impact of cyclic nitramines. International Journal of Quantum Chemistry. 2019;119:e25868.

[428] Zhang C, Shu Y, Huang Y, Zhao X, Dong H. Investigation of correlation between impact sensitivities and nitro group charges in nitro compounds. The Journal of Physical Chemistry B. 2005;109:8978–82.

[429] Bondarchuk SV. Quantification of impact sensitivity based on solid-state derived criteria. The Journal of Physical Chemistry A. 2018;122:5455–63.

[430] Cawkwell M, Manner V. Ranking the drop-weight impact sensitivity of common explosives using Arrhenius chemical rates computed from quantum molecular dynamics simulations. The Journal of Physical Chemistry A. 2019;124:74–81.

[431] Mathieu D. Sensitivity of energetic materials: Theoretical relationships to detonation performance and molecular structure. Industrial & Engineering Chemistry Research. 2017;56:8191–201.

[432] Cho S-G, No K-T, Goh E-M, Kim J-K, Shin J-H, Joo Y-D, Seong S-Y. Optimization of neural networks architecture for impact sensitivity of energetic molecules. Bulletin of the Korean Chemical Society. 2005;26:399–408.

[433] Keshavarz MH, Jaafari M. Investigation of the various structure parameters for predicting impact sensitivity of energetic molecules via artificial neural network. Propellants, Explosives, Pyrotechnics. 2006;31:216–25.

[434] Wang R, Jiang J, Pan Y, Cao H, Cui Y. Prediction of impact sensitivity of nitro energetic compounds by neural network based on electrotopological-state indices. Journal of Hazardous Materials. 2009;166:155–86.

[435] Liu W-H, Liu Q-J, Liu F-S, Liu Z-T. Machine learning approaches for predicting impact sensitivity and detonation performances of energetic materials. Journal of Energy Chemistry. 2025;102:161–71.

[436] Bao S-Y, Liu Q-J, Hong D, Liu W-H, Ma X-J, Liu F-S et al. To explore the relationship between energy transfer rate and impact sensitivity by the first-principle calculation method. Journal of Physics and Chemistry of Solids. 2023;177:111298.

[437] Cawkwell MJ, Davis J, Lease N, Marrs FW, Burch A, Ferreira S et al. Understanding explosive sensitivity with effective trigger linkage kinetics. ACS Physical Chemistry Au. 2022;2(5):448–58.

[438] Duarte JC, da Rocha RD, Borges I. Which molecular properties determine the impact sensitivity of an explosive? A machine learning quantitative investigation of nitroaromatic explosives. Physical Chemistry Chemical Physics. 2023;25(9):6877–90.

[439] Wu Q, Wang X, Yan B, Luo S, Zheng X, Tan L et al. Prediction of impact sensitivity and electrostatic spark sensitivity for energetic compounds by machine learning and density functional theory. Journal of Materials Science. 2024;59(20):8894–910.

[440] Peng H, Hao L, Feng J, Xu W, Wei H. Predictive Models for Sensitivities and Detonation Velocity of Energetic Materials Based on Nonlinear Kernel Machine and Heuristic Algorithms. Processes. 2024;13(1):39.

[441] Deng Q, Hu J, Wang L, Liu Y, Guo Y, Xu T et al. Probing impact of molecular structure on bulk modulus and impact sensitivity of energetic materials by machine learning methods. Chemometrics and Intelligent Laboratory Systems. 2021;215:104331.

[442] Politzer P, Murray JS. Are HOMO–LUMO gaps reliable indicators of explosive impact sensitivity? Journal of Molecular Modeling. 2021;27:1–6.

[443] Marrs FW, Davis JV, Burch AC, Brown GW, Lease N, Huestis PL et al. Chemical descriptors for a large-scale study on drop-weight impact sensitivity of high explosives. Journal of Chemical Information and Modeling. 2023;63(3):753–69.

[444] Pallewela GN, Bettens RP. Theoretical investigation of impact sensitivity of nitrogen rich energetic salts. Computational and Theoretical Chemistry. 2021;1201:113267.

[445] Lotfi S, Ahmadi S, Toropova AP, Toropov AA. Construction of reliable QSPR models for predicting the impact sensitivity of nitroenergetic compounds using correlation weights of the fragments of molecular structures. Scientific Reports. 2025;15(1):11160.

[446] Siqueira Soldaini Oliveira R, Borges Jr I. Correlation between molecular charge properties and impact sensitivity of explosives: nitrobenzene derivatives. Propellants, Explosives, Pyrotechnics. 2021;46(2):309–21.

[447] Guo X-N, Chang X-H, Bai Z-X, Liu Q-J, Liu Z-T. First-Principles Calculation of Impact Sensitivity of Energetic Materials Based on Energy Transfer Rate. Journal of Energetic Materials. 2025:1–12.

[448] Bondarchuk SV. Theory of impact sensitivity revisited: mechanical-to-vibrational energy transfer phenomenon. FirePhysChem. 2022;2(4):334–9.

[449] Liu W-H, Zeng W, Liu F-S, Liu Z-T, Liu Q-J. Probing into the theory of impact sensitivity: propelling the understanding of phonon–vibron coupling coefficients. Physical Chemistry Chemical Physics. 2024;26(9):7695–705.

[450] Bidault X, Chaudhuri S. Can a shock-induced phonon up-pumping model relate to impact sensitivity of molecular crystals, polymorphs and cocrystals? RSC Advances. 2022;12(48):31282–92.

[451] Bao S-Y, Zeng W, Liu F-S, Liu Z-T, Liu Q-J. Theoretical relationship between vibrational properties and impact sensitivity of energetic materials from the phonon upon transition theory. Chemical Physics. 2024;576:112085.

[452] Kamlet MJ. The relationship of impact sensitivity with structure of organic high explosives. I. Polynitroaliphatic explosives. In: Sixth Symposium (International) on Detonation. Coronads, CA. 1976. pp. 69–72.

[453] Kamlet MJ, Adolph HG. The relationship of impact sensitivity with structure of organic high explosives. II. Polynitroaromatic explosives. Propellants, Explosives, Pyrotechnics. 1979;4:30–4.

[454] Mullay J. A relationship between impact sensitivity and molecular electronegativity. Propellants, explosives, pyrotechnics. 1987;12:60–3.

[455] Mullay J. Relationships between impact sensitivity and molecular electronic structure. Propellants, Explosives, Pyrotechnics. 1987;12:121–4.

[456] McNesby K, Coffey C. Spectroscopic determination of impact sensitivities of explosives. The Journal of Physical Chemistry B. 1997;101:3097–104.

[457] Jain S. Energetics of propellants, fuels and explosives; a chemical valence approach. Propellants, explosives, pyrotechnics. 1987;12:188–95.

[458] Zeman S. New aspects of impact reactivity of polynitro compounds. Part IV. Allocation of polynitro compounds on the basis of their impact sensitivities. Propellants, Explosives, Pyrotechnics. 2003;28:308–13.

[459] Keshavarz MH, Pouretedal HR. Simple empirical method for prediction of impact sensitivity of selected class of explosives. Journal of Hazardous Materials. 2005;124:27–33.

[460] Keshavarz MH, Pouretedal HR, Semnani A. Novel correlation for predicting impact sensitivity of nitroheterocyclic energetic molecules. Journal of Hazardous Materials. 2007;141:803–7.

[461] Keshavarz MH. Prediction of impact sensitivity of nitroaliphatic, nitroaliphatic containing other functional groups and nitrate explosives. Journal of Hazardous Materials. 2007;148:648–52.

[462] Keshavarz MH, Zali A, Shokrolahi A. A simple approach for predicting impact sensitivity of polynitroheteroarenes. Journal of Hazardous Materials. 2009;166:1115–9.

[463] Keshavarz MH. Simple relationship for predicting impact sensitivity of nitroaromatics, nitramines, and nitroaliphatics. Propellants, Explosives, Pyrotechnics. 2010;35:175–81.

[464] Storm CB, Stine JR, Kramer JF. Sensitivity Relationships in Energetic Materials. In: Chemistry and Physics of Energetic Materials. Dordrecht: Springer, 1990. pp. 605–39.

[465] Kamlet MJ, Adolph HG. Some comments regarding the sensitivities, thermal stabilities, and explosive performance characteristics of fluorodinitromethyl compounds. In: The Seventh Symposium. (Int.) on Detonation. 1981. pp. 60–7.

[466] Storm CB, Ryan RR, Ritchie JP, Hall JH, Bachrach SM. Structural basis of the impact sensitivities of 1-picryl-1,2,3-triazole, 2-picryl-1,2,3-triazole, 4-nitro-1-picryl-1,2,3-triazole, and 4-nitro-2-picryl-1,2,3-triazole. Journal of Physical Chemistry; (USA). 1989;93.

[467] Keshavarz MH, Esmaeilpour K, Khoshandam H, Keshavarz Z, Atabak HH, Damiri S, Afzali A. A novel method for the prediction of the impact sensitivity of quaternary ammonium-based energetic ionic liquids. Central European Journal of Energetic Materials. 2017;14.

[468] Hafner K, Klapötke TM, Schmid PC, Stierstorfer J. Synthesis and characterization of asymmetric 1,2-dihydroxy-5,5'-bitetrazole and selected nitrogen-rich derivatives. European Journal of Inorganic Chemistry. 2015;2015:2794–803.

[469] Bondarchuk SV, Zhang Z, Chen C, Wen L, Zhang J, Liu Y. Grammar of Impact Sensitivity: An Incremental Theory. The Journal of Physical Chemistry A. 2023;127(49):10506–16.

[470] McNesby K, Coffey C. Spectroscopic determination of impact sensitivities of explosives. The Journal of Physical Chemistry B. 1997;101(16):3097–104.

[471] Coffey C, Jacobs S. Detection of local heating in impact or shock experiments with thermally sensitive films. Journal of Applied Physics. 1981;52(11):6991–3.

[472] Jensen TL, Moxnes JF, Unneberg E, Christensen D. Models for predicting impact sensitivity of energetic materials based on the trigger linkage hypothesis and Arrhenius kinetics. Journal of Molecular Modeling. 2020;26:1–14.

[473] Xiong Y, Zhong K, Zhang C-y. Trigger linkage mechanism: Two or multiple steps initiate the spontaneous decay of energetic materials. Energetic Materials Frontiers. 2022;3(1):38–46.

[474] Zeman S, Valenta P, Zeman V, Jakubko J, Kamensky Z. Electric spark sensitivity of polynitro compounds: a comparison of some authors' results. Chinese Journal of Energetic Materials (HanNeng CaiLiao). 1998;6:118–22.

[475] Friedldl Z, Kočí J. Electric spark sensitivity of nitramines. Part I. Aspects of molecular structure. Central European Journal of Energetic Materials. 2006;3:27–44.

[476] Zeman S, Pelikán W, Majzlík J, Kočí J. Electric Spark Sensitivity of Nitramines. Part II. A Problem of "Hot Spots". Central European Journal of Energetic Materials. 2006;3:45–51.

[477] Zeman V, Koci J, Zeman S. Electric spark sensitivity of polynitro compounds: Part II. A correlation with detonation velocities of some polynitro arenes. Chinese Journal of Energetic Materials (HanNeng CaiLiao). 1999;7:127–32.

[478] Auzenau M, Roux M. Electric spark and ESD sensitivity of reactive solids. Part II: energy transfer mechanism and comprehensive study on E50, Propellants, Explosives. Pyrotechnics. 1995;20:96–101.

[479] Skinner D, Olson D, Block-Bolten A. Electrostatic discharge ignition of energetic materials. Propellants, Explosives, Pyrotechnics. 1998;23:34–42.

[480] Hosoya F, Shiino K, Itabashi K. Electric-spark sensitivity of Heat-Resistant Polynitroaromatic Compounds. Propellants, Explosives, Pyrotechnics. 1991;16:119–22.

[481] Zeman V, Koci J, Zeman S. Electric spark sensitivity of polynitro compounds: Part III. A correlation with detonation velocities of some nitramines. Chinese Journal of Energetic Materials (HanNeng CaiLiao). 1999;7:172–5.

[482] Zeman S. The relationship between differential thermal analysis data and the detonation characteristics of polynitroaromatic compounds. Thermochimica Acta. 1980;41:199–212.

[483] Zeman S, Liu N. A new look on the electric spark sensitivity of nitramines. Defence Technology. 2020;16:10–7.

[484] Zeman S. Influence of the energy content and its outputs on sensitivity of polynitroarenes. Journal of Energetic Materials. 2019;37:445–58.

[485] Tan B, Li Z, Guo X, Li J, Han Y, Long X. Insight into electrostatic initiation of nitramine explosives. Journal of Molecular Modeling. 2017;23:1–9.

[486] Wang G, Xiao H, Ju X, Gong X. Calculation of detonation velocity, pressure, and electric sensitivity of nitro arenes based on quantum chemistry. Propellants, Explosives, Pyrotechnics. 2006;31:361–8.

[487] Wang GX, Xiao HM, Xu XJ, Ju XH. Detonation velocities and pressures, and their relationships with electric spark sensitivities for nitramines. Propellants, Explosives, Pyrotechnics. 2006;31:102–9.

[488] Keshavarz MH. Relationship between the electric spark sensitivity and detonation pressure. Indian Journal of Engineering and Materials Sciences. 2008;15:281–6.

[489] Keshavarz MH, Pouretedal HR, Semnani A. A simple way to predict electric spark sensitivity of nitramines. Indian Journal of Engineering and Materials Sciences. 2008;15:505–9.

[490] Keshavarz MH, Pouretedal HR, Semnani A. Simple way to predict electrostatic sensitivity of nitroaromatic compounds. Khimiya. 2008;17:470–84.

[491] Keshavarz MH, Pouretedal HR, Semnani A. Reliable prediction of electric spark sensitivity of nitramines: A general correlation with detonation pressure. Journal of Hazardous Materials. 2009;167:461–6.

[492] Keshavarz MH, Keshavarz Z. Relation between electric spark sensitivity and impact sensitivity of nitroaromatic energetic compounds. Zeitschrift für anorganische und allgemeine Chemie. 2016;642:335–42.

[493] Türker L. Contemplation on spark sensitivity of certain nitramine type explosives. Journal of Hazardous Materials. 2009;169:454–9.

[494] Zhi C, Cheng X, Zhao F. The correlation between electric spark sensitivity of polynitroaromatic compounds and their molecular electronic properties. Propellants, Explosives, Pyrotechnics. 2010;35:555–60.

[495] Wu Q, Wang X, Yan B, Luo S, Zheng X, Tan L et al. Prediction of impact sensitivity and electrostatic spark sensitivity for energetic compounds by machine learning and density functional theory. Journal of Materials Science. 2024;59(20):8894–910.

[496] Keshavarz MH. Theoretical prediction of electric spark sensitivity of nitroaromatic energetic compounds based on molecular structure. Journal of Hazardous Materials. 2008;153:201–6.

[497] Keshavarz MH, Moghadas MH, Kavosh Tehrani M. Relationship between the electrostatic sensitivity of nitramines and their molecular structure. Propellants, Explosives, Pyrotechnics. 2009;34:136–41.

[498] Zeman S, Koci J. Electric spark sensitivity of polynitro compounds: part IV. A relation to thermal decomposition parameters. Chinese Journal of Energetic Materials (HanNeng CaiLiao). 2000;8:18–26.

[499] Keshavarz MH. A novel approach for the prediction of electric spark sensitivity of polynitroarenes based on the measured data from a new instrument. Central European Journal of Energetic Materials. 2019;16:65–76.

[500] Zeman S, Majzlík J. Electric spark sensitivity of polynitro arenes Part I. A comparison of two instruments. Central European Journal of Energetic Materials. 2007;4:15–24.

[501] Keshavarz MH, Damiri S, Bagheri V. Recent advances for prediction of electric spark and shock sensitivities of organic compounds containing energetic functional groups to assess reliable models. Process Safety and Environmental Protection. 2019;131:9–15.

[502] Keshavarz MH. Two novel correlations for prediction of electric spark sensitivity of nitramines based on the experimental data of the new instrument. Zeitschrift für anorganische und allgemeine Chemie. 2018;644:1607–10.

[503] Zeman S, Pelikán V, Majzlík J, Friedl Z, Kočí J. Electric spark sensitivity of nitramines. Part I. Aspects of molecular structure. Central European Journal of Energetic Materials. 2006;3:27–44.

[504] Nazari B, Keshavarz MH, Jafari M, Jafari F. A novel approach for prediction of sensitivity toward the electrical discharge of quaternary ammonium-based energetic ionic liquids or salts. Zeitschrift für anorganische und allgemeine Chemie. 2018;644:1153–7.

[505] Dippold A. Nitrogen-rich energetic materials based on 1,2,4-triazole derivatives. LMU; 2013.

[506] Keshavarz MH, Bani SH, Bakhtiari R, Hosseini SH. Assessment of electrostatic discharge sensitivity of nitrogen-rich heterocyclic energetic compounds and their salts as high energy-density dangerous compounds: A study of structural variables. Defence Technology. 2024;39:15–22.

[507] Price D. Examination of some proposed relations among HE sensitivity data. Journal of Energetic Materials. 1985;3:239–54.

[508] Tan B, Long X, Peng R, Li H, Jin B, Chu S, Dong H. Two important factors influencing shock sensitivity of nitro compounds: bond dissociation energy of X–NO 2 (X=C, N, O) and Mulliken charges of nitro group. Journal of Hazardous Materials. 2010;183:908–12.

[509] Yang K, Chen L, Liu D-Y, Geng D-S, Lu J-Y, Wu J-Y. Quantitative prediction and ranking of the shock sensitivity of explosives via reactive molecular dynamics simulations. Defence Technology. 2022;18(5):843–54.

[510] Jiang J, Xia Q-Y, Xu S-Y, Zhao F-Q, Ju X-H. Evaluating shock sensitivity and decomposition of energetic materials by ReaxFF molecular dynamics. Journal of Materials Science. 2024;59(1):114–29.

[511] Keshavarz MH, Motamedoshariati H, Pouretedal HR, Tehrani MK, Semnani A. Prediction of shock sensitivity of explosives based on small-scale gap test. Journal of Hazardous Materials. 2007;145:109–12.

[512] Dobratz B. LLNL Explosives Handbook: Properties of Chemical Explosives and Explosives and Explosive Simulants. Lawrence Livermore National Lab., CA (USA); 1981.

[513] Keshavarz MH, Pouretedal HR, Tehrani MK, Semnani A. Predicting shock sensitivity of energetic compounds. Asian Journal of Chemistry. 2008;20:1025.

[514] Dobratz BM, Crawford PC, Handbook LE. Properties of Chemical Explosives and Explosives Simulants. Lawrence Livermore National Lab., CA (USA); 1985.

[515] Cooper PW. Explosives Engineering. John Wiley & Sons; 1996.

[516] Kobylkin IF. Calculation of the critical detonation diameter of explosive charges using data on their shock-wave initiation. Combustion, Explosion and Shock Waves. 2006;42:223–6.

[517] Pepekin VI, Gubina TV. On the critical diameter and detonability of explosives. Russian Journal of Physical Chemistry B, Focus on Physics. 2011;5:813–5.

[518] Kobylkin IF. Critical detonation diameter of highly desensitized low-sensitivity explosive formulations. Combustion, Explosion, and Shock Waves. 2009;45:732.

[519] Kobylkin IF. Critical detonation diameter of industrial explosive charges: Effect of the casing. Combustion, Explosion, and Shock Waves. 2011;47:96.

[520] Keshavarz MH, Klapötke TM. A novel method for prediction of the critical diameter of solid pure and composite high explosives to assess their explosion safety in an industrial setting. Journal of Energetic Materials. 2019;37:331–9.

[521] Matyáš R, Šelešovský J, Musil T. Sensitivity to friction for primary explosives. Journal of Hazardous Materials. 2012;213:236–41.

[522] Jungová M, Zeman S, Husarová A. Friction sensitivity of nitramines. Part II. Comparison with thermal reactivity. Chinese Journal of Energetic Materials (HanNeng CaiLiao). 2011;19:607–9.

[523] Friedl Z, Jungová M, Zeman S, Husarová A. Friction sensitivity of nitramines. Part IV: Links to surface electrostatic potentials. Chinese Journal of Energetic Materials (HanNeng CaiLiao). 2012;19:613–5.

[524] Jungová M, Zeman S, Husarová A. Friction Sensitivity of Nitramines. Part I: Comparison with Impact Sensitivity and Heat of Fusion. Chinese Journal of Energetic Materials (HanNeng CaiLiao). 2012;19:603–6.

[525] Zeman S, Jungová M, Husarová A. Friction sensitivity of nitramines. Part III: Comparison with detonation performance. Chinese Journal of Energetic Materials (HanNeng CaiLiao). 2011;19:610–2.

[526] Le Roux JJ. The dependence of friction sensitivity of primary explosives upon rubbing surface roughness. Propellants, Explosives, Pyrotechnics. 1990;15:243–7.

[527] Wang X, Wang J, Fu Y, Liu R, He Y, Li A et al. The sensitivity determination of energetic materials from laser spark spectrometry based on physical-parameter-corrected statistical methods. Journal of Analytical Atomic Spectrometry. 2021;36(12):2603–11.

[528] Bondarchuk SV. Friction sensitivity of nitramine energetic materials: a prediction based on genetic function approximation. FirePhysChem 2022;2(3):272–8.

[529] Muravyev NV, Meerov DB, Monogarov KA, Kosareva EK, Melnikov IN, Pronkin DK et al. Impact and friction sensitivity of reactive chemicals: from reproducibility study to benchmark data set for modeling. Industrial & Engineering Chemistry Research. 2024;63(15):6504–11.

[530] Muravyev NV, Meerov DB, Monogarov KA, Melnikov IN, Kosareva EK, Fershtat LL et al. Sensitivity of energetic materials: Evidence of thermodynamic factor on a large array of CHNOFCl compounds. Chemical Engineering Journal. 2021;421:129804.

[531] Li H-Y, Zeng W, Liu F-S, Liu Z-T, Bai Z-X, Liu Q-J. Theoretical study of friction sensitivity of energetic materials. Journal of Physics and Chemistry of Solids. 2025;202:112692.

[532] Keshavarz MH, Hayati M, Ghariban-Lavasani S, Zohari N. A new method for predicting the friction sensitivity of nitramines. Central European Journal of Energetic Materials. 2015;12:215–27.

[533] Klapötke TM. New nitrogen-rich high explosives. In: High Energy Density Materials. Springer, 2007. pp. 85–121.

[534] Jafari M, Keshavarz MH, Joudaki F, Mousaviazar A. A simple method for predicting friction sensitivity of quaternary ammonium-based energetic ionic liquids. Propellants, Explosives, Pyrotechnics. 2018;43:568–73.

[535] Dippold AA, Klapötke TM. A study of dinitro-bis-1,2,4-triazole-1, 1′-diol and derivatives: design of high-performance insensitive energetic materials by the introduction of N-oxides. Journal of the American Chemical Society. 2013;135:9931–8.

[536] Pourmortazavi SM, Rahimi-Nasrabadi M, Kohsari I, Hajimirsadeghi SS. Non-isothermal kinetic studies on thermal decomposition of energetic materials: KNF and NTO. Journal of Thermal Analysis and Calorimetry. 2011;110:857–63.

[537] Cusu JP, Musuc AM, Matache M, Oancea D. Kinetics of exothermal decomposition of some ketone-2,4-dinitrophenylhydrazones. Journal of Thermal Analysis and Calorimetry. 2012;110:1259–66.

[538] Lee J-S, Hsu C-K, Chang C-L. A study on the thermal decomposition behaviors of PETN, RDX, HNS and HMX. Thermochimica Acta. 2002;392:173–6.

[539] Chen ZX, Xiao H. Impact sensitivity and activation energy of pyrolysis for tetrazole compounds. International Journal of Quantum Chemistry. 2000;79:350–7.

[540] Sinditskii VP, Smirnov SP, Egorshev VY. Thermal decomposition of NTO: an explanation of the high activation energy. Propellants, Explosives, Pyrotechnics. 2007;32:277.

[541] Zeman S, Dimun M, Truchlik Š. The relationship between kinetic data of the low-temperature thermolysis and the heats of explosion of organic polynitro compounds. Thermochimica Acta. 1984;78:181–209.

[542] Zeman S. Thermal stabilities of polynitroaromatic compounds and their derivatives. Thermochimica Acta. 1979;31:269–83.

[543] Zeman S. Non-isothermal differential thermal analysis in the study of the initial state of the thermal decomposition of polynitroaromatic compounds in the condensed state. Thermochimica Acta. 1980;39:117–24.

[544] Zeman S. The thermoanalytical study of some aminoderivatives of 1,3,5-trinitrobenzene. Thermochimica acta. 1993;216:157–68.

[545] Zeman S. Relationship between the Arrhenius Parameters of the Low-temperature Thermolysis and the 13C and 15N Chemical Shifts of Nitramines. Thermochimica Acta. 1992;202:191–200.

[546] Zeman S. Kinetic compensation effect and thermolysis mechanisms of organic polynitroso and polynitro compounds. Thermochimica Acta. 1997;290:199–217.

[547] Fathollahi M, Sajady H. QSPR modeling of decomposition temperature of energetic cocrystals using artificial neural network. Journal of Thermal Analysis and Calorimetry. 2018;133:1663–72.

[548] Dong J, Yan Q-L, Liu P-J, He W, Qi X-F, Zeman S. The correlations among detonation velocity, heat of combustion, thermal stability and decomposition kinetics of nitric esters. Journal of Thermal Analysis and Calorimetry. 2018;131:1391–403.

[549] Zeman S. Analysis and prediction of the Arrhenius parameters of low-temperature thermolysis of nitramines by means of the 15N NMR spectroscopy. Thermochimica Acta. 1999;333:121–9.

[550] Zeman S, Jalový Z. Heats of fusion of polynitro derivatives of polyazaisowurtzitane. Thermochimica Acta. 2000;345:31–8.

[551] Zeman S, Friedl Z. Relationship between electronic charges at nitrogen atoms of nitro groups and thermal reactivity of nitramines. Journal of Thermal Analysis and Calorimetry. 2004;77:217–24.

[552] Sorescu DC, Rice BM, Thompson DL. Molecular packing and molecular dynamics study of the transferability of a generalized nitramine intermolecular potential to non-nitramine crystals. The Journal of Physical Chemistry A. 1999;103:989–98.

[553] Kissinger HE. Reaction kinetics in differential thermal analysis. Analytical Chemistry. 1957;29:1702–6.

[554] Zeman S. Modified Evans–Polanyi–Semenov relationship in the study of chemical micromechanism governing detonation initiation of individual energetic materials. Thermochimica Acta. 2002;384:137–54.

[555] Keshavarz MH. A new method to predict activation energies of nitroparaffins. Indian Journal of Engineering and Materials Sciences. 2009;16:429–32.

[556] Manelis GB. Problemy Kinetiki Elementarnykh Khimicheskikh Reaktsii (Problems of the Kinetics of Primary Chemical Reactions). Moscow: Nauka; 1973.

[557] Keshavarz MH. Simple method for prediction of activation energies of the thermal decomposition of nitramines. Journal of Hazardous Materials. 2009;162:1557–62.

[558] Keshavarz MH, Pouretedal HR, Shokrolahi A, Zali A, Semnani A. Predicting activation energy of thermolysis of polynitro arenes through molecular structure. Journal of Hazardous Materials. 2008;160:142–7.

[559] Keshavarz MH, Zohari N, Seyedsadjadi SA. Validation of improved simple method for prediction of activation energy of the thermal decomposition of energetic compounds. Journal of Thermal Analysis and Calorimetry. 2013;114:497–510.

[560] Talawar MB, Sivabalan R, Mukundan T, Muthurajan H, Sikder AK, Gandhe BR, Rao AS. Environmentally compatible next generation green energetic materials (GEMs). Journal of Hazardous materials. 2009;161:589–607.

[561] Ando T, Fujimoto Y, Morisaki S. Analysis of differential scanning calorimetric data for reactive chemicals. Journal of Hazardous Materials. 1991;28:251–80.

[562] Wu J-N, Song S-W, Tian X-L, Wang Y, Qi X-J. Machine learning-based prediction and interpretation of decomposition temperatures of energetic materials. Energetic Materials Frontiers. 2023;4(4):254–61.

[563] Zhang Z-X, Cao Y-L, Chen C, Wen L-Y, Ma Y-D, Wang B-Z et al. Machine learning-assisted quantitative prediction of thermal decomposition temperatures of energetic materials and their thermal stability analysis. Energetic Materials Frontiers. 2023;5(4):274–82.

[564] Zhang Z, Chen C, Cao Y, Wen L, He X, Liu Y. Descriptors applicability in machine learning-assisted prediction of thermal decomposition temperatures for energetic materials: Insights from model evaluation and outlier analysis. Thermochimica Acta. 2024;735:179717.

[565] Hossain MM, Roy K. QSPR modeling of thermal stability of reactive and self-reactive chemicals using 2D descriptors: predictions of heat of decomposition, self-accelerating decomposition temperature, and onset temperature. Materials Today Communications. 2025;46:112751.

[566] Zhang X, Liu Q-J, Liu F-S, Liu Z-T. Predicting the thermal decomposition temperature of energetic materials from a simple model. Journal of Molecular Modeling. 2024;30(8):277.

[567] Saraf SR, Rogers WJ, Mannan MS. Prediction of reactive hazards based on molecular structure. Journal of Hazardous Materials. 2003;98:15–29.

[568] Fayet G, Joubert L, Rotureau P, Adamo C. On the use of descriptors arising from the conceptual density functional theory for the prediction of chemicals explosibility. Chemical Physics Letters. 2009;467:407–11.

[569] Fayet G, Rotureau P, Joubert L, Adamo C. On the prediction of thermal stability of nitroaromatic compounds using quantum chemical calculations. Journal of hazardous materials. 2009;171:845–50.

[570] Fayet G, Rotureau P, Joubert L, Adamo C. QSPR modeling of thermal stability of nitroaromatic compounds: DFT vs. AM1 calculated descriptors. Journal of molecular modeling. 2010;16:805–12.

[571] Fayet G, Del Rio A, Rotureau P, Joubert L, Adamo C. Predicting the thermal stability of nitroaromatic compounds using chemoinformatic tools. Molecular Informatics. 2011;30:623–34.

[572] Fayet G, Rotureau P, Joubert L, Adamo C. Development of a QSPR model for predicting thermal stabilities of nitroaromatic compounds taking into account their decomposition mechanisms. Journal of Molecular Modeling. 2011;17:2443–53.

[573] Fayet G, Rotureau P, Adamo C. On the development of QSPR models for regulatory frameworks: The heat of decomposition of nitroaromatics as a test case. Journal of Loss Prevention in the Process Industries. 2013;26:1100–5.

[574] Keshavarz MH, Ghani K, Asgari A. A new method for predicting heats of decomposition of nitroaromatics. Zeitschrift für anorganische und allgemeine Chemie. 2015;641:1818–23.

[575] Lu Y, Ng D, Mannan MS. Prediction of the reactivity hazards for organic peroxides using the QSPR approach. Industrial & Engineering Chemistry Research. 2010;50:1515–22.

[576] Prana V, Rotureau P, Fayet G, André D, Hub S, Vicot P, Rao L, Adamo C. Prediction of the thermal decomposition of organic peroxides by validated QSPR models. Journal of Hazardous Materials. 2014;276:216–24.

[577] Pan Y, Zhang Y, Jiang J, Ding L. Prediction of the self-accelerating decomposition temperature of organic peroxides using the quantitative structure–property relationship (QSPR) approach. Journal of Loss Prevention in the Process Industries. 2014;31:41–9.

[578] Gao Y, Xue Y, Lü Z-G, Wang Z, Chen Q, Shi N, Sun F. Self-accelerating decomposition temperature and quantitative structure–property relationship of organic peroxides. Process Safety and Environmental Protection. 2015;94:322–8.

[579] Zohari N, Keshavarz MH, Dalaei Z. Prediction of decomposition onset temperature and heat of decomposition of organic peroxides using simple approaches. Journal of Thermal Analysis and Calorimetry. 2016;125:887–96.

[580] Wakakura M, Iiduka Y. Trends in chemical hazards in Japan. Journal of Loss prevention in the process industries. 1999;12:79–84.

[581] Keshavarz MH, Pouretedal HR, Semnani A. Relationship between thermal stability and molecular structure of polynitro arenes. Indian Journal of Engineering and Materials Sciences. 2009;16:61–4.

[582] Keshavarz MH, Mousaviazar A, Hayaty M. A novel approach for assessment of thermal stability of organic azides through prediction of their temperature of maximum mass loss. Journal of Thermal Analysis and Calorimetry. 2017;129:1659–65.

[583] Kumari D, Anjitha SG, Pant CS, Patil M, Singh H, Banerjee S. Synthetic approach to novel azido esters and their utility as energetic plasticizers. RSC Advances. 2014;4:39924–33.

[584] Fedoroff BT, Sheffield OE, Reese EF, Sheffield OE, Clift GD, Dunkle CG, Walter H, Mclean DC. In: Encyclopedia of Explosives and Related Items, Part 2700, Picatinny Arsenal. Dower, NJ. 1960.

[585] Keshavarz MH, Moradi S, Saatluo BE, Rahimi H, Madram AR. A simple accurate model for prediction of deflagration temperature of energetic compounds. Journal of Thermal Analysis and Calorimetry. 2013;112:1453–63.

[586] Drees D, Löffel D, Messmer A, Schmid K. Synthesis and characterization of azido plasticizer. Propellants, Explosives, Pyrotechnics. 1999;24:159–62.

[587] Keshavarz MH, Nazari B, Jafari M, Yazdani Z. A novel and simple approach for predicting activation energy of thermolysis of some selected ionic liquids. Journal of Thermal Analysis and Calorimetry. 2018;134:2383–90.

[588] Cao Y, Mu T. Comprehensive investigation on the thermal stability of 66 ionic liquids by thermogravimetric analysis. Industrial & Engineering Chemistry Research. 2014;53:8651–64.

[589] Keshavarz MH, Pouretedal HR, Saberi E. A new method for predicting decomposition temperature of imidazolium-based energetic ionic liquids. Zeitschrift für anorganische und allgemeine Chemie. 2017;643:171–9.

[590] Schneider S, Hawkins T, Rosander M, Mills J, Vaghjiani G, Chambreau S. Liquid azide salts and their reactions with common oxidizers IRFNA and N_2O_4. Inorganic Chemistry. 2008;47:6082–9.

[591] Zohari N, Abrishami F, Zeynali V. Prediction of decomposition temperature of azole-based energetic compounds in order to assess of their thermal stability. Journal of Thermal Analysis and Calorimetry. 2019;141:1453–63.

[592] Kumar D, Mitchell LA, Parrish DA, Jean'Ne MS. Asymmetric N, N′-ethylene-bridged azole-based compounds: Two way control of the energetic properties of compounds. Journal of Materials Chemistry A. 2016;4:9931–40.

[593] Keshavarz MH, Mousavi S, Drikvand M. Assessment of the Thermal Decomposition Temperature of High-Energy Heterocyclic Aromatic Compounds in Order to Increase Their Safety during Storage, Handling and Application. Central European Journal of Energetic Materials. 2022;19(1).

[594] Koci Jí, Zeman V, Zeman S. Electric spark sensitivity of polynitro compounds. Part V. A relationship between electric spark and impact Zohari N, Keshavarz MH, Seyedsadjadi SA. A link between impact sensitivity of energetic sensitivities of energetic materials. Chinese Journal of Energetic Materials (HanNeng CaiLiao). 2001;9:60–5.

[595] Zohari N, Keshavarz MH, Seyedsadjadi SA. A novel method for risk assessment of electrostatic sensitivity of nitroaromatics through their activation energies of thermal decomposition. Journal of Thermal Analysis and Calorimetry. 2014;115:93–100.

[596] Keshavarz MH, Hayati M, Ghariban-Lavasani S, Zohari N. Relationship between activation energy of thermolysis and friction sensitivity of cyclic and acyclic nitramines. Zeitschrift für anorganische und allgemeine Chemie. 2016;642:182–8.

[597] Keshavarz MH, Zohari N, Seyedsadjadi SA. Relationship between electric spark sensitivity and activation energy of the thermal decomposition of nitramines for safety measures in industrial processes. Journal of Loss Prevention in the Process Industries. 2013;26:1452–6.

[598] Zohari N, Seyed-Sadjadi SA, Marashi-Manesh S. The Relationship between Impact Sensitivity of Nitroaromatic Energetic Compounds and their Electrostatic Sensitivity. Central European Journal of Energetic Materials. 2016;13:427–43.

[599] Wu Y-Q, Huang F-L. A microscopic model for predicting hot-spot ignition of granular energetic crystals in response to drop-weight impacts. Mechanics of Materials. 2011;43:835–52.

[600] Keshavarz MH, Ghaffarzadeh M, Omidkhah MR, Farhadi K. New correlation between electric spark and impact sensitivities of nitramine energetic compounds for assessment of their safety. Zeitschrift für anorganische und allgemeine Chemie. 2017;643:1227–31.

[601] Ferdowsi M, Yazdani F, Omidkhah MR, Keshavarz MH. A general relationship between electric spark and impact sensitivities of nitroaromatics and nitramines. Zeitschrift für anorganische und allgemeine Chemie. 2018;644:1623–8.

[602] Östmark H, Bergman H, Ekvall K, Langlet A. A study of the sensitivity and decomposition of 1,3,5-trinitro-2-oxo-1,3,5-triazacyclo-hexane. Thermochimica Acta. 1995;260:201–16.

[603] Keshavarz MH, Ghaffarzadeh M, Omidkhah MR, Farhadi K. Correlation between Shock Sensitivity of Nitramine Energetic Compounds based on Small-scale Gap Test and Their Electric Spark Sensitivity. Zeitschrift für anorganische und allgemeine Chemie. 2017;643:2158–62.

[604] Tan B, Long X, Peng R, Li H, Jin B, Chu S. On the shock sensitivity of explosive compounds with small-scale gap test. The Journal of Physical Chemistry A. 2011;115:10610–6.

[605] Ferdowsi M, Yazdani F, Omidkhah MR, Keshavarz MH. Reliable prediction of shock sensitivity of energetic compounds based on small-scale gap test through their electric spark sensitivity. Zeitschrift für anorganische und allgemeine Chemie. 2018;644:888–92.

[606] Agrawal JP, Dodke VS. Some novel high energy materials for improved performance. Zeitschrift für anorganische und allgemeine Chemie. 2021;647(19):1856–82.

[607] Agrawal JP, Dodke VS. Validation of Approaches (Salt Formation & Introduction of Amino Group/s) for Imparting/Improving Thermal Stability of Explosives, Part I. Propellants, Explosives, Pyrotechnics. 2022;47(3):e202100265.

[608] Li C, Li H, Zong H-H, Huang Y, Gozin M, Sun CQ et al. Strategies for achieving balance between detonation performance and crystal stability of high-energy-density materials. iScience. 2020;23(3).

[609] Jiao F, Xiong Y, Li H, Zhang C. Alleviating the energy & safety contradiction to construct new low sensitivity and highly energetic materials through crystal engineering. CrystEngComm. 2018;20(13):1757–68.

[610] Chen B, Lu H, Chen J, Chen Z, Yin S-F, Peng L et al. Recent progress on nitrogen-rich energetic materials based on tetrazole skeleton. Topics in Current Chemistry. 2023;381(5):25.

[611] Patil VB, Zeman S. Progress in energy–safety balanced cocrystallization of four commercially attractive nitramines. Crystal Growth & Design. 2024;24(17):7361–88.

[612] O'Sullivan OT, Zdilla MJ. Properties and promise of catenated nitrogen systems as high-energy-density materials. Chemical Reviews. 2020;120(12):5682–744.

[613] Wozniak DR, Piercey DG. Review of the current synthesis and properties of energetic pentazolate and derivatives thereof. Engineering. 2020;6(9):981–91.

[614] Yao Y, Lin Q, Zhou X, Lu M. Recent research on the synthesis pentazolate anion cyclo-N5−. FirePhysChem. 2021;1(1):33–45.

[615] Gong L, Chen G, Liu Y, Wang T, Zhang J, Yi X et al. Energetic metal–organic frameworks achieved from furazan and triazole ligands: synthesis, crystal structure, thermal stability and energetic performance. New Journal of Chemistry. 2021;45(47):22299–305.

[616] Gruhne MS, Lenz T, Rösch M, Lommel M, Wurzenberger MH, Klapötke TM et al. Nitratoethyl-5 H-tetrazoles: improving the oxygen balance through application of organic nitrates in energetic coordination compounds. Dalton Transactions. 2021;50(31):10811–25.

[617] Wurzenberger MH, Gruhne MS, Lommel M, Stierstorfer J. 1-Amino-5-methyltetrazole in energetic 3 d transition metal complexes–ligand design for future primary explosives. Propellants, Explosives, Pyrotechnics. 2021;46(2):207–13.

[618] Hanafi S, Trache D, He W, Xie W-X, Mezroua A, Yan Q-L. Thermostable energetic coordination polymers based on functionalized go and their catalytic effects on the decomposition of AP and RDX. The Journal of Physical Chemistry C. 2020;124(9):5182–95.

[619] Wurzenberger MH, Gruhne MS, Lommel M, Szimhardt N, Klapötke TM, Stierstorfer J. Comparison of 1-ethyl-5H-tetrazole and 1-azidoethyl-5H-tetrazole as ligands in energetic transition metal complexes. Chemistry–An Asian Journal. 2019;14(11):2018–28.

[620] Xu Z, Hou T, Zhang X, Yang F, Zhang L, Jiang S et al. Self-assembly and controllable regulation of 1, 5′-bitetrazolate-2 N-oxide-based EMOFs toward high dimensionality. Crystal Growth & Design. 2023;24(1):461–70.

[621] Zhong K, Zhang C. Review of the decomposition and energy release mechanisms of novel energetic materials. Chemical Engineering Journal. 2024:149202.

[622] Zhang S, Gao Z, Lan D, Jia Q, Liu N, Zhang J et al. Recent advances in synthesis and properties of nitrated-pyrazoles based energetic compounds. Molecules. 2020;25(15):3475.

[623] He T, Kong X-J, Li J-R. Chemically stable metal–organic frameworks: rational construction and application expansion. Accounts of Chemical Research. 2021;54(15):3083–94.

[624] Hou T, Leng H, Chen M, Luo J, Zhang C, Zhang Y et al. Integrating three types of structure reinforcements abounding in heat-resistant explosives to construct a 3D solvent-free EMOF with superb stability. New Journal of Chemistry. 2024;48(28):12697–705.

[625] Schock M, Bräse S. Reactive & efficient: Organic azides as cross-linkers in material sciences. Molecules. 2020;25(4):1009.

[626] Rajak R, Kumar P, Ghule VD, Dharavath S. Polytetrazole containing thermally stable and insensitive alkali metal-based 3D energetic metal–organic frameworks. Inorganic Chemistry. 2023;62(21):8389–96.

[627] Li S, Wang Y, Qi C, Zhao X, Zhang J, Zhang S et al. 3D energetic metal–organic frameworks: synthesis and properties of high energy materials. Angew Chem, Int Ed. 2013;52(52):14031–5.

[628] Dong Y, Peng P, Hu B, Su H, Li S, Pang S. High-density energetic metal–organic frameworks based on the 5, 5'-dinitro-2 H, 2' H-3, 3'-bi-1, 2, 4-triazole. Molecules. 2017;22(7):1068.

[629] Dalirandeh Z, Jafari M, Mousaviazar A. A novel and reliable method for prediction of the density of energetic metal–organic frameworks. Chemical Papers. 2024;78(4):2323–38.

[630] Dalirandeh Z, Jafari M, Mousaviazar A. The use of simple structural parameters of energetic metal-organic frameworks to assess their density. Journal of Solid State Chemistry. 2024;333:124608.

[631] Qu X, Zhai L, Wang B, Wei Q, Xie G, Chen S et al. Copper-based energetic MOFs with 3-nitro-1 H-1, 2, 4-triazole: solvent-dependent syntheses, structures and energetic performances. Dalton Transactions. 2016;45(43):17304–11.

[632] Dalirandeh Z, Jafari M, Mousaviazar A. Simple method for predicting the heats of formation of energetic metal–organic frameworks. Russian Journal of Physical Chemistry A. 2024;98(3):395–405.

[633] Zhang J, Zhu Z, Zhou M, Zhang J, Hooper JP, Shreeve Jn M. Superior high-energy-density biocidal agent achieved with a 3D metal–organic framework. ACS Applied Materials & Interfaces. 2020;12(36):40541–7.

[634] Mousaviazar A, Shirazi Z, Keshavarz MH, Mansouri N. A novel approach for prediction of exothermic decomposition temperature of energetic complexes through additive and non-additive descriptors. Journal of Thermal Analysis and Calorimetry. 2022;147(22):12907–17.

[635] Wurzenberger MH, Gruhne MS, Lommel M, Szimhardt N, Klapötke TM, Stierstorfer J. Comparison of 1-ethyl-5H-tetrazole and 1-azidoethyl-5H-tetrazole as ligands in energetic transition metal complexes. Chemistry – An Asian Journal. 2019;14(11):2018–28.

[636] Damavarapu R, Klapötke TM, Stierstorfer J, Tarantik KRJP. Barium salts of tetrazole derivatives – synthesis and characterization. Propellants, Explosives, Pyrotechnics. 2010;35(4):395–406.

[637] Sutton GP, Biblarz O. Rocket Propulsion Elements. John Wiley & Sons; 2011.

[638] Yu C, Wei Z. An Introduction to Propellants. Nova Science Publishers, Inc.; 2020.

[639] Penner SS, Ducarme J. The chemistry of propellants: a meeting organised by the AGARD combustion and propulsion panel. Elsevier; 2016.

[640] Wen Y, Mo H, Tan B, Lu X, Wang B, Liu N. Progress in synthesis and properties of oxetane-based energetic polymers. European Polymer Journal. 2023;194:112161.

[641] Hildebrand JH. A critique of the theory of solubility of non-electrolytes. Chemical Reviews. 1949;44(1):37–45.

[642] Walden DM, Bundey Y, Jagarapu A, Antontsev V, Chakravarty K, Varshney J. Molecular simulation and statistical learning methods toward predicting drug–polymer amorphous solid dispersion miscibility, stability, and formulation design. Molecules. 2021;26(1):182.

[643] Bergin SD, Sun Z, Rickard D, Streich PV, Hamilton JP, Coleman JN. Multicomponent solubility parameters for single-walled carbon nanotube–solvent mixtures. ACS Nano. 2009;3(8):2340–50.

[644] Hansen CM. 50 Years with solubility parameters – past and future. Progress in Organic Coatings. 2004;51(1):77–84.

[645] Hansen CM. Hansen Solubility Parameters: A User's Handbook. Second ed. CRC press; 2007.

[646] Stefanis E, Panayiotou C. A new expanded solubility parameter approach. International Journal of Pharmaceutics. 2012;426(1–2):29–43.

[647] Chen Y, Wang Q, Liu Z, Li Z, Chen W, Zhou L, Qin J, Meng Y, Mu T. Vaporization enthalpy, long-term evaporation and evaporation mechanism of polyethylene glycol-based deep eutectic solvents. New Journal of Chemistry. 2020;44(22):9493–501.

[648] Carvalho SP, Lucas EF, González G, Spinelli LS. Determining Hildebrand solubility parameter by ultraviolet spectroscopy and microcalorimetry. Journal of the Brazilian Chemical Society. 2013;24:1998–2007.

[649] Forster A, Hempenstall J, Tucker I, Rades T. Selection of excipients for melt extrusion with two poorly water-soluble drugs by solubility parameter calculation and thermal analysis. International Journal of Pharmaceutics. 2001;226(1–2):147–61.

[650] Brandrup J, Immergut EH, Grulke EA, Abe A, Bloch DR. Polymer Handbook. Wiley New York; 1999.

[651] Li X, Winnik MA, Guillet JE. A fluorescence method to determine the solubility parameters δH of soluble polymers at infinite dilution. Cyclization dynamics of polymers. 11. Macromolecules. 1983;16(6):992–5.

[652] Albahri TA. Accurate prediction of the solubility parameter of pure compounds from their molecular structures. Fluid Phase Equilibria. 2014;379:96–103.

[653] Stefanis E, Panayiotou C. Prediction of Hansen solubility parameters with a new group-contribution method. International Journal of Thermophysics. 2008;29:568–85.

[654] Code JE, Holder AJ, Eick JD. Direct and indirect quantum mechanical QSPR Hildebrand solubility parameter models. QSAR & Combinatorial Science. 2008;27(7):841–9.

[655] Bagheri M, Golbraikh A. Rank-based ant system method for non-linear QSPR analysis: QSPR studies of the solubility parameter. SAR and QSAR in Environmental Research. 2012;23(1–2):59–86.

[656] Chi M, Gargouri R, Schrader T, Damak K, Maâlej R, Sierka M. Atomistic descriptors for machine learning models of solubility parameters for small molecules and polymers. Polymers. 2021;14(1):26.

[657] Koç Dİ, Koç ML. A genetic programming-based QSPR model for predicting solubility parameters of polymers. Chemometrics and Intelligent Laboratory Systems. 2015;144:122–7.

[658] Keshavarz MH, Shafiee M, Jazi BN. Simple approach for reliable prediction of solubility of polymers in environmentally compatible solvents. Industrial & Engineering Chemistry Research. 2022;61(6):2425–33.

[659] Van Krevelen DW, Te Nijenhuis K. Properties of Polymers: Their Correlation with Chemical Structure; Their Numerical Estimation and Prediction from Additive Group Contributions. Fourth ed. Elsevier; 2009.

[660] Bak J, Yoo B. Intrinsic viscosity of binary gum mixtures with xanthan gum and guar gum: Effect of NaCl, sucrose, and pH. International Journal of Biological Macromolecules. 2018;111:77–81.

[661] Kulicke W-M, Clasen C. Viscosimetry of Polymers and Polyelectrolytes. Springer; 2004.

[662] Huggins ML. The viscosity of dilute solutions of long-chain molecules. IV. Dependence on concentration. Journal of the American Chemical Society. 1942;64(11):2716–8.

[663] Kraemer EO. Molecular weights of celluloses and cellulose derivates. Industrial & Engineering Chemistry. 1938;30(10):1200–3.

[664] Lovell PA. Dilute solution viscometry. In: Price C, Booth C, editors. Comprehensive Polymer Science and Supplements. Oxford: Pergamon Press, 1989. pp. 173–97.

[665] Esau A. Characterization of industrial flocculants through intrinsic viscosity measurements: University of British Columbia; 2008.

[666] Fuoss RM. Viscosity function for polyelectrolytes. Journal of Polymer Science. 1948;3(4):603–4.

[667] Fuoss RM, Strauss UP. The viscosity of mixtures of polyelectrolytes and simple electrolytes. Annals of the New York Academy of Sciences. 1949;51(4):836–51.

[668] Lopez CG, Richtering W. Viscosity of semidilute and concentrated nonentangled flexible polyelectrolytes in salt-free solution. The Journal of Physical Chemistry B. 2019;123(26):5626–34.

[669] Colby RH. Structure and linear viscoelasticity of flexible polymer solutions: comparison of polyelectrolyte and neutral polymer solutions. Rheologica Acta. 2010;49:425–42.

[670] Afantitis A, Melagraki G, Sarimveis H, Koutentis PA, Markopoulos J, Igglessi-Markopoulou O. Prediction of intrinsic viscosity in polymer–solvent combinations using a QSPR model. Polymer. 2006;47(9):3240–8.

[671] Gharagheizi F. QSPR analysis for intrinsic viscosity of polymer solutions by means of GA-MLR and RBFNN. Computational Materials Science. 2007;40(1):159–67.

[672] Wang S, Cheng M, Zhou L, Dai Y, Dang Y, Ji X. QSPR modelling for intrinsic viscosity in polymer–solvent combinations based on density functional theory. SAR and QSAR in Environmental Research. 2021;32(5):379–93.

[673] Danielsen SP, Beech HK, Wang S, El-Zaatari BM, Wang X, Sapir L, Ouchi T, Wang Z, Johnson PN, Hu Y. Molecular characterization of polymer networks. Chemical Reviews. 2021;121(8):5042–92.

[674] Öhrn OE. Preliminary report on the influence of adsorption on capillary dimensions of viscometers. Journal of Polymer Science. 1955;17(83):137–40.

[675] Boris DC, Colby RH. Rheology of sulfonated polystyrene solutions. Macromolecules. 1998;31(17):5746–55.

[676] Keshavarz MH, Shafiee M, Jazi BN. A simple correlation for reliable prediction of intrinsic viscosity (limiting viscosity number) of different polymer-solvent combinations. Fluid Phase Equilibria. 2022;557:113422.

[677] Krause S, Gormley JJ, Roman N, Shetter JA, Watanabe WH. Glass temperatures of some acrylic polymers. Journal of Polymer Science Part A: General Papers. 1965;3(10):3573–86.

[678] Kratochvíl J, Šturcová A, Sikora A, Dybal J. Note on the glass transition temperature of poly (vinylphenol). European Polymer Journal. 2009;45(6):1851–6.

[679] Calleja G, Jourdan A, Ameduri B, Habas J-P. Where is the glass transition temperature of poly (tetrafluoroethylene)? A new approach by dynamic rheometry and mechanical tests. European Polymer Journal. 2013;49(8):2214–22.

[680] Monkos K. Determination of the glass-transition temperature of proteins from a viscometric approach. International journal of biological macromolecules. 2015;74:1–4.

[681] Otsuka S, Kuwajima I, Hosoya J, Xu Y, Yamazaki M, editors. PoLyInfo: Polymer database for polymeric materials design. In: 2011 International Conference on Emerging Intelligent Data and Web Technologies. IEEE, 2011.

[682] Xie R, Weisen AR, Lee Y, Aplan MA, Fenton AM, Masucci AE, Kempe F, Sommer M, Pester CW, Colby RH. Glass transition temperature from the chemical structure of conjugated polymers. Nature Communications. 2020;11(1):893.

[683] Camacho-Zuniga C, Ruiz-Trevino FA. A new group contribution scheme to estimate the glass transition temperature for polymers and diluents. Industrial & Engineering Chemistry Research. 2003;42(7):1530–4.

[684] Chen X, Sztandera L, Cartwright HM. A neural network approach to prediction of glass transition temperature of polymers. International Journal of Intelligent Systems. 2008;23(1):22–32.

[685] Yu X. Support vector machine-based QSPR for the prediction of glass transition temperatures of polymers. Fibers and Polymers. 2010;11:757–66.

[686] Liu W, Cao C. Artificial neural network prediction of glass transition temperature of polymers. Colloid and Polymer Science. 2009;287:811–8.

[687] Kim C, Chandrasekaran A, Jha A, Ramprasad R. Active-learning and materials design: the example of high glass transition temperature polymers. Mrs Communications. 2019;9(3):860–6.

[688] Alesadi A, Cao Z, Li Z, Zhang S, Zhao H, Gu X, Xia W. Machine learning prediction of glass transition temperature of conjugated polymers from chemical structure. Cell Reports Physical Science. 2022;3(6).

[689] Pilania G, Iverson CN, Lookman T, Marrone BL. Machine-learning-based predictive modeling of glass transition temperatures: a case of polyhydroxyalkanoate homopolymers and copolymers. Journal of Chemical Information and Modeling. 2019;59(12):5013–25.

[690] Toropov AA, Toropova AP, Kudyshkin VO, Bozorov NI, Rashidova SS. Applying the Monte Carlo technique to build up models of glass transition temperatures of diverse polymers. Structural Chemistry. 2020;31:1739–43.

[691] Afantitis A, Melagraki G, Makridima K, Alexandridis A, Sarimveis H, Iglessi-Markopoulou O. Prediction of high weight polymers glass transition temperature using RBF neural networks. Journal of Molecular Structure: THEOCHEM. 2005;716(1–3):193–8.

[692] Tao L, Varshney V, Li Y. Benchmarking machine learning models for polymer informatics: an example of glass transition temperature. Journal of Chemical Information and Modeling. 2021;61(11):5395–413.

[693] Khan P, Roy K. QSPR modelling for prediction of glass transition temperature of diverse polymers. SAR and QSAR in Environmental Research. 2018;29(12):935–56.

[694] Karuth A, Alesadi A, Xia W, Rasulev B. Predicting glass transition of amorphous polymers by application of cheminformatics and molecular dynamics simulations. Polymer. 2021;218:123495.

[695] Shahrousvand E, Hamadanian M, Keshavarz MH. A general method for assessment of glass transition temperature of polymeric materials only from various structural factors in their repeating unit structure. Materials Today Communications. 2024;38:108405.

[696] Afzal MAF, Cheng C, Hachmann J. Combining first-principles and data modeling for the accurate prediction of the refractive index of organic polymers. The Journal of Chemical Physics. 2018;148(24).

[697] Thomas SW, Khan RR, Puttananjegowda K, Serrano-Garcia W. Conductive polymers and metal oxide polymeric composites for nanostructures and nanodevices. In: Advances in Nanostructured Materials and Nanopatterning Technologies. Elsevier, 2020. pp. 243–71.

[698] Xu J, Du K, Peng F, Sun Z, Zhong Z, Feng W, Ying L. Highly conductive polymer electrodes for polymer light-emitting diodes. npj Flexible Electronics. 2024;8(1):38.

[699] Qiu J, Hu G, Wang Y, Wang Y, Luo M, Hu X. A high refractive index resist for UV-nanoimprint soft lithography based on titanium-containing elemental polymer oligomers. Journal of Materials Chemistry C. 2022;10(1):219–26.

[700] Cerna JR, Alejandro AS, Matla M. Analysis of the influence of glycidyl methacrylate on molecular weight and refractive index in styrene-methylmethacrylate-glycidyl methacrylate copolymers through mixture design of experiments. Journal of applied polymer science. 2009;114(3):1935–41.

[701] Kasarova SN, Sultanova NG, Ivanov CD, Nikolov ID. Analysis of the dispersion of optical plastic materials. Optical Materials. 2007;29(11):1481–90.

[702] Bobbitt JM, Mendivelso-Perez D, Smith EA. Scanning angle Raman spectroscopy: A nondestructive method for simultaneously determining mixed polymer fractional composition and film thickness. Polymer. 2016;107:82–8.

[703] Khan PM, Rasulev B, Roy K. QSPR modeling of the refractive index for diverse polymers using 2D descriptors. ACS Omega. 2018;3(10):13374–86.

[704] Erickson ME, Ngongang M, Rasulev B. A refractive index study of a diverse set of polymeric materials by QSPR with quantum-chemical and additive descriptors. Molecules. 2020;25(17):3772.

[705] Schustik SA, Cravero F, Ponzoni I, Díaz MF. Polymer informatics: Expert-in-the-loop in QSPR modeling of refractive index. Computational Materials Science. 2021;194:110460.

[706] Wills TJ, Polshakov DA, Robinson MC, Lee AA. Impact of chemist-in-the-loop molecular representations on machine learning outcomes. Journal of Chemical Information and Modeling. 2020;60(10):4449–56.

[707] Zhang J, Wang Z, Wang Z, Wei L. Advanced multi-material optoelectronic fibers: a review. Journal of Lightwave Technology. 2020;39(12):3836–45.

[708] Gepreel KA, Zayed EM, Alngar ME, Biswas A, Guggilla P, Khan S, Yıldırım Y, Alzahrani AK, Belic MR. Optical solitons with Kudryashov's arbitrary form of refractive index and generalized non-local nonlinearity. Optik. 2021;243:166723.

[709] Li J. A review: Development of novel fiber-optic platforms for bulk and surface refractive index sensing applications. Sensors and Actuators Reports. 2020;2(1):100018.

[710] Hamadanian M, Keshavarz MH, Shahrousvand E. The reliable predicting refractive index for diverse polymers only from structural moieties in repeating unit structures. Materials Today Communications. 2023;35:105823.

[711] Keshavarz MH, Klapötke TM, Jaafari M, Ebadpour R, Dalirandeh Z, Moghayadnia R. Energetic Materials Designing Bench (EMDB). In: http://www.emdb-software.com/ (keshavarz7@gmail.com). 2025.

[712] Wahler S, Klapötke TM. Research output software for energetic materials based on observational modelling 2.1 (RoseBoom2. 1©). Materials Advances. 2022;3(21):7976–86.

[713] Wahler S. RoseBoom 2.3, RoseEnergetic GmbH, Munich, 2023.

[714] Wahler S, Klapötke TM. Comparison of the Implemented Detonation Velocity Predictions in the Research Output Software for Energetic Materials Based on Observational Modelling (RoseBoom©) with 30 Experimental Values. Central European Journal of Energetic Materials. 2023;20(1).

[715] Silva AL, Almeida AR, Ribeiro da Silva MD, Reinhardt J, Klapötke TM. On the Enthalpy of Formation and Enthalpy of Sublimation of Dihydroxylammonium 5, 5′-bitetrazole-1, 1′-dioxide (TKX-50). Propellants, Explosives, Pyrotechnics. 2023;48(7):e202200361.

[716] Silva AL, León GP, Ribeiro da Silva MD, Klapötke TM, Reinhardt J. Enthalpy of Formation of the Nitrogen-Rich Salt Guanidinium 5, 5′-Azotetrazolate (GZT) and a Simple Approach for Estimating the Enthalpy of Formation of Energetic C, H, N, O Salts. Thermo. 2023;3(4):549–65.

[717] Bathelt H, Volk F, editors. The ICT Thermodynamic Code (ICT-Code). Fraunhofer-Institut für Chemische Technologie (Internationale Jahrestagung) 1996; 1996.

[718] Kelzenberg S, Kempa PB, Wurster S, Herrmann M, Fischer T, editors. New version of the ICT-thermodynamic code. In: 45th International Annual Conference of ICT, 2014.

[719] Sućeska M, editor. Calculation of detonation parameters by EXPLO5 computer program. In: Materials Science Forum. Trans Tech Publ, 2004.

[720] Sućeska M. EXPLO5, Version 8.01, Brodarski Institute, Zagreb, 2025.

[721] Willer RL. Calculation of the density and detonation properties of C, H, N, O and F compounds: use in the design and synthesis of new energetic materials. Journal of the Mexican Chemical Society. 2009;53(3):108–19.

[722] Jenkins HDB, Roobottom HK, Passmore J, Glasser L. Relationships among ionic lattice energies, molecular (formula unit) volumes, and thermochemical radii. Inorganic Chemistry. 1999;38(16):3609–20.

[723] Byrd EF, Rice BM. A comparison of methods to predict solid phase heats of formation of molecular energetic salts. The Journal of Physical Chemistry A. 2009;113(1):345–52.

[724] Gutowski KE, Rogers RD, Dixon DA. Accurate thermochemical properties for energetic materials applications. II. Heats of formation of imidazolium-, 1, 2, 4-triazolium-, and tetrazolium-based energetic salts from isodesmic and lattice energy calculations. The Journal of Physical Chemistry B. 2007;111(18):4788–800.

[725] Sinditskii VP, Filatov SA, Kolesov VI, Kapranov KO, Asachenko AF, Nechaev MS et al. Combustion behavior and physico-chemical properties of dihydroxylammonium 5, 5′-bistetrazole-1, 1′-diolate (TKX-50). Thermochimica Acta. 2015;614:85–92.

[726] Lal S, Gao H, Shreeve JM. 1, 3, 5, 6, 7-pentanitro-3, 6, 7-triazabicyclo-[3.1. 1]-heptane (UIX): A powerful potential explosive with zero oxygen balance. Chem Asian J. 2025, 0, e70003: doi.org/10.1002/asia.202500290.

[727] Curtiss LA, Raghavachari K, Redfern PC, Pople JA. Assessment of Gaussian-2 and density functional theories for the computation of enthalpies of formation. The Journal of Chemical Physics. 1997;106(3):1063–79.

[728] Byrd EF, Rice BM. Improved prediction of heats of formation of energetic materials using quantum mechanical calculations. The Journal of Physical Chemistry A. 2006;110(3):1005–13.

[729] Rice BM, Pai SV, Hare J. Predicting heats of formation of energetic materials using quantum mechanical calculations. Combustion and flame. 1999;118(3):445–58.

[730] Hobbs ML, Baer MR. Calibrating the BKW-EOS with a large product species data base and measured CJ properties. In: Proc. of the 10th Symp. (International) on Detonation, ONR, 1993. pp. 409.

[731] Pedley JB. Thermochemical Data of Organic Compounds. Springer; 2012.

[732] Vadhe PP, Pawar RB, Sinha RK, Asthana SN, Rao AS. Cast aluminized explosives (review). Combustion, Explosion, and Shock Waves. 2008;44:461–77.

[733] Manaa MR, Fried LE, Kuo I-FW. Determination of enthalpies of formation of energetic molecules with composite quantum chemical methods. Chemical Physics Letters. 2016;648:31–5.

[734] Frisch MJ, Trucks GW, Schlegel HB, Scuseria GE, Robb MA, Cheeseman JR, Scalmani G, Barone V, Mennucci B, Petersson GA, Nakatsuji H, Caricato M, Li X, Hratchian HP, Izmaylov AF, Bloino J, Zheng G, Sonnenberg JL, Hada M, Ehara M, Toyota K, Fukuda R, Hasegawa J, Ishida M, Nakajima T, Honda Y, Kitao O, Nakai H, Vreven T, Montgomery JA, Peralta JE, Ogliaro F, Bearpark M, Heyd JJ, Brothers E, Kudin KN, Staroverov VN, Kobayashi R, Normand J, Raghavachari K, Rendell A, Burant JC, Iyengar SS, Tomasi J, Cossi M, Rega N, Millam JM, Klene M, Knox JE, Cross JB, Bakken V, Adamo C, Jaramillo J, Gomperts R, Stratmann RE, Yazyev O, Austin AJ, Cammi R, Pomelli C, Ochterski JW, Martin RL, Morokuma K, Zakrzewski VG, Voth GA, Salvador P, Dannenberg JJ, Dapprich S, Daniels AD, Farkas O, Foresman JB, Ortiz JV, Cioslowski J, Fox DJ. GAUSSIAN 09 (Revision A. 1). Wallingford: Gaussian, Inc., 2009.

[735] Stevens WR, Ruscic B, Baer T. Heats of formation of C6H5•, C6H5+, and C6H5NO by threshold photoelectron photoion coincidence and active thermochemical tables analysis. The Journal of Physical Chemistry A. 2010;114:13134–45.

[736] Simmie JM. A database of formation enthalpies of nitrogen species by compound methods (CBS-QB3, CBS-APNO, G3, G4). The Journal of Physical Chemistry A. 2015;119:10511–26.

[737] Verevkin SP, Emel'yanenko VN, Zaitsau DH, Heintz A, Muzny CD, Frenkel M. Thermochemistry of imidazolium-based ionic liquids: experiment and first-principles calculations. Physical Chemistry Chemical Physics. 2010;12:14994–5000.

[738] Emel'yanenko VN, Verevkin SP, Heintz A. The gaseous enthalpy of formation of the ionic liquid 1-butyl-3-methylimidazolium dicyanamide from combustion calorimetry, vapor pressure measurements, and ab initio calculations. Journal of the American Chemical Society. 2007;129:3930–7.

[739] Zhang Z-H, Tan Z-C, Sun L-X, Jia-Zhen Y, Lv X-C, Shi Q. Thermodynamic investigation of room temperature ionic liquid: The heat capacity and standard enthalpy of formation of EMIES. Thermochimica Acta. 2006;447:141–6.

[740] Ye C, Xiao J-C, Twamley B, Shreeve JnM. Energetic salts of azotetrazolate, iminobis(5-tetrazolate) and 5,5′-bis(tetrazolate). Chemical communications. 2005;2750–2.

[741] Gutowski KE, Rogers RD, Dixon DA. Accurate thermochemical properties for energetic materials applications. II. Heats of formation of imidazolium-, 1,2,4-triazolium-, and tetrazolium-based energetic salts from isodesmic and lattice energy calculations. The Journal of Physical Chemistry B. 2007;111:4788–800.

[742] Pedley JB, Naylor RD, Kirby SP. Thermochemical Data of Organic Compounds. 2nd ed. New York: Chapman and Hall Ltd.; 1986.

[743] Darwich C, Klapötke TM, Welch JM, Suceska M. Synthesis and characterization of 3,4,5-triamino-1,2,4-triazolium 5-nitrotetrazolate. Propellants, Explosives, Pyrotechnics. 2007;32:235–43.

[744] Darwich C, Klapötke TM, Sabaté CM. 1,2,4-Triazolium-cation-based energetic salts. Chemistry–A European Journal. 2008;14:5756–71.

[745] Xue H, Shreeve JM. Energetic Ionic Liquids from Azido Derivatives of 1,2,4-Triazole. Advanced Materials. 2005;17:2142–6.

[746] Xue H, Twamley B, Jean'ne MS. Energetic salts of substituted 1,2,4-triazolium and tetrazolium 3,5-dinitro-1,2,4-triazolates. Journal of Materials Chemistry. 2005;15:3459–65.

Index

https://doi.org/10.1515/9783112206768-023